THE COMPLETE GUIDE TO HOME PLUMBING

A Comprehensive Manual, from Basic Repairs to Advanced Projects

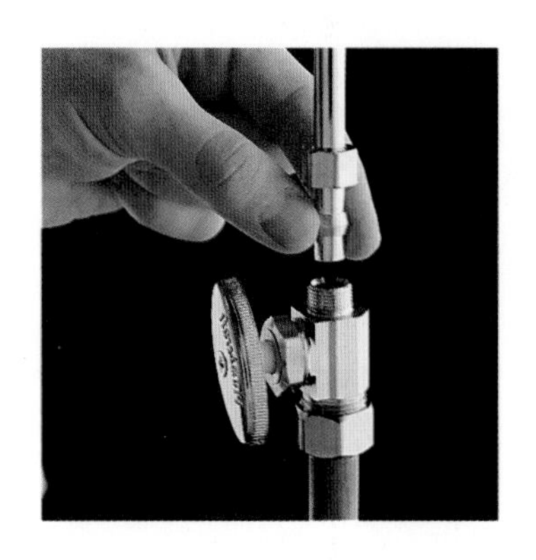

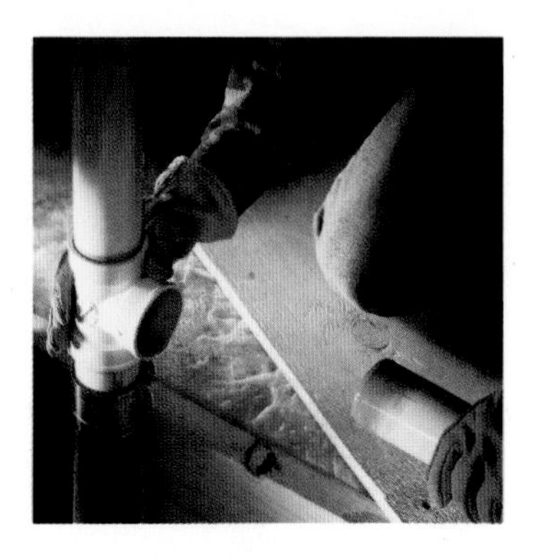

MINNETONKA, MINNESOTA

Contents

Planning & Installing New Plumbing

Creative Publishing International, Inc.
5900 Green Oak Drive
Minnetonka, Minnesota 55343
1-800-328-3895

Printed in U.S.A.

President/CEO: David D. Murphy
Vice President/Editorial: Patricia K. Jacobsen
Vice President/Retail Sales & Marketing: Richard M. Miller

THE COMPLETE GUIDE TO HOME PLUMBING
Created by: The Editors of Creative Publishing international, Inc., in cooperation with Black & Decker.

Printed on American paper by:
R.R. Donnelley & Sons
10 9 8 7 6

Replacing & Repairing Old Plumbing

Library of Congress
Cataloging-in-Publication Data

The complete guide to home plumbing : a comprehensive manual, from basic repairs to advanced projects.

p. cm.
At head of title: Black & Decker.
ISBN 0-86573-775-4 (softcover)
1. Plumbing--Amateurs' manuals. 2. Dwellings--Remodeling--Amateurs' manuals. I. Black & Decker Corporation (Towson, Md.) II. Creative Publishing International.
TH6124.C66 1998
646'.1--dc21 98-34999 CIP

Portions of *The Complete Guide to Home Plumbing* are from the Black & Decker® books *Home Plumbing Projects & Repairs* and *Advanced Home Plumbing.* Other books in the Black & Decker® Home Improvement Library™ include:

Everyday Home Repairs, Decorating With Paint & Wallcovering, Carpentry: Tools • Shelves • Walls • Doors, Building Decks, Basic Wiring & Electrical Repairs, Workshop Tips & Techniques, Advanced Home Wiring, Carpentry: Remodeling, Landscape Design & Construction, Bathroom Remodeling, Built-In Projects for the Home, Refinishing & Finishing Wood, Exterior Home Repairs & Improvements, Home Masonry Repairs & Projects, Building Porches & Patios, Flooring Projects & Techniques, Advanced Deck Building, Advanced Home Plumbing, Remodeling Kitchens

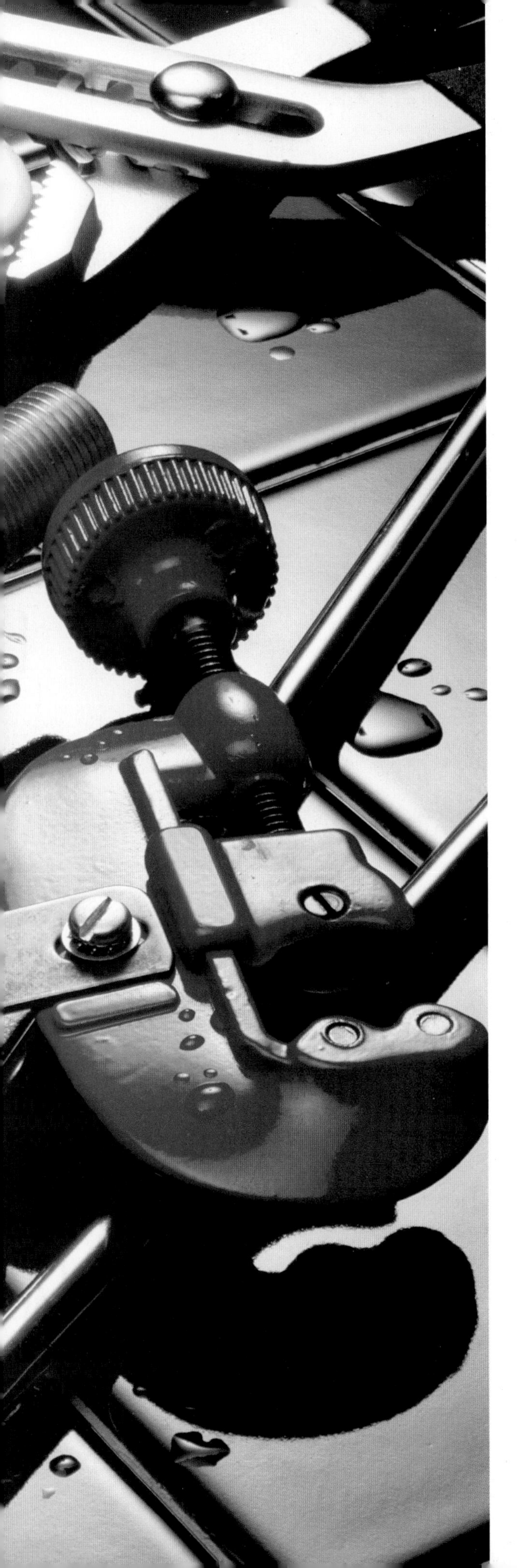

Introduction

It is difficult to overstate the importance of the plumbing system on the quality of modern life. A well-functioning plumbing system is essential to your safety, health, and cleanliness. To keep your home working smoothly, you should know how your plumbing system works. When plumbing problems arise, you will need to know how to solve them, and quickly. Over time, faucets develop leaks, clogs block drain lines, and appliances wear out and need to be replaced. As a homeowner, you can either call a professional plumber, or save money by making your own repairs.

Home plumbing tasks may seem intimidating, but with *The Complete Guide to Home Plumbing*, you will be able to handle them confidently. This book will help you understand how plumbing works. It also provides the information you need for making repairs to your home plumbing and for installing supply lines, drain lines, and plumbing fixtures.

Some of the projects in *The Complete Guide to Home Plumbing* are easy, requiring only basic skills and tools; others are more complex. Many plumbing projects require basic carpentry and electrical techniques. If you are unsure of your basic do-it-yourself skills, you should have good books on these subjects, such as *Bathroom Remodeling*, *Carpentry: Remodeling*, *Kitchen Remodeling*, and *The Complete Guide to Home Wiring,* all from the Black & Decker® Home Improvement Library™. Before starting a project, read through the directions to decide if you are prepared to complete the project with the equipment and knowledge you already have, or whether you should get help from a professional plumber.

This book is arranged in five sections, and is designed to be easy for you to use. With clear directions, expert advice from professional plumbers, cross references, and hundreds of detailed color photographs, this book will walk you through the most common repair and installation procedures, one step at a time.

Introduction

The first section, which includes this introduction, provides a glossary of terms you will need to know in order to make the best use of this book, plus an explanation of the water cycle and a description of home plumbing principles. To help you gain an understanding of how plumbing works, the opening pages give you a dramatic look at an entire home plumbing system, with all pipes color-coded for easy identification. Included are detailed descriptions of each part of the plumbing system to help you diagnose problems and plan possible repairs.

Planning Your Project

The next section, "Planning Your Project," provides important background information. Read this section carefully before you attempt the plumbing projects described later in the book. In this section, you'll learn the basic mechanics of the overall plumbing network and how to inspect and map your existing plumbing system. This section also explains the principles that make your plumbing system work and tells you how to make sure your project will comply with the Plumbing Code.

Plumbing Tools & Materials

A third section, "Plumbing Tools & Materials," shows what you will need to begin and complete a plumbing project successfully. It includes photographs and descriptions of the common hand tools, specialty plumbing tools, power tools, and rental tools that are used in this book. This section also shows the many different kinds of pipes and fittings available, and how to cut, fit, repair, and replace each kind. After reviewing this section, you will have the basic knowledge you need to begin most common plumbing projects.

Installing New Plumbing

The fourth section describes the process of adding new supply lines and drains, and includes sample home plumbing projects for bathrooms and kitchens. Some of these projects —particularly those for installing showers, bathtubs, whirlpools, and dishwashers—require carpentry skills. Whirlpools, dishwashers, and food disposers also involve electrical work. You should be familiar with these skills before taking on any of these projects. This section concludes with an overview of outdoor plumbing.

Repairing & Replacing Old Plumbing

Many home plumbing projects involve existing plumbing lines and fixtures instead of new ones. The fifth section, "Repairing & Replacing Old Plumbing," will help you decide whether to fix what you have, change parts of it, or tear out the old and start over. This section includes directions for running new plumbing pipes when walls and floors are finished and for replacing plumbing pipes between your water meter and individual fixtures. It also provides instructions for repairing faucets, fixing toilets, maintaining drains, fixing tub and shower plumbing, and repairing or replacing both electric and gas water heaters. At the end of this section, you will find advice on how to prevent or repair some of the most common plumbing annoyances: burst, frozen, or noisy pipes.

The Complete Guide to Home Plumbing takes the mystery out of your plumbing system and its individual parts. Creative Publishing international is proud to offer this solid, comprehensive reference book, and we're confident that it will be an important addition to your home improvement library.

Glossary of Terms

Access panel:
Opening in a wall or ceiling that provides access to the plumbing system

Appliance:
Powered device that uses water, such as a water heater, dishwasher, washing machine, whirlpool, or water softener

Auger:
Flexible tool used for clearing obstructions in drain lines

Ballcock:
Valve that controls the water supply entering a toilet tank

Blow bag:
Expanding rubber device that attaches to a garden hose; used for clearing floor drains

Branch drain line:
Pipe that connects additional lines to a drain system

Branch line:
Pipe that connects additional lines to a water supply system

Cleanout:
Cover in a waste pipe or trap that provides access for cleaning

Closet auger:
Flexible rod used to clear obstructions in toilets

Closet bend:
Curved fitting that fits between a closet flange and a toilet drain

Closet flange:
Ring at the opening of a toilet drain, used as the base for a toilet

Coupling:
Fitting that connects two pieces of pipe

DWV:
Drain, waste, and vent; the system for removing water from a house

DWV stack:
Pipe that connects house drain system to a sewer line at the bottom and vents air to outside of house at the top

Elbow:
Angled fitting that changes the direction of a pipe

Fixture:
Device that uses water, such as a sink, tub, shower, sillcock, or toilet

Flapper (tank ball):
Rubber seal that controls the flow of water from a toilet tank to a toilet bowl

Flux (soldering paste):
Paste applied to metal joints before soldering to increase joint strength

Hand auger (snake):
Hand tool with flexible shaft, used for clearing clogs in drain lines

Hose bib:
Any faucet spout that is threaded to accept a hose

I.D.:
Inside diameter; plumbing pipes are classified by I.D.

Loop vent:
A special type of vent configuration used in kitchen sink island installations

Main shutoff valve:
Valve that controls water supply to an entire house; usually next to the water meter

Motorized auger:
Power tool with flexible shaft, used for clearing tree roots from sewer lines

Nipple:
Pipe with threaded ends

O.D.:
Outside diameter

Plumber's putty:
A soft material used for sealing joints between fixtures and supply or drain parts

Reducer:
A fitting that connects pipes of different sizes

Riser:
Assembly of water supply fittings and pipes that distributes water upward

Run:
Assembly of pipes that extends from water supply to fixture, or from drain to stack

Saddle valve:
Fitting clamped to copper supply pipe, with hollow spike that punctures the pipe to divert water to another device, usually a dishwasher or refrigerator icemaker

Sanitary fitting:
Fitting that joins DWV pipes; allows solid material to pass through without clogging

Shutoff valve:
Valve that controls the water supply for one fixture or appliance

Sillcock:
Compression faucet used on the outside of a house

Soil stack:
Main vertical drain line, which carries waste from all branch drains to a sewer line

Solder:
Metal alloy used for permanently joining metal (usually copper) pipes

T-fitting:
Fitting shaped like the letter T used for creating or joining branch lines

Trap:
Curved section of drain, filled with standing water, that prevents sewer gases from entering a house

Union:
Fitting that joins two sections of pipe, but can be disconnected without cutting

Vacuum breaker:
Attachment for outdoor and below-ground fixtures that prevents waste water from entering supply lines if water supply pressure drops

Wet vent:
Pipe that serves as a drain for one fixture and as a vent for another

Y-fitting:
Fitting shaped like the letter Y used for creating or joining branch lines

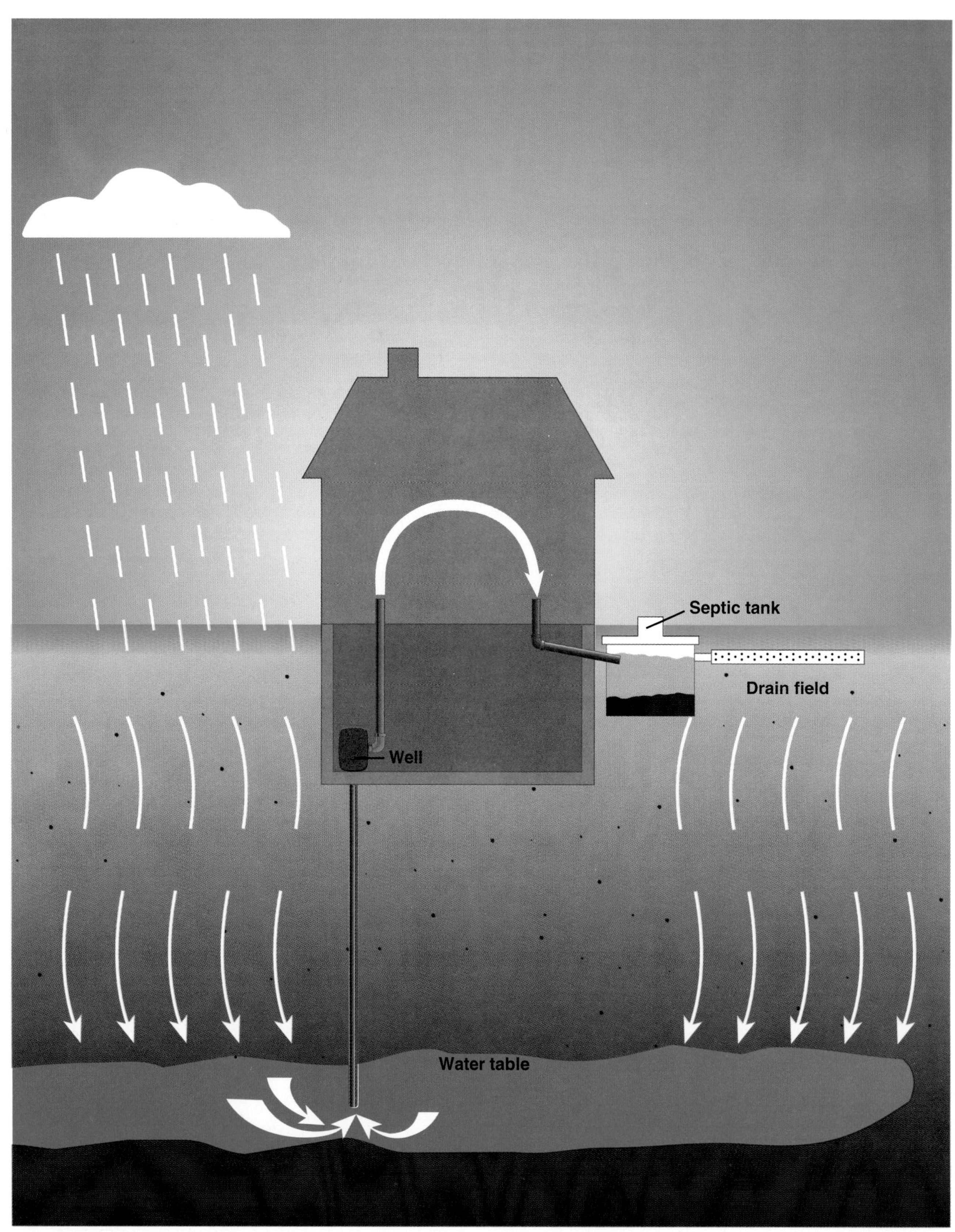

In a basic plumbing cycle, which is still used in many rural homes, a well pumps fresh water from the water table. A system of in-house plumbing pipes then distributes the water to various fixtures throughout the house. After use, the waste water travels through a system of drain pipes to a septic system, which separates solid wastes and returns the water to the soil. In metropolitan areas, the methods for supplying fresh water and handling waste water are more complex (see pages 10 to 11).

Understanding the Water Cycle

Each day in the United States, roughly 500 billion gallons of water are withdrawn from surface and underground water supply sources for use in homes and businesses. In private homes, water is used for drinking and food preparation, for cleaning and washing, and for irrigating landscapes and gardens. Finally, we depend on water to carry septic wastes safely away from our homes.

Just as fresh water is essential to basic human existence, a well-designed plumbing system is essential to a healthy life and productive society. The science of transporting water to and from a home through a system of pipes originated some 5,000 years ago in ancient Sumer and is still evolving today.

Most homeowners are familiar with the in-home portion of the water cycle—the plumbing system with its network of faucets, toilets, and other plumbing fixtures. Many homeowners can make minor repairs to the fixtures and pipes. However, few homeowners fully understand the beginning and ending stages of this sequence—called the hydrologic cycle. Knowing where the fresh water supply originates can help you make decisions on how to manage the water entering your home; and understanding how waste water is recycled may change your decisions on what materials to flush down the drain.

Distributing fresh water. The fresh water that enters your home originates in one of two sources: surface water or groundwater. Surface water supplies include lakes, streams, rivers, and artificial storage reservoirs. Groundwater comes from natural underground caverns or from aquifers—porous, water-saturated layers of gravel, sand, and silt. This water is either pumped from the ground through a well, or rises to the surface through natural springs.

In about 25% of homes, mostly rural residences, a well delivers fresh water directly from groundwater sources with no chemical treatment. This is done because groundwater is generally regarded as pure enough to drink. In recent years, however, it has been found that some groundwater supplies have unhealthy levels of natural substances, such as nitrates, as well as toxic pollutants and dangerous microorganisms. If your water supply comes from a well, it is a good idea to have a sample tested by a laboratory. If a test finds that your water supply has harmful contaminants, consider installing an in-home purification or filtration system.

In addition, groundwater pumped from a well often has a high mineral content. This "hard" water may prematurely age your plumbing pipes as the minerals build up on the inside surfaces of the pipes. Installing a water softener helps reduce mineral levels and can extend the life of your plumbing system.

In metropolitan areas and many smaller communities, fresh water distribution is a function of a public utility. The water pumped from the ground or taken from surface sources is first cleansed at a central treatment plant before it is distributed to homes. Public utilities generally do an adequate job of purifying the water supply, but in large industrial areas, especially those that derive their water supply from surface water, it is not uncommon for very low levels of potentially harmful substances to be present. Your local division of the Environment Protection Agency should have information on the water quality in your area.

Recycling waste water. About 78% of all water that enters a home or business eventually finds its way back to a groundwater or surface water source. Because this waste water eventually becomes part of the fresh water supply again, it must be purified before it is released into the environment.

In rural areas, this purification is accomplished by simple home septic systems that separate solid wastes and transport the water back into the soil. As this water makes its slow journey back to the water table, it is filtered pure by many layers of rock and soil.

In most cities and towns, waste water purification is accomplished by a system of sewer utility pipes that carry raw sewage to a central treatment facility. After solid wastes are removed, the water is purified and released.

Keep in mind that individual septic systems and urban sewage treatment plants are designed to recycle waste water and organic solids only. Flushing any synthetic materials or chemicals down your drain can jeopardize the system's function—transforming waste liquid into pure fresh water for the next user.

Man-made reservoirs created by constructing dams across river valleys are common in arid western regions. By capturing water from melting snows in the mountains, communities can store enough water to supply their needs through the dry summer months. In other regions, rivers and lakes provide ongoing sources of fresh water. Many man-made reservoirs are hydroelectric projects used to generate electricity, as well as to store fresh water.

Water towers and enclosed tanks store the water pumped from treatment plants. They are usually positioned on high ground, so gravity can produce the downward pressure necessary to force water through the water mains. Every 2.3 feet of height produces 1 pound of water pressure.

Distributing Fresh Water

Although the specific methods for collecting, treating, and distributing fresh water vary from community to community, the process is generally the same no matter where you live. Water is pumped from an underground source or is channeled from a lake, river, or reservoir into a large controlled storage facility.

From the storage reservoir, the water is pumped to a water treatment plant, where it is cleansed and purified. Water treatment plants use a variety of physical and chemical procedures. In a typical plant, water is strained and filtered to remove solids, is aerated with sprayers to remove dissolved gases, and is then disinfected with chlorine to kill organisms. In most communities, fluoride is added to the water supply to help reduce tooth decay.

Finally, the treated fresh water is pumped to enclosed storage towers or tanks, where a network of distribution pipes carries the water to homes and businesses in the community.

Recycling Waste Water

The systems used to purify waste water after it leaves the home range from simple home septic tanks to enormous sewage treatment plants that handle millions of gallons of raw sewage each day. But no matter what system is used, the process of cleansing waste water is simple in principle.

All waste treatment systems use a combination of physical, biological, and chemical processes to remove contaminants from water. However, the success of any waste water treatment system can be compromised by improper use. Bleaches, fertilizers, and phosphorus detergents flushed into the waste system may interfere with the natural consumption of solid wastes by microorganisms. And pesticides and other chemicals poured down sink drains by homeowners may eventually find their way to the fresh water supply.

Lagoon systems are growing in popularity, especially for small communities. Originally, this design was used simply to hold untreated waste water until it evaporated, but modern lagoons are state-of-the-art treatment facilities in which aquatic plants are used to break down solid wastes. The plant material is periodically harvested and dried to create fertilizer, and the water is used to irrigate farming lands or urban landscapes.

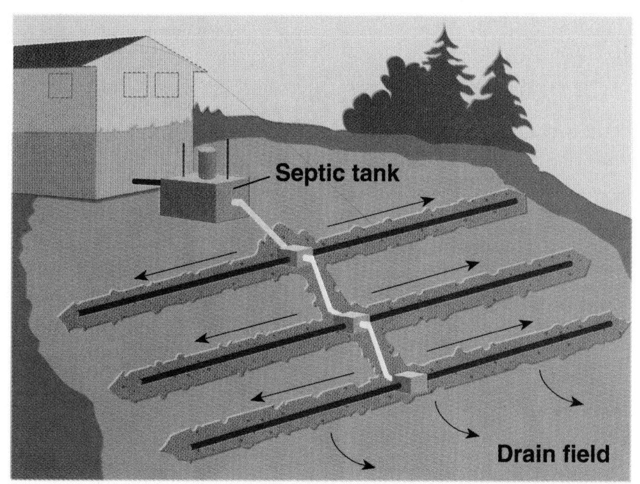

Private septic system consists of an underground tank and a system of pipes fanning out from the tank. When sewage reaches the tank, the solid wastes settle to the bottom, where they are consumed by microorganisms. As the tank fills, the water flows out of the tank through porous drain pipes that distribute the water into the soil. The water is filtered clean as it drains down through thick layers of soil and rock on its return to the water table. Used correctly, a septic system requires only that residual solid wastes be pumped out every few years.

Sewer treatment plant is the popular urban solution for treating billions of gallons of waste water that drain from our homes and businesses. Sewage, a combination of water and solid wastes, is first directed to settling tanks, where the solid wastes precipitate to the bottom. This is followed by a biological treatment, in which microorganisms digest remaining organic material in the water. The water is then filtered, disinfected with chlorine, and discharged to irrigation canals, streams, or lakes. The residual solid waste is often processed as agricultural fertilizer.

The Home Plumbing System

Because most of a plumbing system is hidden inside walls and floors, it may seem to be a complex maze of pipes and fittings. In fact, home plumbing is simple and straightforward. Understanding how home plumbing works is an important first step toward doing routine maintenance and money-saving repairs.

A typical home plumbing system includes three basic parts: a water supply system, fixtures and appliances, and a drain system. These three parts can be seen clearly in the photograph of the cutaway house on the opposite page.

Fresh water enters a home through a main supply line **(1)**. This fresh water source is provided by either a municipal water company or a private underground well. If the source is a municipal supplier, the water passes through a meter **(2)** that registers the amount of water used. A family of four uses about four hundred gallons of water each day.

Immediately after the main supply enters the house, a branch line splits off **(3)** and is joined to a hot water heater **(4)**. From the water heater, a hot water line runs parallel to the cold water line to bring the water supply to fixtures and appliances throughout the house. Fixtures include sinks, bathtubs, showers, and laundry tubs. Appliances include water heaters, dishwashers, clothes washers, and water softeners. Toilets and exterior sillcocks are examples of fixtures that require only a cold water line.

The water supply to fixtures and appliances is controlled with faucets and valves. Faucets and valves have moving parts and seals that eventually may wear out or break, but they are easily repaired or replaced.

Waste water then enters the drain system. It first must flow past a trap **(5)**, a U-shaped piece of pipe that holds standing water and prevents sewer gases from entering the home. Every fixture must have a drain trap.

The drain system works entirely by gravity, allowing waste water to flow downhill through a series of large-diameter pipes. These drain pipes are attached to a system of vent pipes. Vent pipes **(6)** bring fresh air to the drain system, preventing suction that would slow or stop drain water from flowing freely. Vent pipes usually exit the house at a roof vent **(7)**.

All waste water eventually reaches a main waste and vent stack **(8)**. The main stack curves to become a sewer line **(9)** that exits the house near the foundation. In a municipal system, this sewer line joins a main sewer line located near the street. Where sewer service is not available, waste water empties into a septic system.

Water meter and main shutoff valve are located where the main water supply pipe enters the house. The water meter is the property of your local municipal water company. If the water meter leaks, or if you suspect it is not functioning properly, call your water company for repairs.

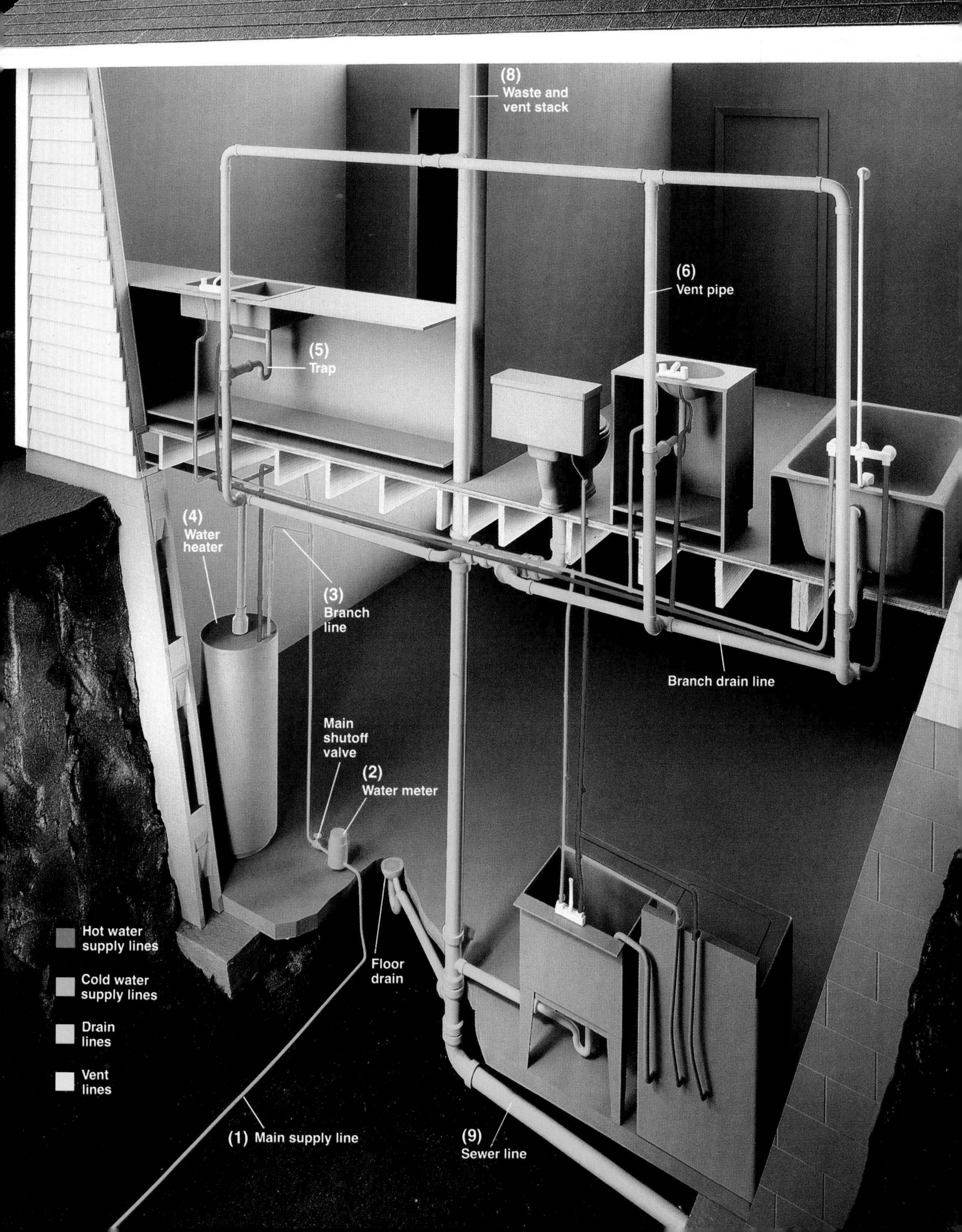
(7)
Roof vent
(8)
Waste and
vent stack
(6)
Vent pipe
(5)
Trap
(4)
Water
heater
(3)
Branch
line
Branch drain line
Main
shutoff
valve
(2)
Water meter
Floor
drain
Hot water
supply lines
Cold water
supply lines
Drain
lines
Vent
lines
(1) Main supply line
(9)
Sewer line

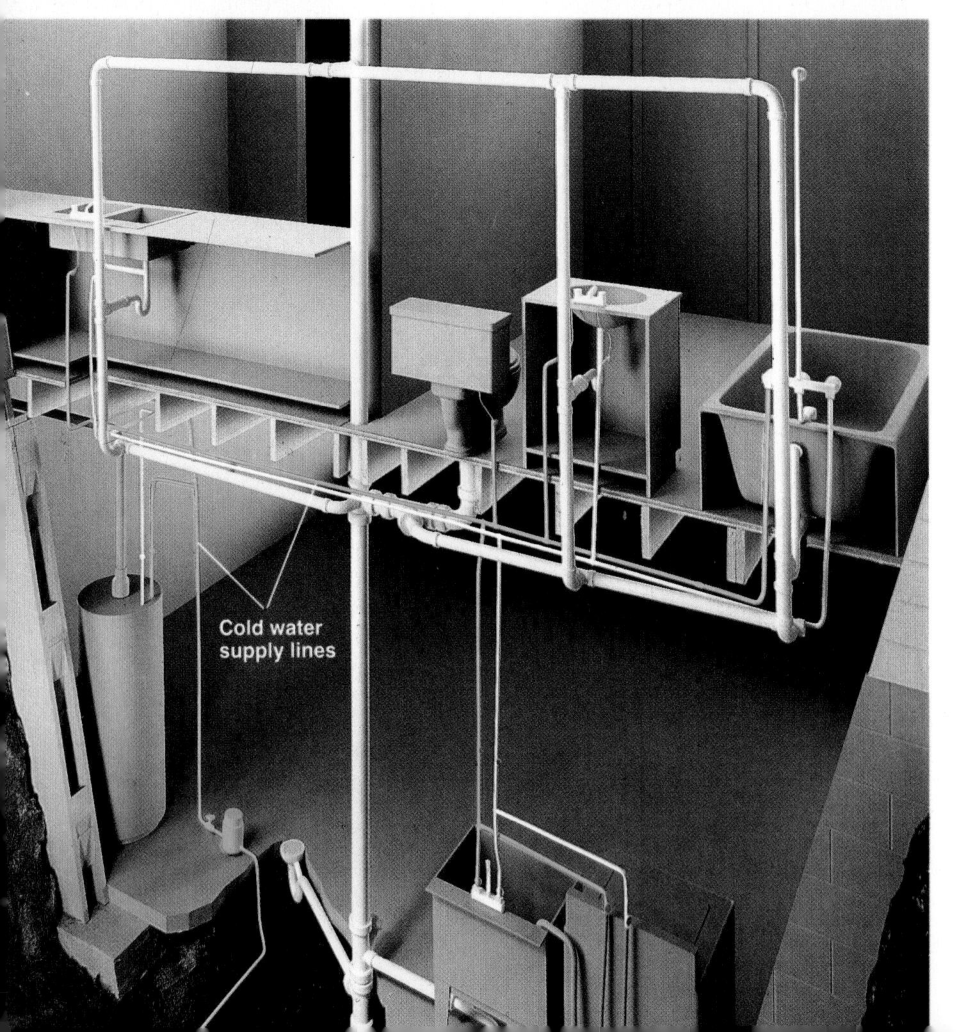

Water Supply System

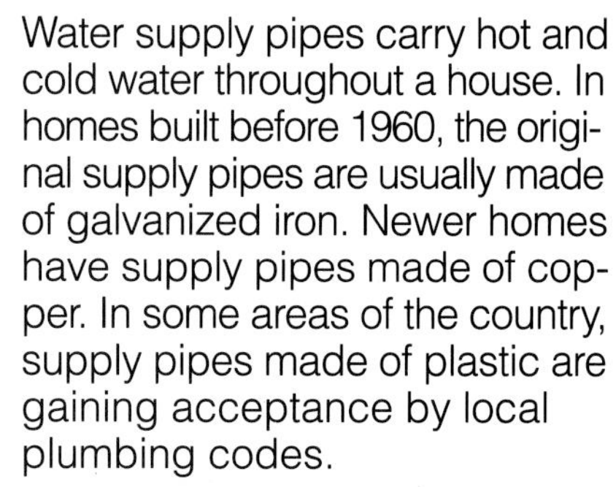

Water supply pipes carry hot and cold water throughout a house. In homes built before 1960, the original supply pipes are usually made of galvanized iron. Newer homes have supply pipes made of copper. In some areas of the country, supply pipes made of plastic are gaining acceptance by local plumbing codes.

Water supply pipes are made to withstand the high pressures of the water supply system. They have small diameters, usually ½" to 1", and are joined with strong, watertight fittings. The hot and cold lines run in tandem to all parts of the house. Usually, the supply pipes run inside wall cavities or are strapped to the undersides of floor joists.

Hot and cold water supply pipes are connected to fixtures or appliances. Fixtures include sinks, tubs, and showers. Some fixtures, such as toilets or hose bibs, are supplied only by cold water. Appliances include dishwashers and clothes washers. A refrigerator icemaker uses only cold water. Tradition says that hot water supply pipes and faucet handles are found on the left-hand side of a fixture. Cold water is on the right.

Because of high water pressure, leaks are the most common problems for the water supply system. This is especially true of galvanized iron pipe, which has limited resistance to corrosion.

Drain, Waste, Vent System

Drain pipes use gravity to carry waste water away from fixtures, appliances, and other drains. This waste water is carried out of the house to a municipal sewer system or septic tank.

Drain pipes are usually plastic or cast iron. In some older homes, drain pipes may be made of copper or lead. Because they are not part of the supply system, lead drain pipes pose no health hazard. However, lead pipes are no longer manufactured for home plumbing systems.

Drain pipes have diameters ranging from 1¼" to 4" . These large diameters allow waste water to pass easily.

Traps are an important part of the drain system. These curved sections of drain pipe hold standing water, and they are usually found near any drain opening. The standing water of a trap prevents sewer gases from backing up into the home. Each time a drain is used, the standing trap water is flushed away and is replaced by new water.

In order to work properly, the drain system requires air. Air allows waste water to flow freely down drain pipes.

To allow air into the drain system, drain pipes are connected to vent pipes. All drain systems must include vents, and the entire system is called the drain, waste, vent (DWV) system. One or more vent stacks, located on the roof, provide the air needed for the DWV system to work.

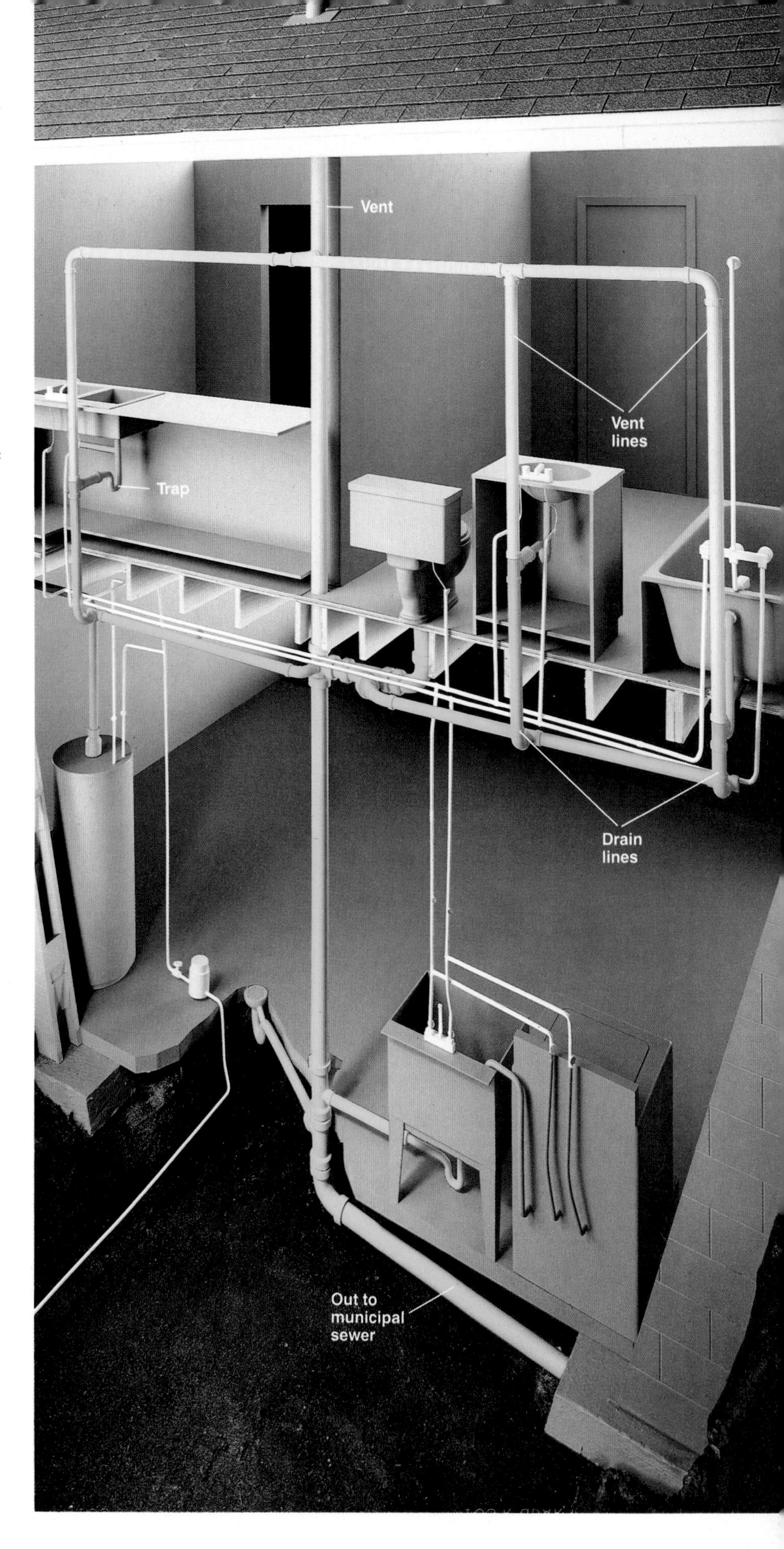

Planning Your Project

Mapping Your Plumbing System

Find supply lines inside walls or beneath floors by listening for running water with a stethoscope or a drinking glass while a helper runs water from the fixture. Use fixture locations on the floors above and below to find the general location of pipes.

Mapping your home's plumbing system is a good way to familiarize yourself with the plumbing layout and can help you when planning plumbing renovation projects. With a good map, you can envision the best spots for new fixtures and plan new pipe routes more efficiently. Maps also help in emergencies, when you need to locate burst or leaking pipes quickly.

Draw a plumbing map for each floor on tracing paper, so you can overlay floors and still read the information below. Make your drawings to scale and have all plumbing fixtures marked. Fixture templates and tracing paper are available at drafting supply stores.

Tips for Mapping Your Plumbing System

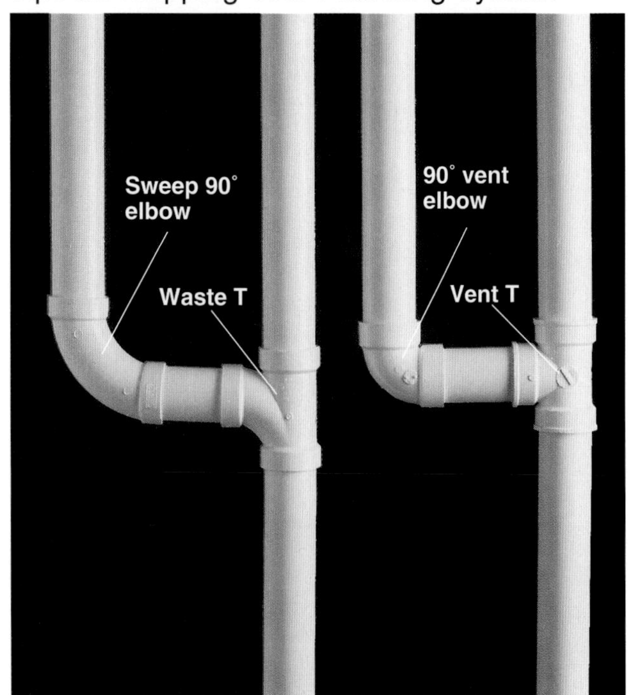

Identify drains and vents by the shape of their fittings. Drain pipes (left) require gradual changes in direction, requiring the use of Y-fittings, waste T-fittings, and sweep 90° elbows. Vents (right) can use fittings with abrupt changes in direction, such as vent T-fittings and vent elbows.

Pinpoint the location of the main stack when all interior walls are finished by jiggling a hand auger down the roof vent while a helper listens to walls from inside the house. Always exercise caution when working on a roof.

Use floor plans of your house to create your plumbing map. Convert the general outlines for each story to tracing paper. The walls can be drawn larger than scale to fit all the plumbing symbols you will map, but keep overall room dimensions and plumbing fixtures to scale. Be sure to make diagrams for basements and attic spaces as well.

Locate and map all valves throughout the supply lines. This will allow you to shut off only the necessary branches when making repairs, while maintaining service to the rest of the house. Use the correct symbols (right) to identify different valve types (page 40).

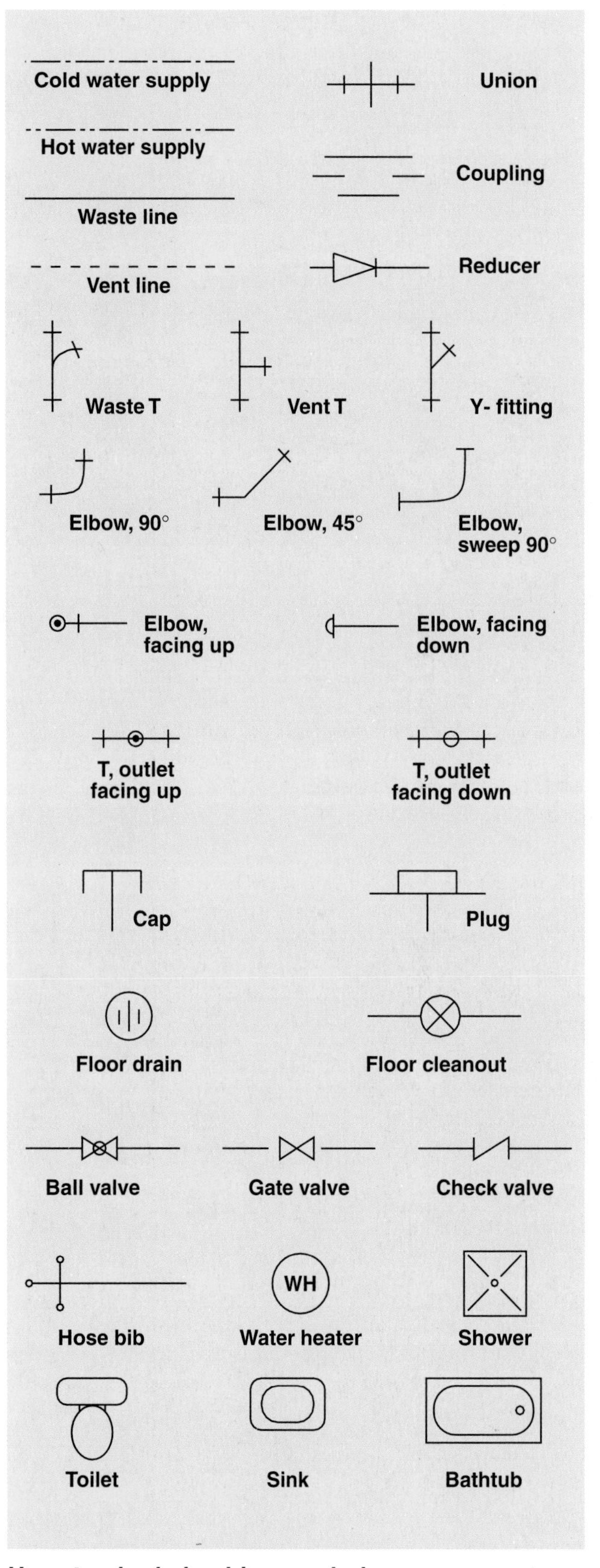

Use standard plumbing symbols on your map to identify the components of your plumbing system. These symbols will help you and your building inspector follow connections and transitions more easily.

How to Map Water Supply Pipes

1 Locate the water meter, usually found along a basement wall. The meter is the first main fitting in the supply line. Mark its location on your basement diagram.

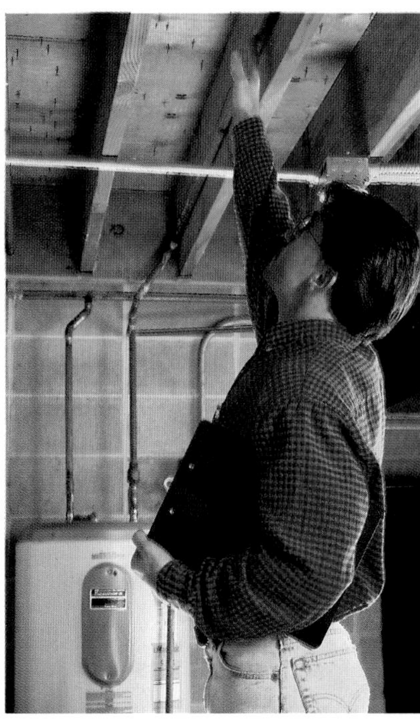

2 Follow the cold water distribution pipe past the main shutoff valve to the water heater, generally the first appliance to receive water. Map the valve and water heater locations on the basement diagram.

3 Locate cold water branch pipes leading to sillcocks, which supply water to hose bibs outside the house. Indicate these branches on the basement map.

4 Return to the water heater and map the location of hot and cold supply lines running to basement utility fixtures, such as a washing machine and utility sink.

5 Map the routes to any remaining basement plumbing fixtures on your basement diagram. Pipe runs that serve both basement and first-floor plumbing should be marked on both the basement and first-story maps.

6 Find where vertical supply pipe risers extend up into the floors above. Supply lines generally rise straight to the first-floor fixtures. If there is doubt, measure from the nearest outside wall to the supply line riser, and do the same at the respective first-floor fixture. If measurements are not the same, there is a hidden offset in the pipe route.

7 Map the supply routes to all first-story fixtures by laying the first-floor diagram over the basement map and transferring the locations of vertical supply risers. Indicate any jogs in the supply lines occurring between floors.

8 Overlay the second-story diagram over the first-floor map, and mark the location of supply pipes—generally they will extend directly up from fixtures below. If first-story and second-story fixtures are not closely aligned, the supply pipes follow an offset route in wall or floor cavities. By overlaying the maps, you can see the relation and distance between fixtures and accurately estimate pipe routes. If no obvious path exists for supply pipes, try locating the pipes by listening in likely areas with a stethoscope (page 18).

How to Map DWV Pipes

1 From the basement, locate the main waste-vent stack and any fixtures that drain directly into it, such as a basement toilet.

2 Determine the path of the main drain under the basement slab by following the main stack to the cleanout hub on the basement floor. The cleanout is usually located near a basement wall facing the street.

3 Note any auxiliary waste-vent stacks that enter the basement floor. These are typically 2"-diameter pipes, compared to 3"- or 4"-diameter main waste-vent stacks. Auxiliary waste-vent stacks are often located near basement utility sinks or below a kitchen located far from the main stack.

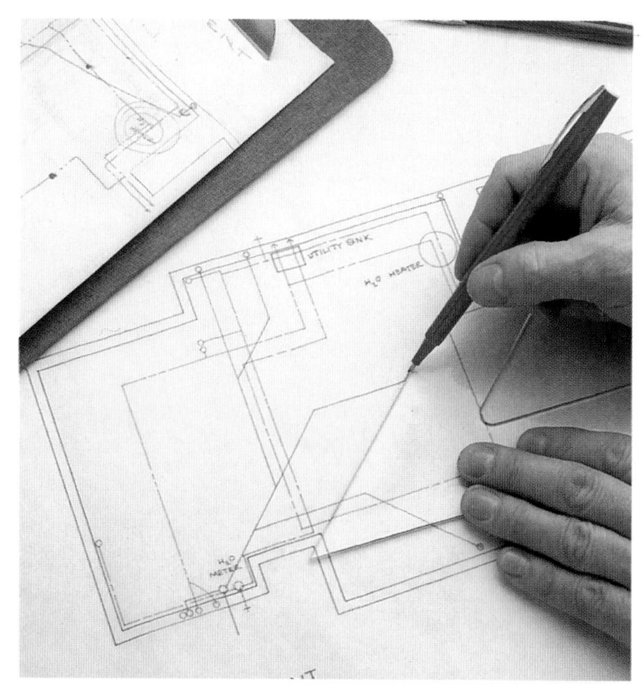

4 On your basement diagram, map the location of the main waste-vent stack, the cleanout hub, and the horizontal main drain pipe. Also note the location of auxiliary stacks, and estimate the path of the horizontal drain pipe connecting the auxiliary stacks to the main drain.

5 From the basement, note the location of horizontal drain pipes running overhead, and the points where vertical drain pipes extend up into the floors above. Overlay your first-story diagram onto the basement map, then transfer the location of the vertical waste-vent stacks. Mark the location of all horizontal drain pipes running below the floor.

6 Overlay the second-story diagram over the basement map, transfer the location of the vertical waste-vent stacks, and mark the location of any horizontal drain pipes running beneath the floor. Since the floor spaces between the first and second story are usually finished, you may need to estimate their location. These horizontal drain pipes will usually drain into the nearest waste-vent stack.

7 Finally, map the location of vent pipes as accurately as you can. If possible, look in your attic to determine where vent pipes emerge from the story below. Indicate whether the individual vent pipes connect to a waste-vent stack or extend through the roofline.

Understanding Plumbing Codes

The Plumbing Code is the set of regulations that building officials and inspectors use to evaluate your project plans and the quality of your work. Codes vary from region to region, but most are based on the National Uniform Plumbing Code, the authority we used in the development of this book.

Code books are available for reference at bookstores and government offices. However, they are highly technical, difficult-to-read manuals. More user-friendly for do-it-yourselfers are the variety of Code handbooks available at bookstores and libraries. These handbooks are based on the National Uniform Plumbing Code, but are easier to read and include many helpful diagrams and photos.

Plumbing Code handbooks sometimes discuss three different plumbing "zones" in an effort to accommodate variations in regulations from state to state. The states included in each zone are listed below.

Zone 1: Washington, Oregon, California, Nevada, Idaho, Montana, Wyoming, North Dakota, South Dakota, Minnesota, Iowa, Nebraska, Kansas, Utah, Arizona, Colorado, New Mexico, Indiana, parts of Texas.

Zone 2: Alabama, Arkansas, Louisiana, Tennessee, North Carolina, Mississippi, Georgia, Florida, South Carolina, parts of Texas, parts of Maryland, parts of Delaware, parts of Oklahoma, parts of West Virginia.

Zone 3: Virginia, Kentucky, Missouri, Illinois, Michigan, Ohio, Pennsylvania, New York, Connecticut, Massachusetts, Vermont, New Hampshire, Rhode Island, New Jersey, parts of Delaware, parts of West Virginia, parts of Maine, parts of Maryland, parts of Oklahoma.

Remember that your local Plumbing Code always supersedes the national Code. On some issues, the local Code may be less demanding than the national Code, but on other issues it may be more restrictive. Your local building inspector is a valuable source of information and may provide you with a convenient summary sheet of the regulations that apply to your project.

The plumbing inspector is the final authority when it comes to evaluating your work. By visually examining and testing your new plumbing, the inspector ensures that your work is safe and functional.

Getting a Permit

To ensure public safety, your community requires that you obtain a permit for most plumbing projects, including all the projects demonstrated in this book.

When you visit your city Building Inspection office to apply for a permit, the building official will want to review three drawings of your plumbing project: a site plan, a water supply diagram, and a drain-waste-vent diagram. These drawings are described on this page. If the official is satisfied that your project meets Code requirements, he will issue you a plumbing permit, which is your legal permission to begin work. The building official will also specify an inspection schedule for your project. As your project nears completion, you will be asked to arrange for an inspector to visit your home while the pipes are exposed and review the installation to ensure its safety.

Although do-it-yourselfers often complete complex plumbing projects without obtaining a permit or having the work inspected, we strongly urge you to comply with the legal requirements in your area. A flawed plumbing system can be dangerous, and it can potentially threaten the value of your home.

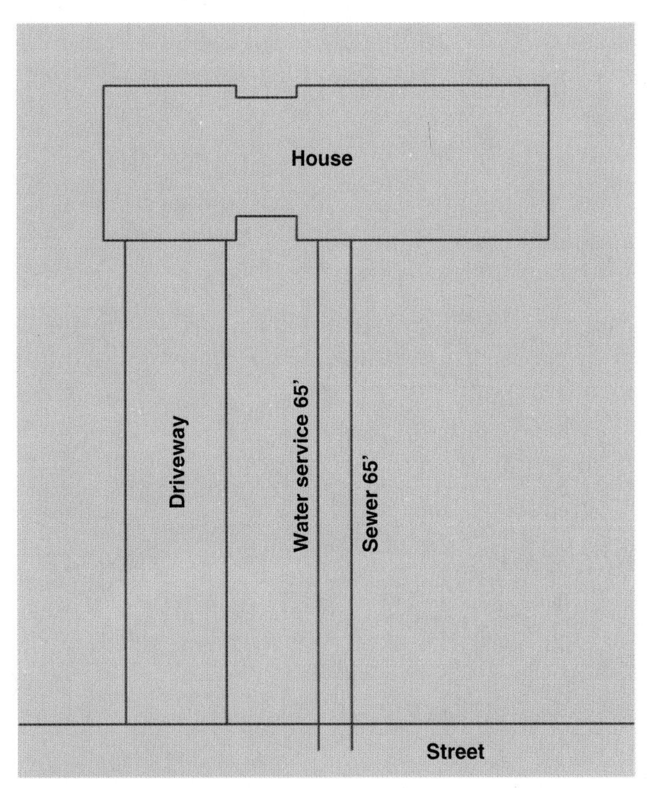

The site plan shows the location of the water main and sewer main with respect to your yard and home. The distances from your foundation to the water main and from the foundation to the main sewer should be indicated on the site plan.

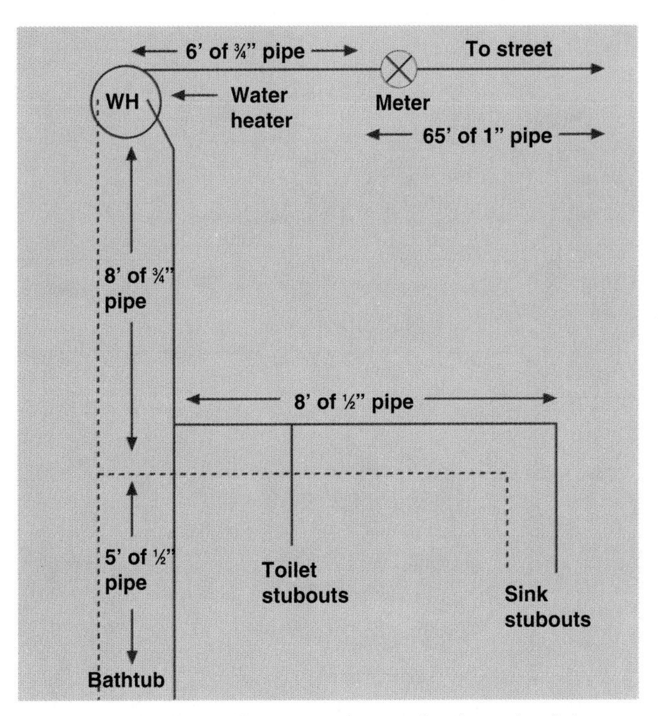

The supply riser diagram shows the length of the hot and cold water pipes and the relation of the fixtures to one another. The inspector will use this diagram to determine the proper size for the new water supply pipes in your system.

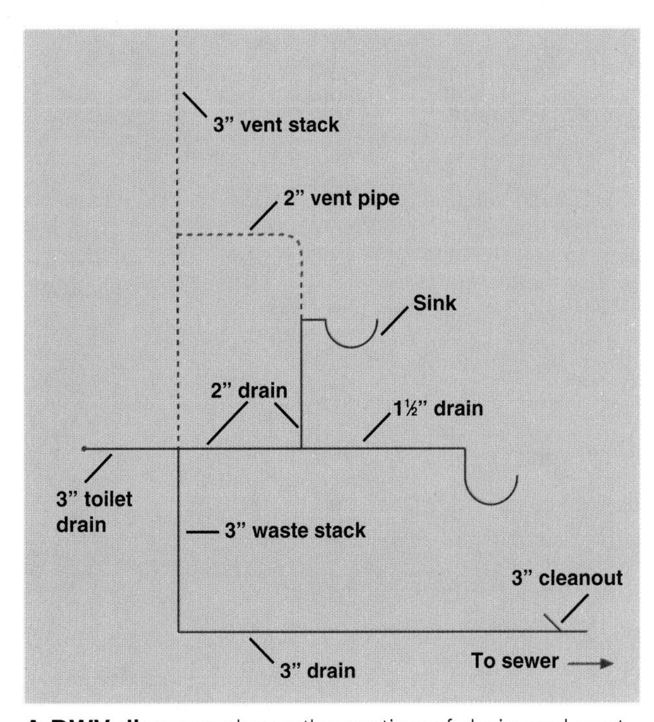

A DWV diagram shows the routing of drain and vent pipes in your system. Make sure to indicate the lengths of drain pipes and the distances between fixtures. The inspector will use this diagram to determine if you have properly sized the drain traps, drain pipes, and vent pipes in your project.

Sizing for Water Distribution Pipes

Fixture	Unit rating
Toilet	3
Vanity sink	1
Shower	2
Bathtub	2
Dishwasher	2
Kitchen sink	2
Clothes washer	2
Utility sink	2
Sillcock	3

Size of service pipe from street	Size of distribution pipe from water meter	Maximum length (ft.)—total fixture units					
		40	60	80	100	150	200
¾"	½"	9	8	7	6	5	4
¾"	¾"	27	23	19	17	14	11
¾"	1"	44	40	36	33	28	23
1"	1"	60	47	41	36	30	25
1"	1¼"	102	87	76	67	52	44

Water distribution pipes are the main pipes extending from the water meter throughout the house, supplying water to the branch pipes leading to individual fixtures. To determine the size of the distribution pipes, you must first calculate the total demand in "fixture units" (above, left) and the overall length of the water supply lines, from the street hookup through the water meter and to the most distant fixture in the house. Then, use the second table (above, right) to calculate the minimum size for the water distribution pipes. Note that the fixture unit capacity depends partly on the size of the street-side pipe that delivers water to your meter.

Sizes for Branch Pipes & Supply Tubes

Fixture	Min. branch pipe size	Min. supply tube size
Toilet	½"	⅜"
Vanity sink	½"	⅜"
Shower	½"	½"
Bathtub	½"	½"
Dishwasher	½"	½"
Kitchen sink	½"	½"
Clothes washer	½"	½"
Utility sink	½"	½"
Sillcock	¾"	N.A.
Water heater	¾"	N.A.

Branch pipes are the water supply lines that run from the distribution pipes toward the individual fixtures. **Supply tubes** are the vinyl, chromed copper, or mesh tubes that carry water from the branch pipes to the fixtures. Use the chart above as a guide when sizing branch pipes and supply tubes.

Valve Requirements

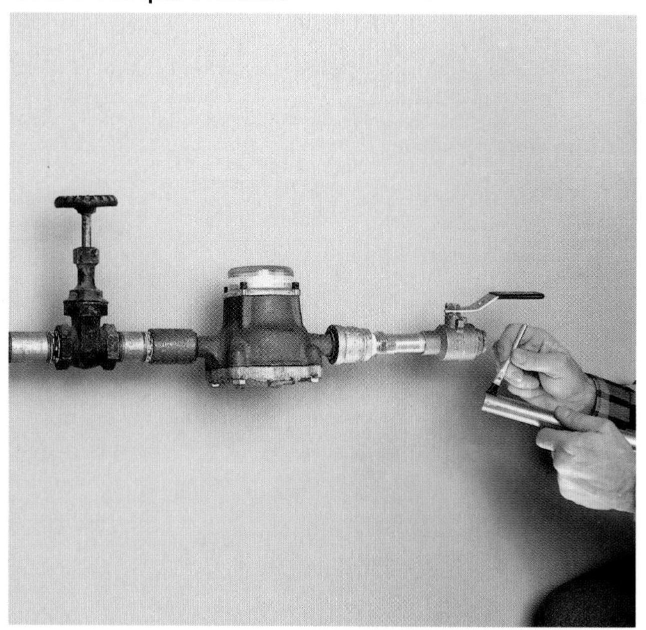

Full-bore gate valves or ball valves are required in the following locations: on both the street side and house side of the water meter; on the inlet pipes for water heaters and heating system boilers. Individual fixtures should have accessible shutoff valves, but these need not be full-bore valves. All sillcocks must have individual control valves located inside the house.

Modifying Water Pressure

Pressure-reducing valve (shown above) is required if the water pressure coming into your home is greater than 80 pounds per square inch (psi). The reducing valve should be installed near the point where the water service enters the building. A **booster pump** may be required if the water pressure in your home is below 40 psi.

Preventing Water Hammer

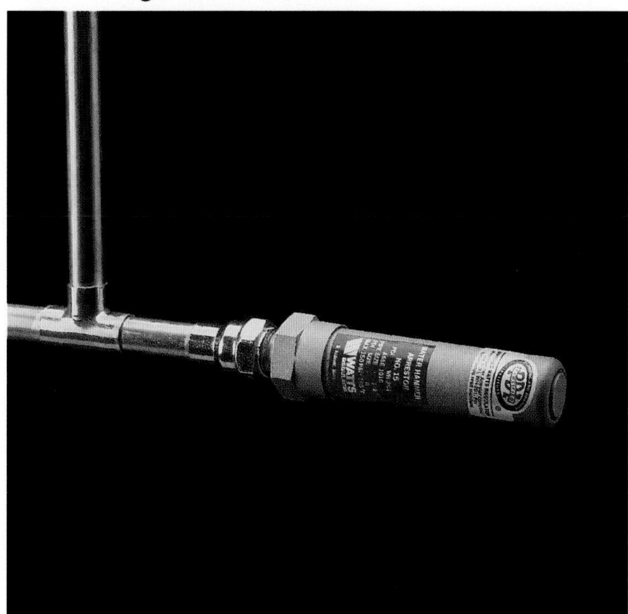

Water hammer arresters may be required by Code. Water hammer is a problem that may occur when the fast-acting valves on washing machines or other appliances cause pipes to vibrate against framing members. The arrester works as a shock absorber, with a watertight diaphragm inside. It is mounted to a T-fitting installed near the appliance.

Anti-siphon Devices

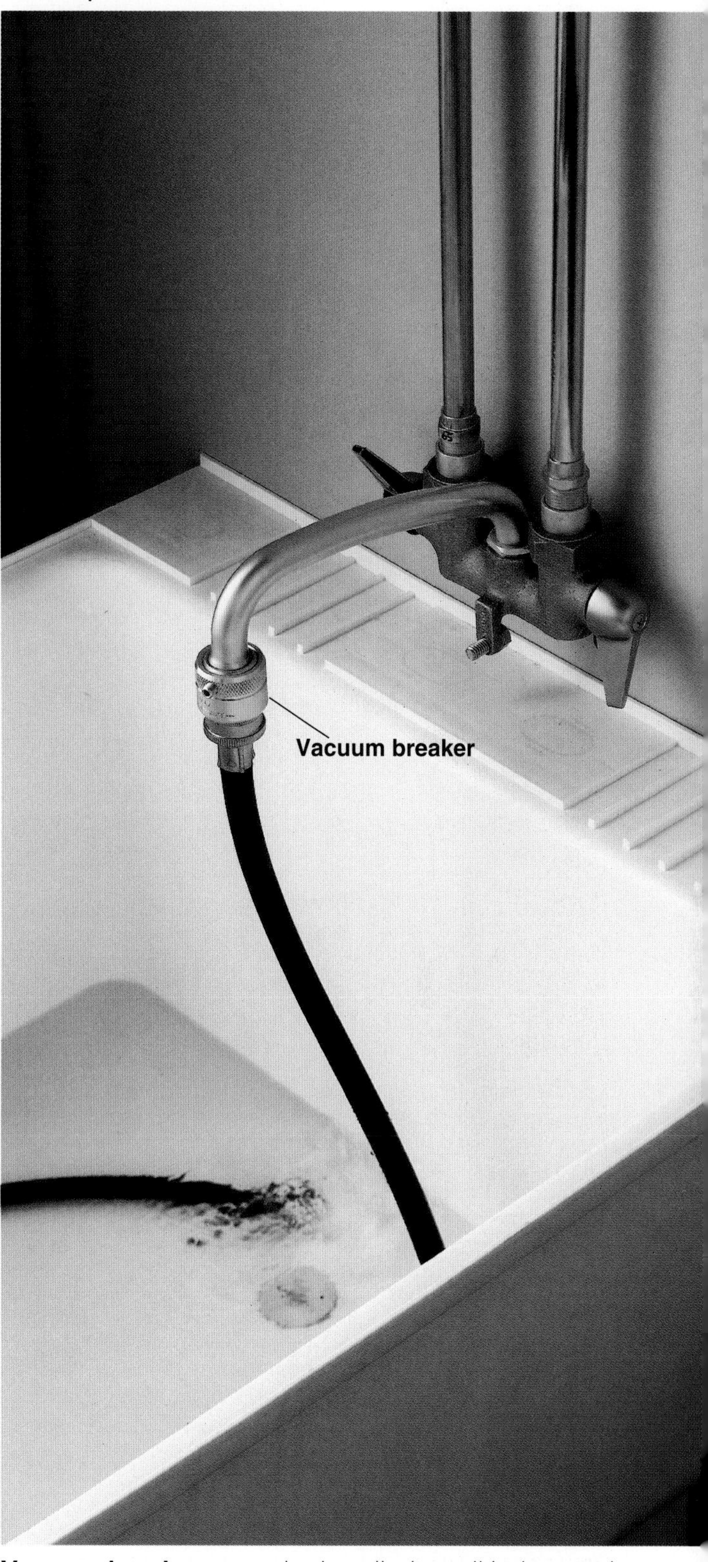

Vacuum breakers must be installed on all indoor and outdoor hose bibs and any outdoor branch pipes that run underground (page 145, step 7). Vacuum breakers prevent contaminated water from being drawn into the water supply pipes in the event of a sudden drop in water pressure in the water main. When a drop in pressure produces a partial vacuum, the breaker prevents siphoning by allowing air to enter the pipes.

Drain cleanouts make your DWV system easier to service. In most areas, the Plumbing Code requires that you place cleanouts at the end of every horizontal drain run. Where horizontal runs are not accessible, removable drain traps will suffice as cleanouts.

Pipe Support Intervals

Type of pipe	Vertical support interval	Horizontal support interval
Copper	6 ft.	10 ft.
ABS	4 ft.	4 ft.
CPVC	3 ft.	3 ft.
PVC	4 ft.	4 ft.
Steel	12 ft.	15 ft.
Iron	5 ft.	15 ft.

Minimum intervals for supporting pipes are determined by the type of pipe and its orientation in the system. See page 40 for acceptable pipe support materials. Remember that the measurements shown above are minimum requirements; many plumbers install pipe supports at closer intervals.

Fixture Units & Minimum Trap Size

Fixture	Fixture units	Min. trap size
Shower	2	2"
Vanity sink	1	1¼"
Bathtub	2	1½"
Dishwasher	2	1½"
Kitchen sink	2	1½"
Kitchen sink*	3	1½"
Clothes washer	2	1½"
Utility sink	2	1½"
Floor drain	1	2"

*Kitchen sink with attached food disposer

Minimum trap size for fixtures is determined by the drain fixture unit rating, a unit of measure assigned by the Plumbing Code. NOTE: Kitchen sinks rate 3 units if they include an attached food disposer, 2 units otherwise.

Sizes for Horizontal & Vertical Drain Pipes

Pipe size	Maximum fixture units for horizontal branch drain	Maximum fixture units for vertical drain stacks
1¼"	1	2
1½"	3	4
2"	6	10
2½"	12	20
3"	20	30
4"	160	240

Drain pipe sizes are determined by the load on the pipes, as measured by the total fixture units. Horizontal drain pipes less than 3" in diameter should slope ¼" per foot toward the main drain. Pipes 3" or more in diameter should slope ⅛" per foot. NOTE: Horizontal or vertical drain pipes for a toilet must be 3" or larger.

Vent Pipe Sizes, Critical Distances

Size of fixture drain	Minimum vent pipe size	Maximum trap-to-vent distance
1¼"	1¼"	2½ ft.
1½"	1¼"	3½ ft.
2"	1½"	5 ft.
3"	1½"	6 ft.
4"	3"	10 ft.

Vent pipes are usually one pipe size smaller than the drain pipes they serve. Code requires that the distance between the drain trap and the vent pipe fall within a maximum "critical distance," a measurement that is determined by the size of the fixture drain. Use this chart to determine both the minimum size for the vent pipe and the maximum critical distance.

Vent Pipe Orientation to Drain Pipe

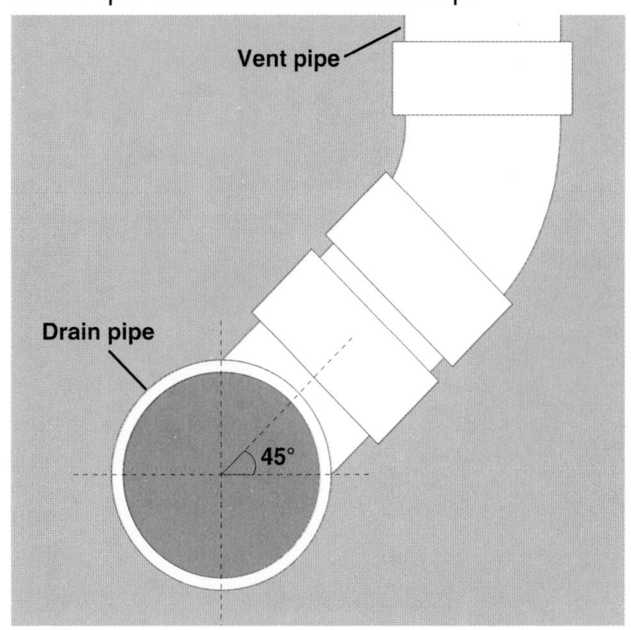

Vent pipes must extend in an upward direction from drains, no less than 45° from horizontal, This ensures that waste water cannot flow into the vent pipe and block it. At the opposite end, a new vent pipe should connect to an existing vent pipe or main waste-vent stack at a point at least 6" above the highest fixture draining into the system.

Wet Venting

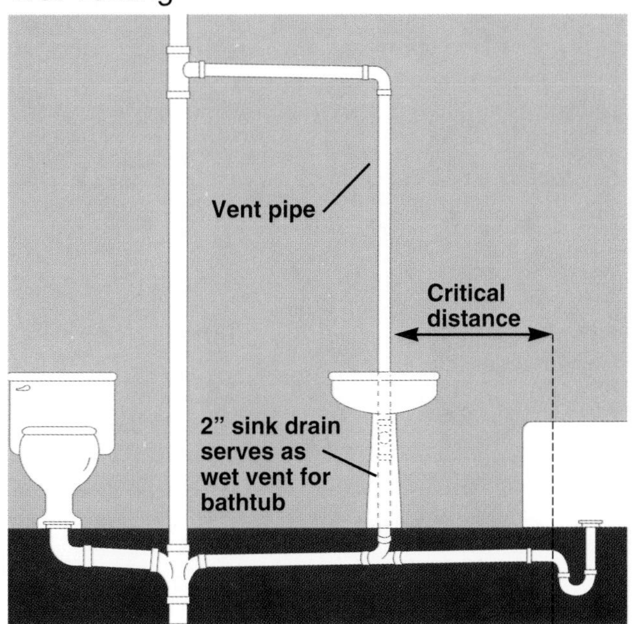

Wet vents are pipes that serve as a vent for one fixture and a drain for another. The sizing of a wet vent is based on the total fixture units it supports (opposite page): a 3" wet vent can serve up to 12 fixture units; a 2" wet vent is rated for 4 fixture units; a 1½" wet vent, for only 1 fixture unit. NOTE: The distance between the wet-vented fixture and the wet vent itself must be no more than the maximum critical distance (above, left).

Auxiliary Venting

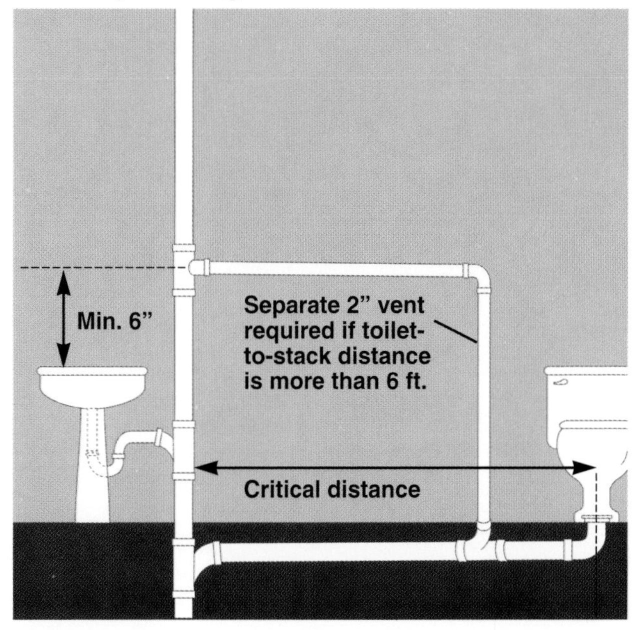

Fixtures must have auxiliary vents if the distance to the main waste-vent stack exceeds the critical distance (above, left). A toilet, for example, should have a separate vent pipe if it is located more than 6 ft. from the main waste-vent stack. This secondary vent pipe should connect to the stack or an existing vent pipe at a point at least 6" above the highest fixture on the system.

Testing New Plumbing Pipes

A **pressure gauge and air pump** are used to test DWV lines. The system is first blocked off at each fixture and at points near where the new drain and vent pipes connect to the main stack. Air is then pumped into the system to a pressure of 5 pounds per square inch (psi). To pass inspection, the system must hold this pressure for 15 minutes.

When the building inspector comes to review your new plumbing, he may require that you perform a pressure test on the DWV and water supply lines as he watches. The inspection and test should be performed after the system is completed, but before the new pipes are covered with wallboard. To ensure that the inspection goes smoothly, it is a good idea to perform your own pretest, so you can locate and repair any problems before the inspector makes his visit.

The DWV system is tested by blocking off the new drain and vent pipes, then pressuring the system with air to see if it leaks. At the fixture stub-outs, the DWV pipes can be capped off or plugged with test balloons designed for this purpose. The air pump, pressure gauge, and test balloons required to test the DWV system can be obtained at tool rental centers.

Testing the water supply lines is a simple matter of turning on the water and examining the joints for leaks. If you find a leak, you will need to drain the pipes, then resolder the faulty joints.

How to Test New DWV Pipes

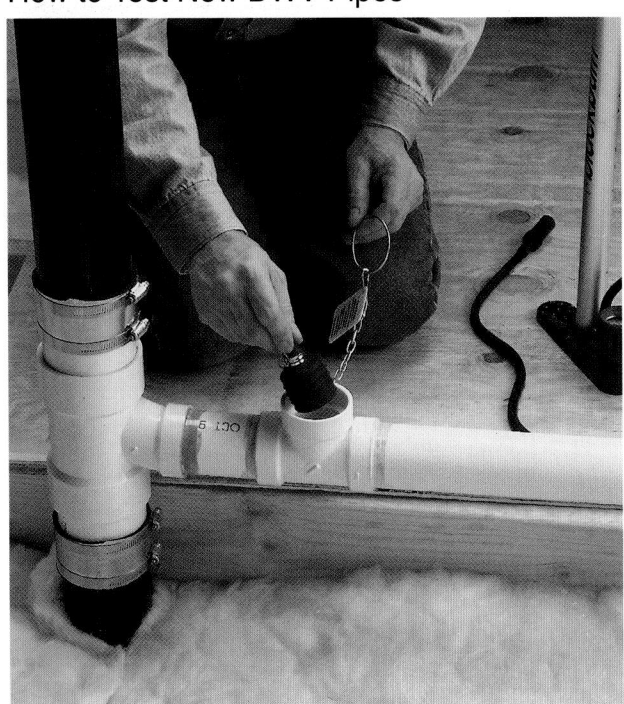

1 Insert a test balloon into the test T-fittings at the top and bottom of the new DWV line, blocking the pipes entirely. NOTE: Ordinary T-fittings installed near the bottom of the drain line and near the top of the vent line are generally used for test fittings.

2 Block toilet drains with a test balloon designed for a toilet bend. Large test balloons may need to be inflated with an air pump.

3 Cap off the remaining fixture drains by solvent-gluing test caps onto the stub-outs. After the system is tested, these caps are simply knocked loose with a hammer.

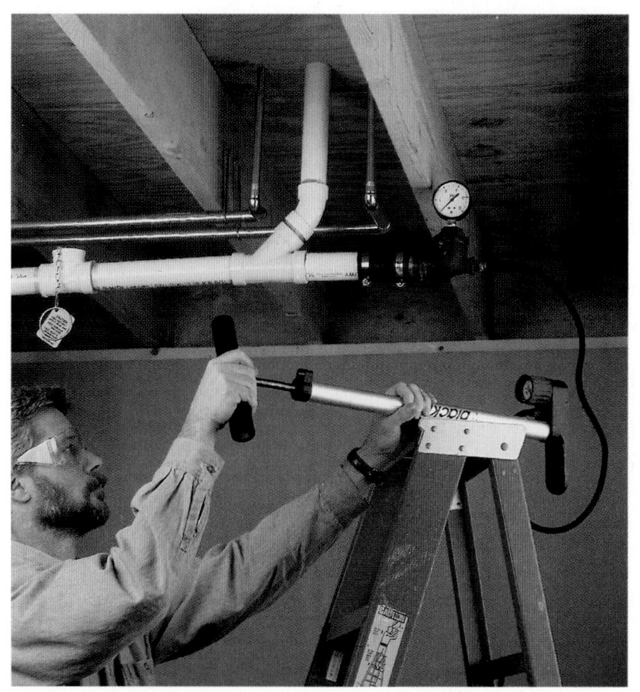

4 At a cleanout fitting, insert a *weenie*—a special test balloon with an air gauge and inflation valve. Attach an air pump to the valve on the weenie, and pressurize the pipes to 5 psi. Watch the pressure gauge for 15 minutes to ensure that the system does not lose pressure.

5 If the DWV system loses air when pressurized, check each joint for leaks by rubbing soapy water over the fittings and looking for active bubbles. When you identify a problem joint, cut away the existing fitting and solvent-glue a new fitting in place, using couplings and short lengths of pipe.

6 After the DWV system has been inspected and approved by a building official, remove the test balloons and close the test T-fittings by solvent-gluing caps onto the open inlets.

Plumbing Tools & Materials

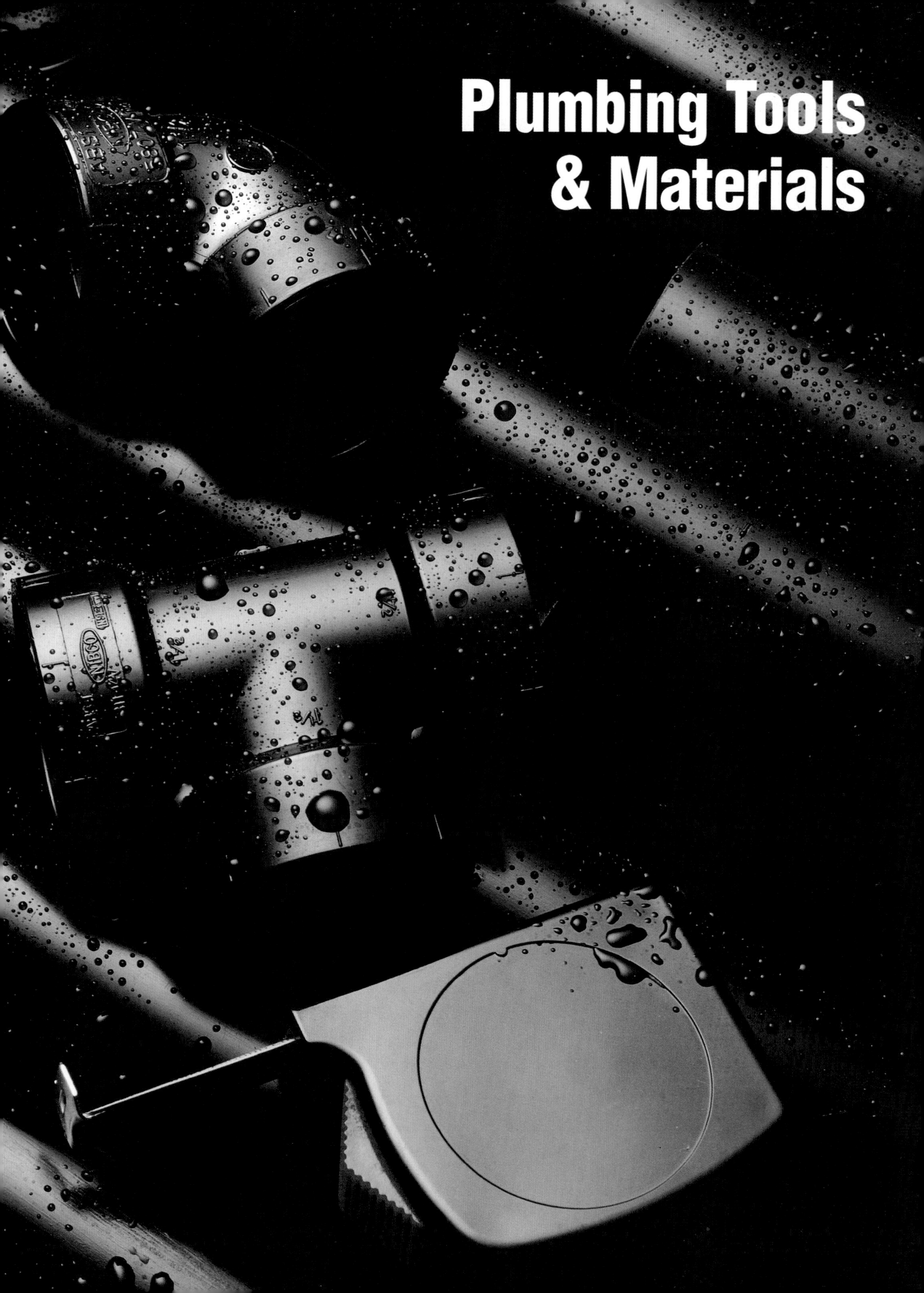

Tools for Plumbing

Many plumbing projects and repairs can be completed with basic hand tools you probably already own. Adding a few simple plumbing tools will prepare you for all the projects in this book. Specialty tools, such as a snap cutter or appliance dolly, are available at rental centers. When buying tools, invest in quality products.

Always care for tools properly. Clean tools after using them, wiping them free of dirt and dust with a soft rag. Prevent rust on metal tools by wiping them with a rag dipped in household oil. If a metal tool gets wet, dry it immediately, and then wipe it with an oiled rag. Keep tool boxes and cabinets organized. Make sure all tools are stored securely.

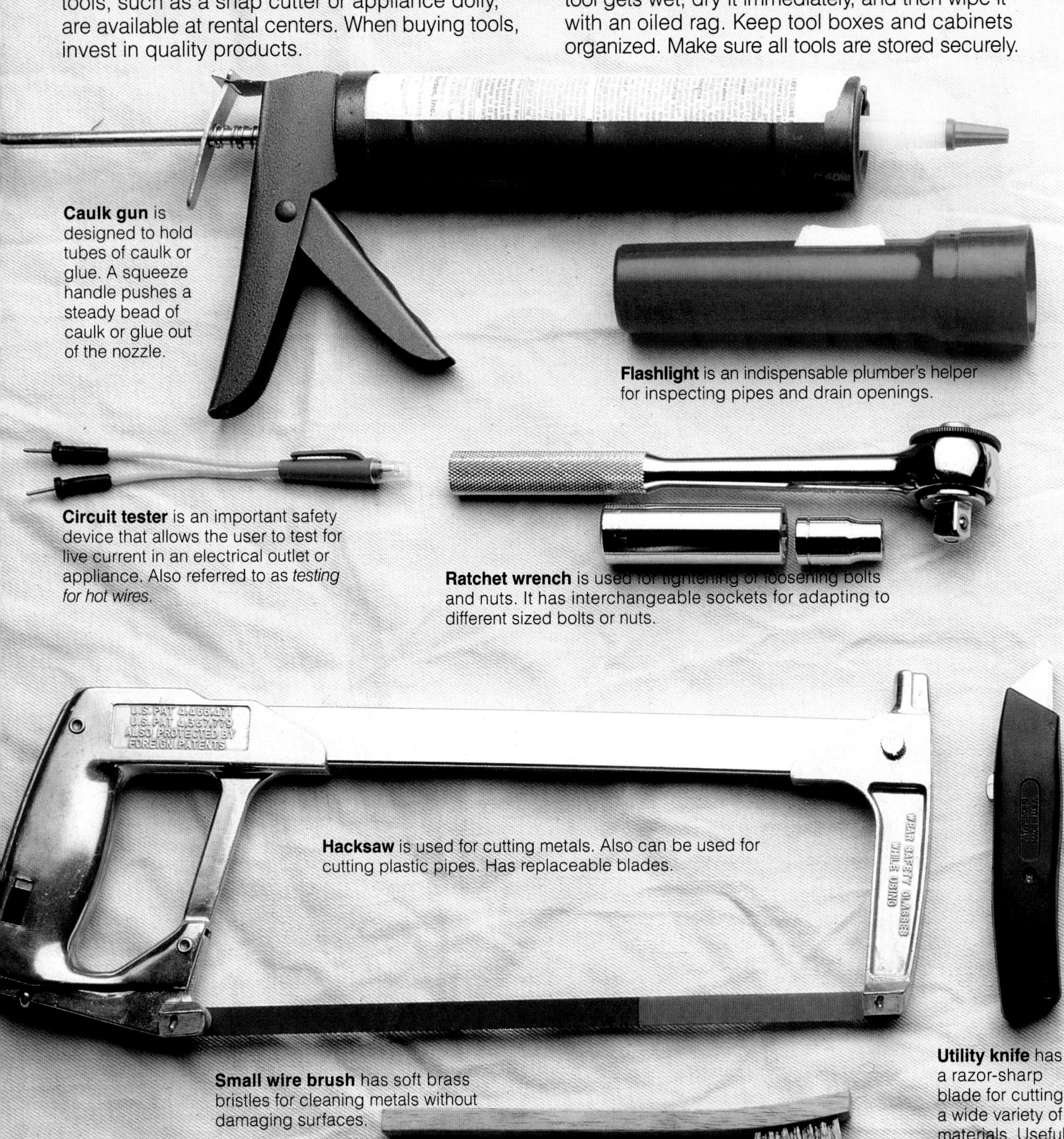

Caulk gun is designed to hold tubes of caulk or glue. A squeeze handle pushes a steady bead of caulk or glue out of the nozzle.

Flashlight is an indispensable plumber's helper for inspecting pipes and drain openings.

Circuit tester is an important safety device that allows the user to test for live current in an electrical outlet or appliance. Also referred to as *testing for hot wires.*

Ratchet wrench is used for tightening or loosening bolts and nuts. It has interchangeable sockets for adapting to different sized bolts or nuts.

Hacksaw is used for cutting metals. Also can be used for cutting plastic pipes. Has replaceable blades.

Utility knife has a razor-sharp blade for cutting a wide variety of materials. Useful for trimming ends of plastic pipes. For safety, the utility knife should have a retractable blade.

Small wire brush has soft brass bristles for cleaning metals without damaging surfaces.

Cold chisel is used with a *ball peen hammer* to cut or chip ceramic tile, mortar, or hardened metals.

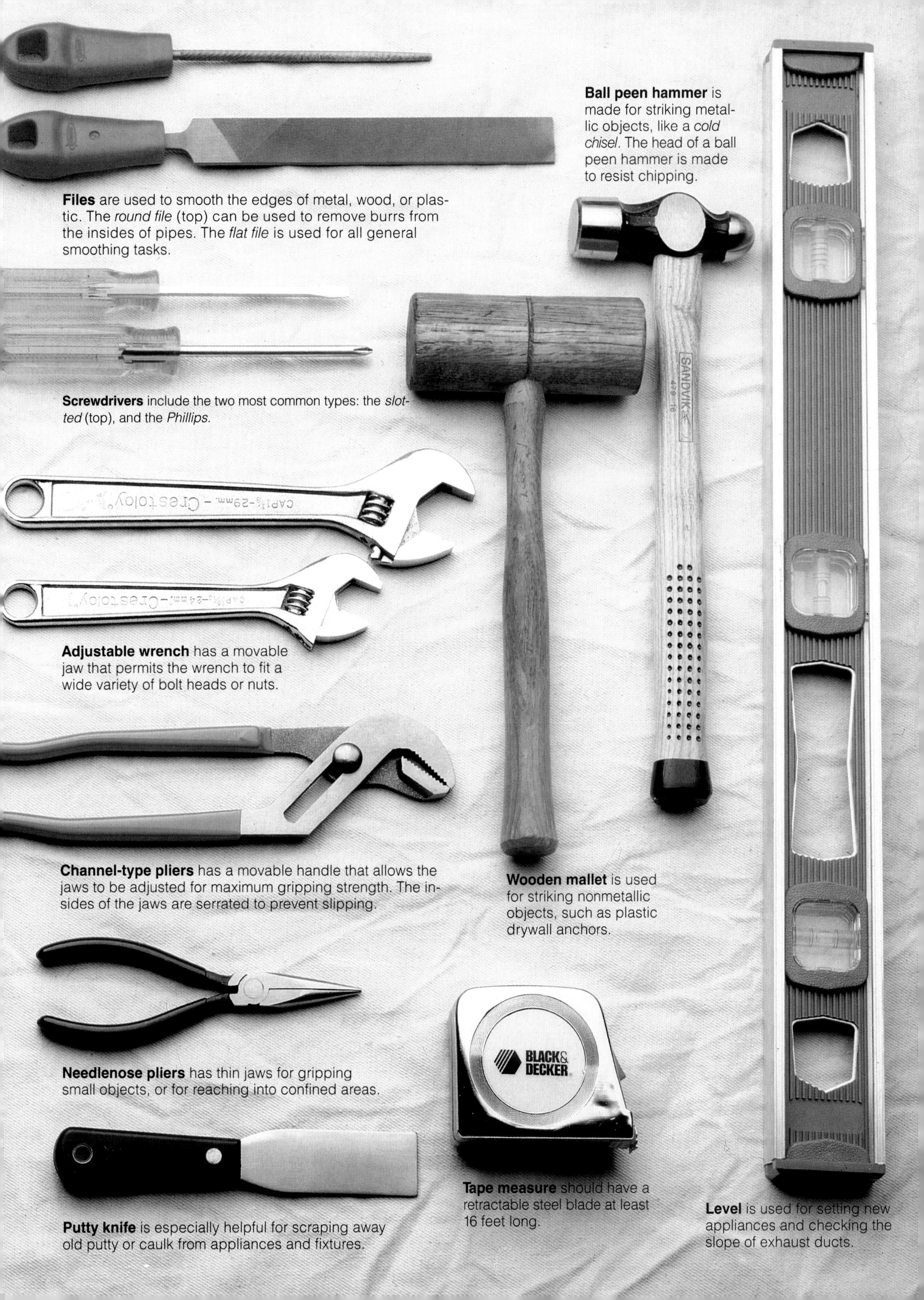

Files are used to smooth the edges of metal, wood, or plastic. The *round file* (top) can be used to remove burrs from the insides of pipes. The *flat file* is used for all general smoothing tasks.

Ball peen hammer is made for striking metallic objects, like a *cold chisel*. The head of a ball peen hammer is made to resist chipping.

Screwdrivers include the two most common types: the *slotted* (top), and the *Phillips.*

Adjustable wrench has a movable jaw that permits the wrench to fit a wide variety of bolt heads or nuts.

Channel-type pliers has a movable handle that allows the jaws to be adjusted for maximum gripping strength. The insides of the jaws are serrated to prevent slipping.

Wooden mallet is used for striking nonmetallic objects, such as plastic drywall anchors.

Needlenose pliers has thin jaws for gripping small objects, or for reaching into confined areas.

Putty knife is especially helpful for scraping away old putty or caulk from appliances and fixtures.

Tape measure should have a retractable steel blade at least 16 feet long.

Level is used for setting new appliances and checking the slope of exhaust ducts.

Tubing cutter makes straight, smooth cuts in plastic and copper pipe. A tubing cutter usually has a dull, triangular blade, called a *reaming tip*, for removing burrs from the insides of pipes.

Closet auger is used to clear toilet clogs. It is a slender tube with a crank handle on one end of a flexible auger cable. A special bend in the tube allows the auger to be positioned in the bottom of the toilet bowl. The bend is usually protected with a rubber sleeve to prevent scratching the toilet.

Plastic tubing cutter works like a gardener's pruners to cut flexible plastic (PB) pipes quickly.

Spud wrench is specially designed for removing or tightening large nuts that are 2" to 4" in diameter. Hooks on the ends of the wrench grab onto the *lugs* of large nuts for increased leverage.

Plunger clears drain clogs with water and air pressure. The *flanged plunger* (shown) is used for toilet bowls. The flange usually can be folded up into the cup for use as a *standard plunger*. Use a standard plunger to clear clogs in sink, tub, shower, and floor drains.

Blow bag, sometimes called an *expansion nozzle*, is used to clear drains. It attaches to a garden hose and removes clogs with powerful spurts of water. The blow bag is best used on floor drains.

Hand auger, sometimes called a *snake*, is used to clear clogs in drain lines. A long, flexible steel cable is stored in the disk-shaped crank. A pistol-grip handle allows the user to apply steady pressure on the cable.

Propane torch (left) is used for soldering fittings to copper pipes. Light the torch quickly and safely using a **spark lighter** (above).

Pipe wrench has a movable jaw that adjusts to fit a variety of pipe diameters. Pipe wrench is used for tightening and loosening pipes, pipe fittings, and large nuts. Two pipe wrenches often are used together to prevent damage to pipes and fittings.

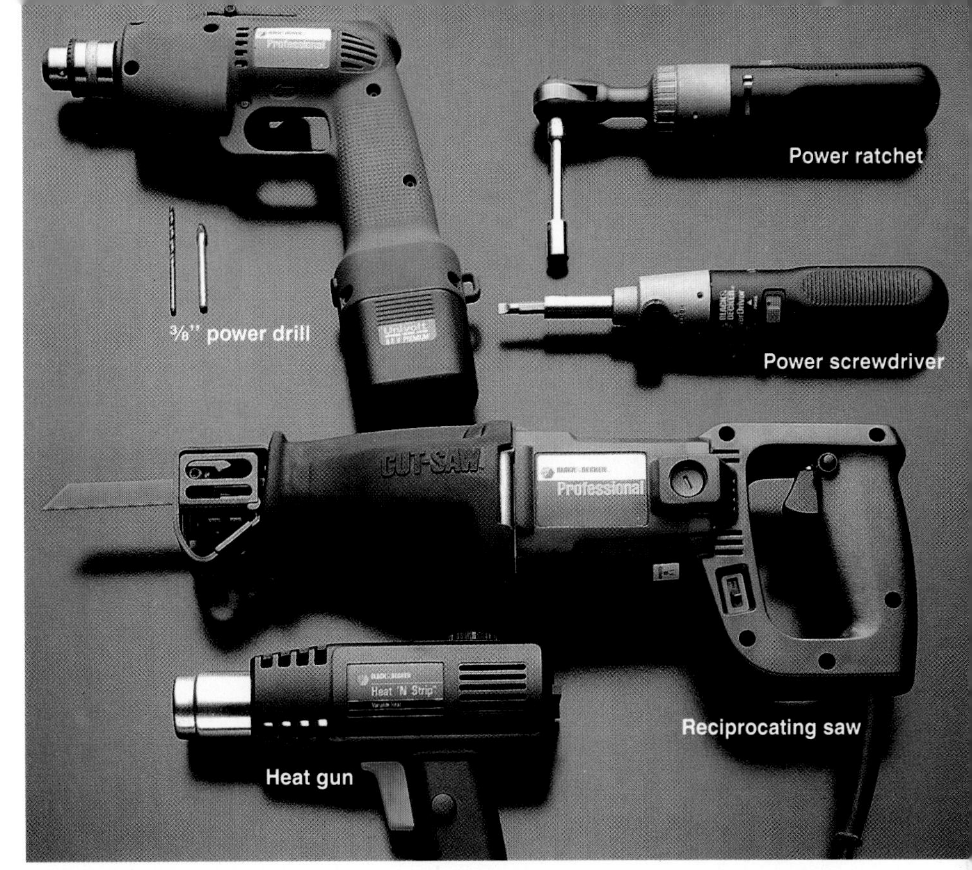

Power hand tools can make any job faster, easier, and safer. Cordless powers tools offer added convenience. Use a cordless 3/8" **power drill** for virtually any drilling task. A cordless **power rachet** makes it easy to turn small nuts or hex-head bolts. The cordless reversible **power screwdriver** drives a wide variety of screws and fasteners. A **reciprocating saw** uses interchangeable blades to cut wood, metal, or plastic. Thaw frozen pipes quickly with a **heat gun.**

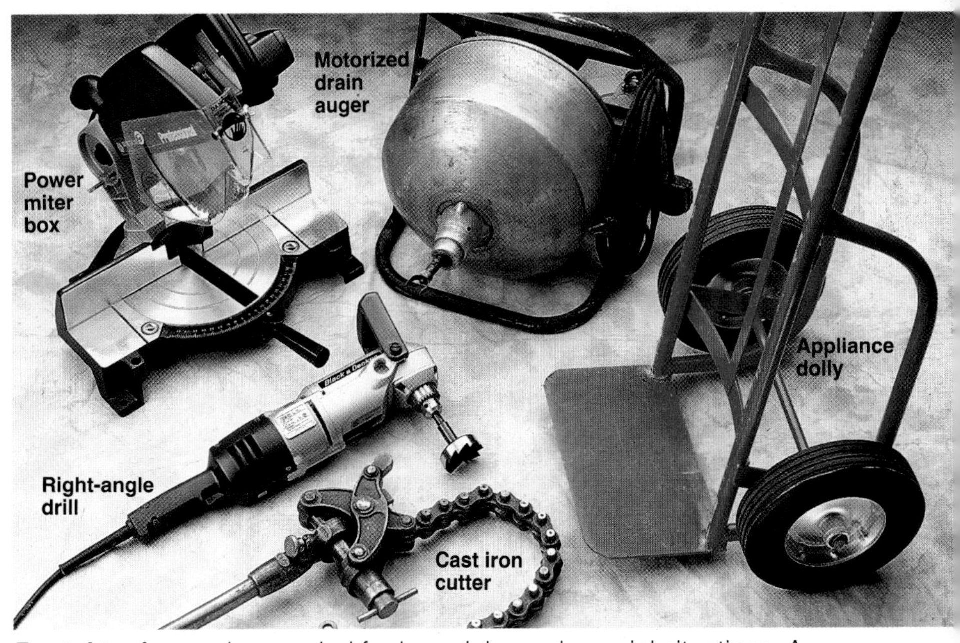

Rental tools may be needed for large jobs and special situations. A **power miter box** makes fast, accurate cuts in a wide variety of materials, including plastic pipes. A **power auger** clears tree roots from sewer service lines. Use an **appliance dolly** to move heavy objects like water heaters. A **snap cutter** is designed to cut tough cast-iron pipes. The **right-angle drill** is useful for drilling holes in hard-to-reach areas.

Plumbing Materials

Check local Plumbing Code for materials allowed in your area. All diameters specified are the interior diameters (I.D.) of pipes.

Benefits & Characteristics
Cast iron is very strong, but is difficult to cut and fit. Repairs and replacements should be made with plastic pipe, if allowed by local Code.
ABS (acrylonitrile butadiene styrene) was the first rigid plastic approved for use in home drain systems. Some local plumbing codes now restrict the use of ABS in new installations.
PVC (polyvinyl chloride) is a modern rigid plastic that is highly resistant to damage by heat or chemicals. It is the best material for drain-waste-vent pipes.
Galvanized iron is very strong, but gradually will corrode. Not advised for new installation. Because galvanized iron is difficult to cut and fit, large jobs are best left to a professional.
CPVC (chlorinated polyvinyl chloride) rigid plastic is chemically formulated to withstand the high temperatures and pressures of water supply systems. Pipes and fittings are inexpensive.
PB (polybutylene) flexible plastic is easy to fit. It bends easily around corners and requires fewer fittings than CPCV. Another type of flexible plastic is polyethylene (PE).
Rigid copper is the best material for water supply pipes. It resists corrosion, and has smooth surfaces that provide good water flow. Soldered copper joints are very durable.
Chromed copper has an attractive shiny surface, and is used in areas where appearance is important. Chromed copper is durable and easy to bend and fit.
Flexible copper tubing is easy to shape, and will withstand a slight frost without rupturing. Flexible copper bends easily around corners, so it requires fewer fittings than rigid copper.
Brass is heavy and durable. **Chromed brass** has an attractive shiny surface, and is used for drain traps where appearance is important.

Common Uses	Lengths	Diameters	Fitting Methods	Tools Used for Cutting
Main drain-waste-vent pipes	5 ft., 10 ft.	3", 4"	Joined with banded neoprene couplings	Snap cutter or hacksaw
Drain & vent pipes; drain traps	10 ft., 20 ft.; or sold by linear ft.	1¼", 1½", 2", 3", 4"	Joined with solvent glue and plastic fittings	Tubing cutter, miter box, or hacksaw
Drain & vent pipes; drain traps	10 ft., 20 ft.; or sold by linear ft.	1¼", 1½", 2", 3", 4"	Joined with solvent glue and plastic fittings	Tubing cutter, miter box, or hacksaw
Drains; hot & cold water supply pipes	1" to 1-ft. nipples; custom lengths up to 20 ft.	½", ¾", 1", 1½", 2"	Joined with galvanized threaded fittings	Hacksaw or reciprocating saw
Hot & cold water supply pipes	10 ft.	⅜", ½", ¾", 1"	Joined with solvent glue and plastic fittings, or with grip fittings	Tubing cutter, miter box, or hacksaw
Hot & cold water supply, where allowed by code	25-ft., 100-ft. coils; or sold by linear ft.	⅜", ½", ¾"	Joined with plastic grip fittings	Flexible plastic tubing cutter, sharp knife, or miter box
Hot & cold water supply pipes	10 ft., 20 ft.; or sold by linear ft.	⅜", ½", ¾", 1"	Joined with metal solder or compression fittings	Tubing cutter, hacksaw, or jig saw
Supply tubing for plumbing fixtures	12", 20", 30"	⅜"	Joined with brass compression fittings	Tubing cutter or hacksaw
Gas tubing; hot & cold water supply tubing	30-ft., 60-ft. coils; or sold by linear ft.	¼", ⅜", ½", ¾", 1"	Joined with brass flare fittings, compression fittings, or metal solder	Tubing cutter or hacksaw
Valves & shutoffs; chromed drain traps	Lengths vary	¼", ½", ¾"; *for drain traps:* 1¼", 1½"	Joined with compression fittings, or with metal solder	Tubing cutter, hacksaw, or reciprocating saw

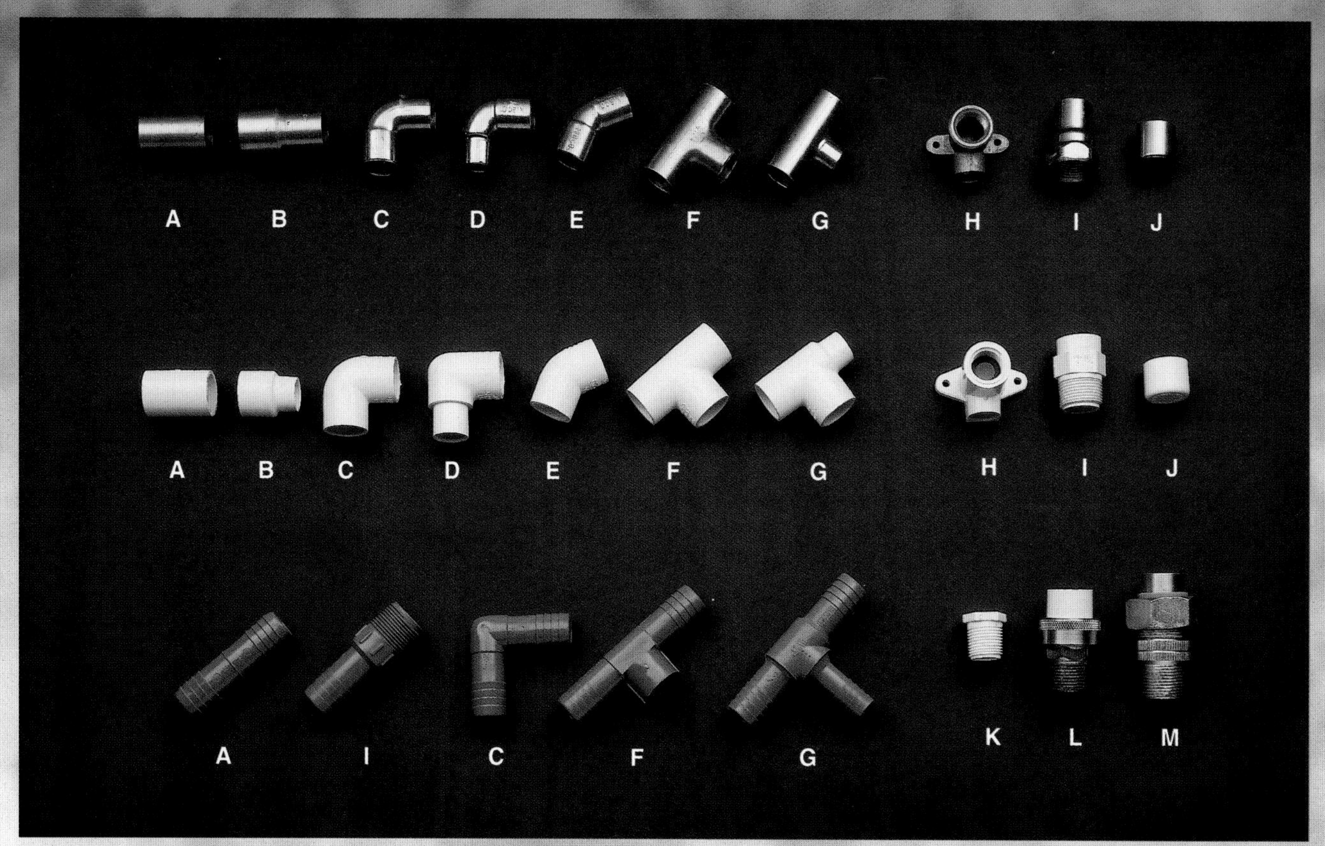

Water supply fittings are available in copper (top), CPVC plastic (center), and PVC plastic (bottom). PVC water supply fittings are gray with barbed sleeves, and are used only with cold water PE pipe. Fittings for each material are available in many shapes, including: unions (A), reducers (B), 90° elbows (C), reducing elbows (D), 45° elbows (E), T-fittings (F), reducing T-fittings (G), drop ear elbows (H), threaded adapters (I), caps (J), plug (K), CPVC to copper transition (L), and copper to steel transition (M).

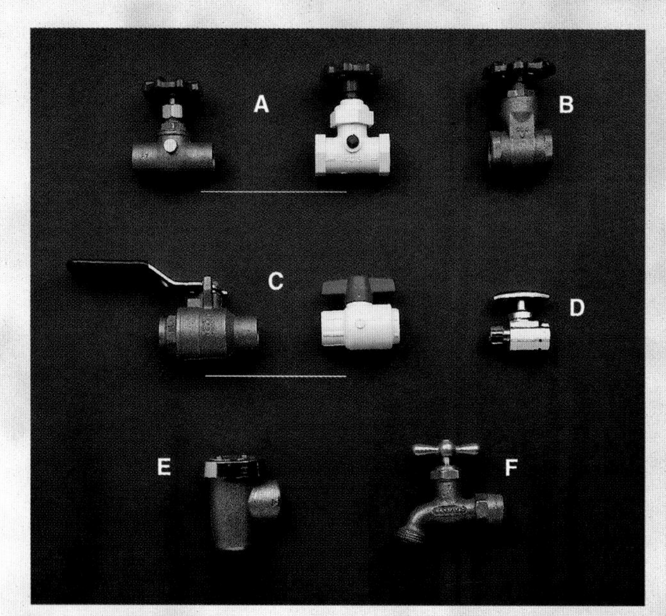

Water supply valves are available in bronze or plastic and in a variety of styles, including: drain-and-waste valves (A), gate valve (B), full-bore ball valves (C), fixture shutoff valve (D), vacuum breaker (E), and hose bib (F).

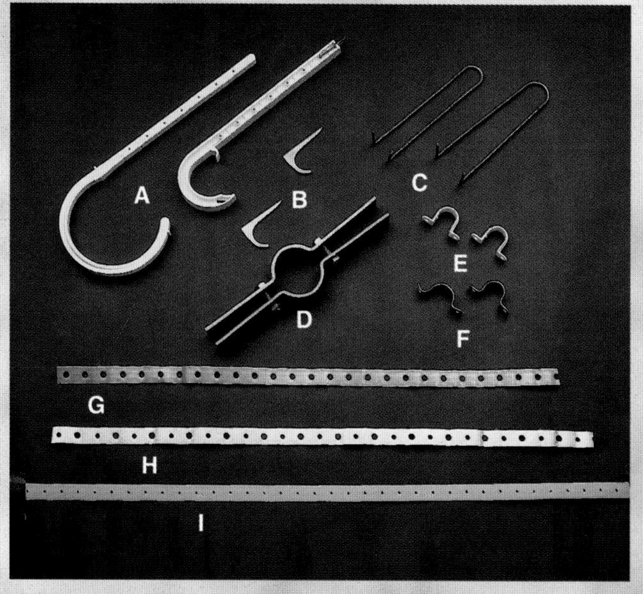

Support materials for pipes include: plastic pipe hangers (A), copper J-hooks (B), copper wire hangers (C), riser clamp (D), copper pipe straps (E), plastic pipe straps (F), flexible copper, steel, and plastic pipe strapping (G, H, I). Do not mix metal types when supporting metal pipes: use copper support materials for copper pipe, steel for steel and cast-iron pipes.

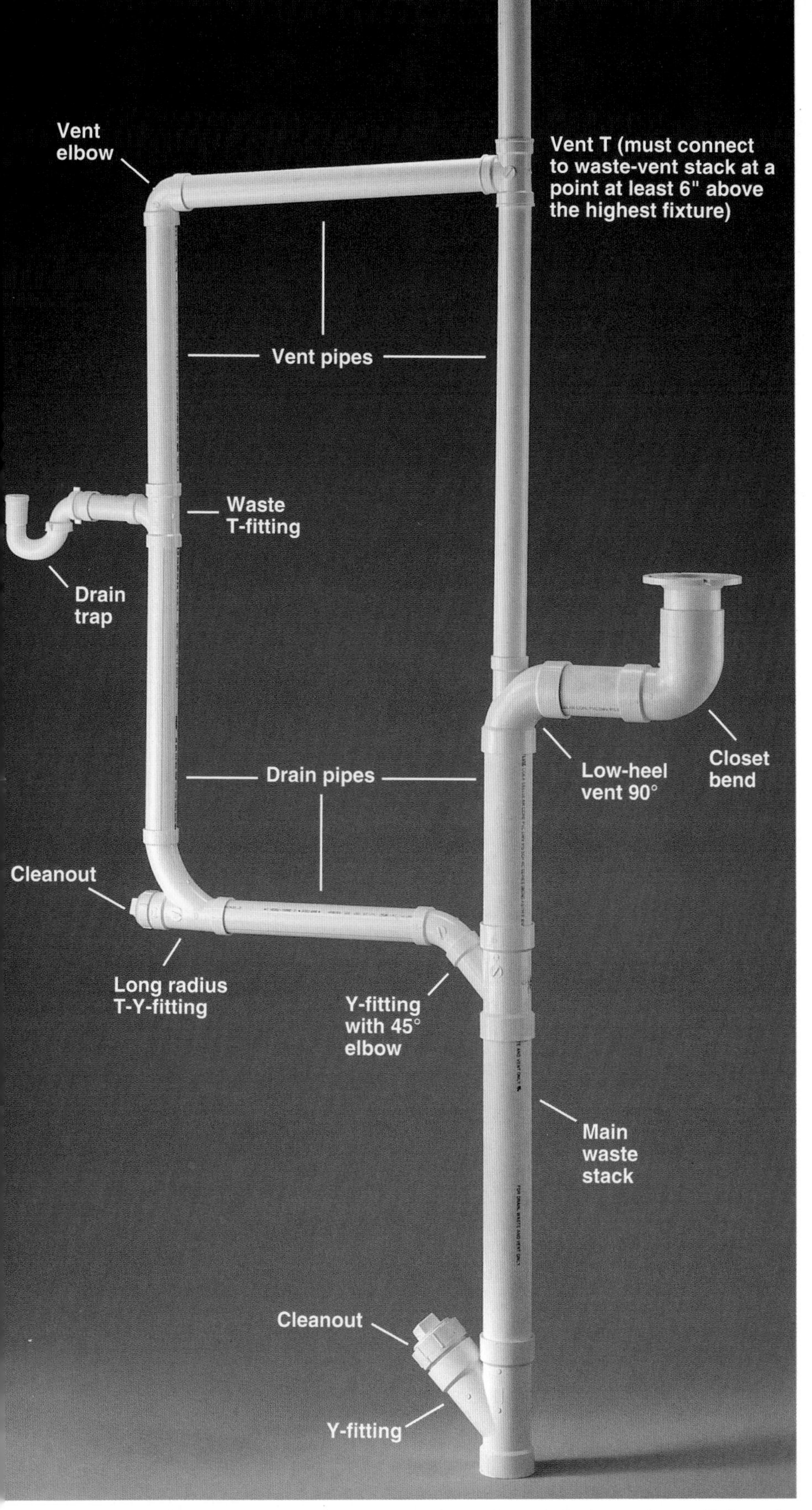

Basic DWV tree shows the correct orientation of drain and vent fittings in a plumbing system. Bends in vent pipes can be very sharp, but drain pipes should use fittings with a noticeable sweep. Fittings used to direct falling waste water from a vertical to a horizontal pipe should have bends that are even more sweeping. Your local Plumbing Code may require that you install cleanout fittings where vertical drain pipes meet horizontal runs.

DWV Fittings

Use the photos on these pages to identify the DWV fittings specified in the project how-to directions found later in this book. Each fitting shown is available in a variety of sizes to match your needs. Always use fittings made from the same material as your DWV pipes.

DWV fittings come in a variety of shapes to serve different functions within the plumbing system.

Vents: In general, the fittings used to connect vent pipes have very sharp bends with no sweep. Vent fittings include the vent T and vent 90° elbow. Standard drain pipe fittings can also be used to join vent pipes.

Horizontal-to-vertical drains: To change directions in a drain pipe from the horizontal to the vertical, use fittings with a noticeable sweep. Standard fittings for this use include waste T-fittings and 90° elbows. Y-fittings and 45° and 22° elbows can also be used for this purpose.

Vertical-to-horizontal drains: To change directions from the vertical to the horizontal, use fittings with a very pronounced, gradual sweep. Common fittings for this purpose include the combination Y-fitting with 45° elbow (often called a combo) and the long radius T-Y-fitting.

Horizontal offsets in drains: Y-fittings, 45° elbows, 22° elbows, and long sweep 90° elbows are used when changing directions in horizontal pipe runs. Whenever possible, horizontal drain pipes should use gradual, sweeping bends rather than sharp turns.

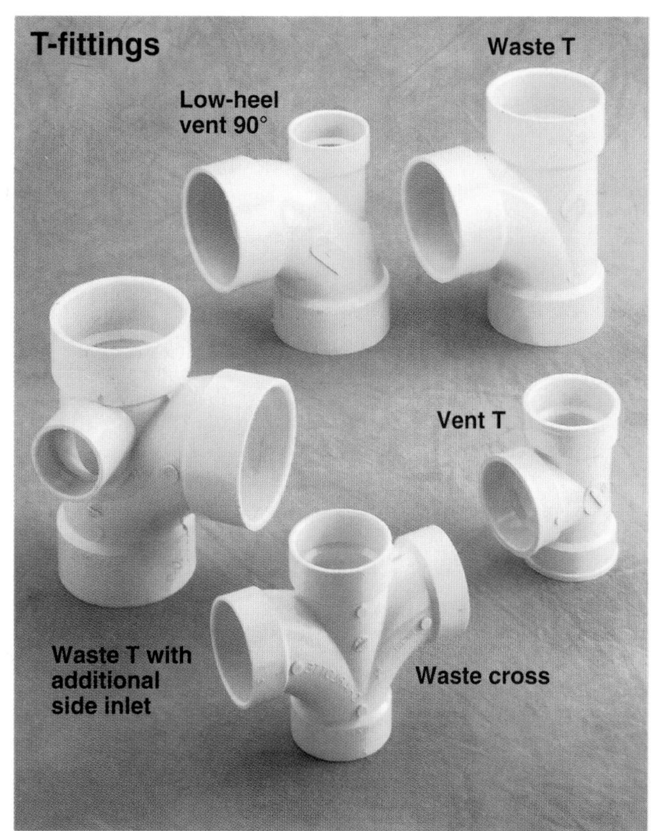

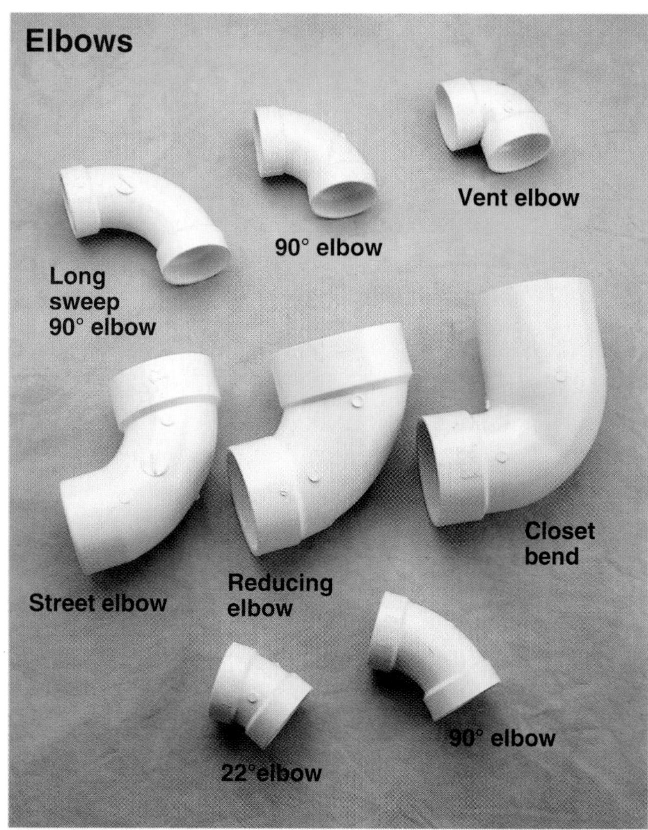

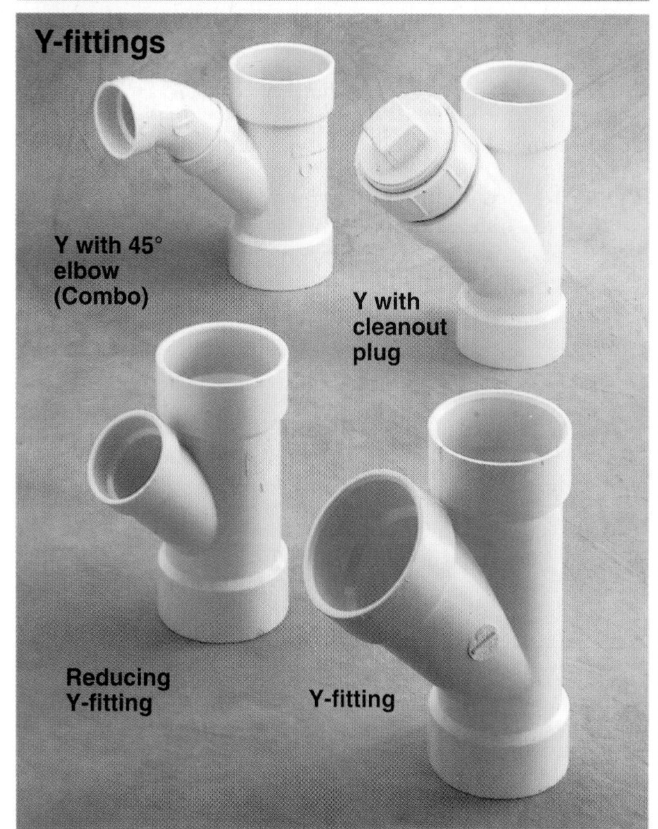

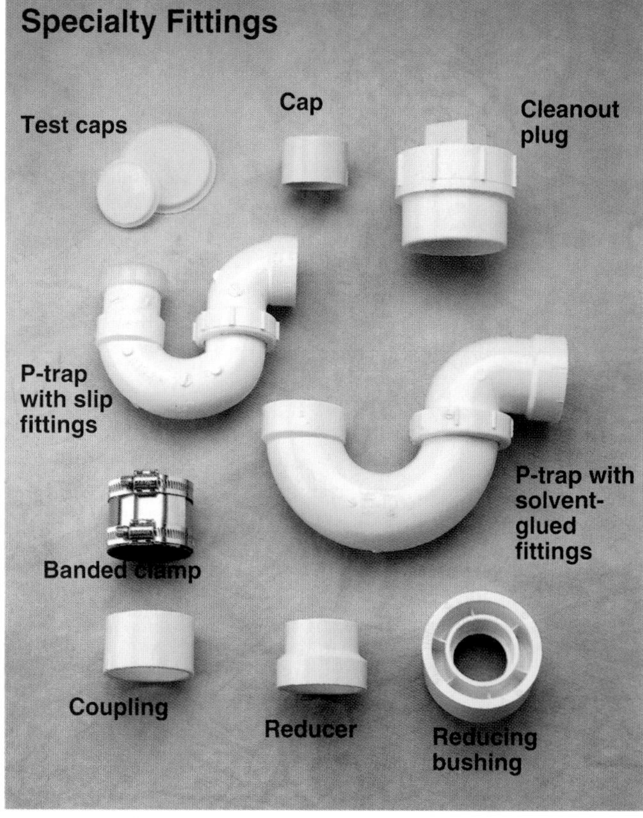

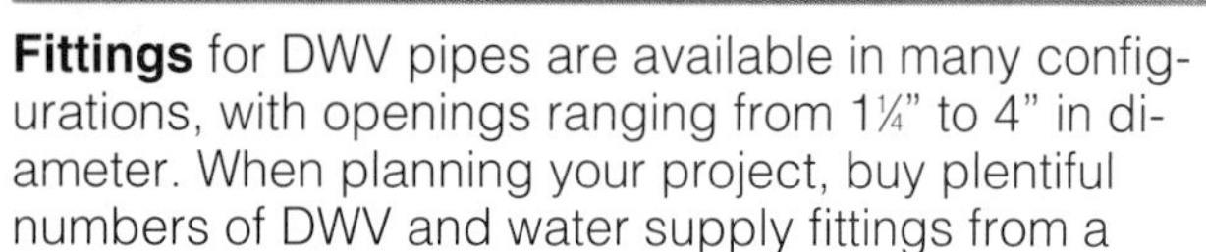

Fittings for DWV pipes are available in many configurations, with openings ranging from 1¼" to 4" in diameter. When planning your project, buy plentiful numbers of DWV and water supply fittings from a reputable retailer with a good return policy. It is much more efficient to return leftover materials after you complete your project than it is to interrupt your work each time you need to shop for a missing fitting.

How to Use Transition Fittings

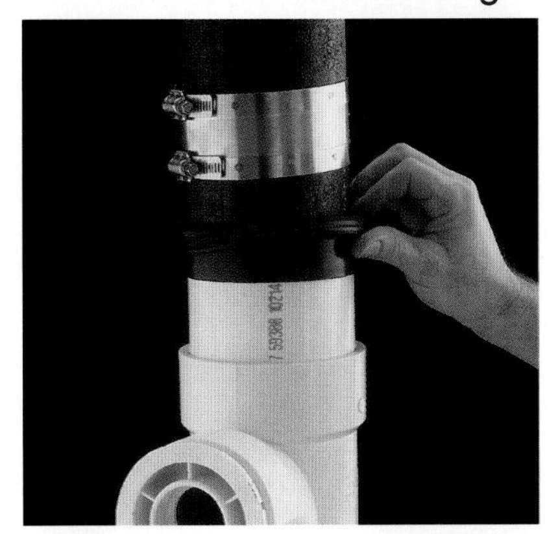

Connect plastic to cast iron with banded couplings (pages 68 to 71). Rubber sleeves cover ends of pipes and ensure a watertight joint.

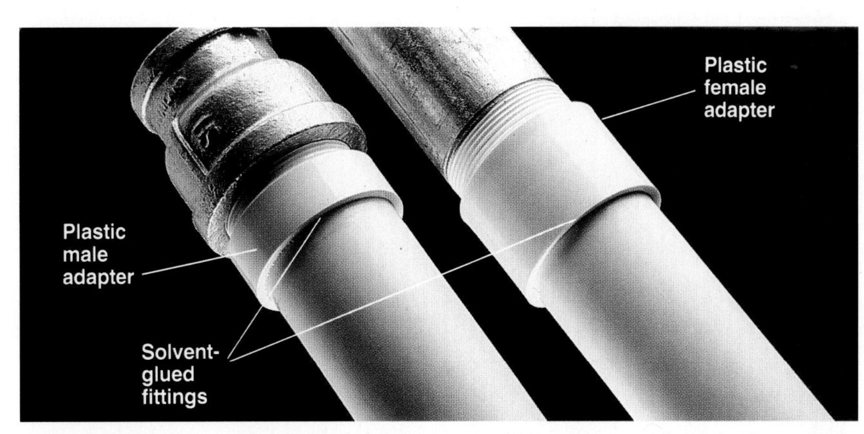

Connect plastic to threaded metal pipes with male and female threaded adapters. Plastic adapter is solvent-glued to plastic pipe. Threads of pipe should be wrapped with Teflon tape. Metal pipe is then screwed directly to the adapter.

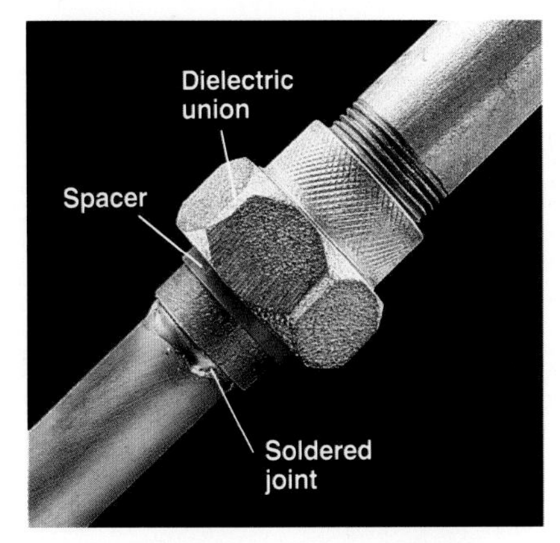

Connect copper to galvanized iron with a dielectric union. Union is threaded onto iron pipe, and is soldered to copper pipe. A dielectric union has plastic spacer that prevents corrosion caused by electrochemical reaction between metals.

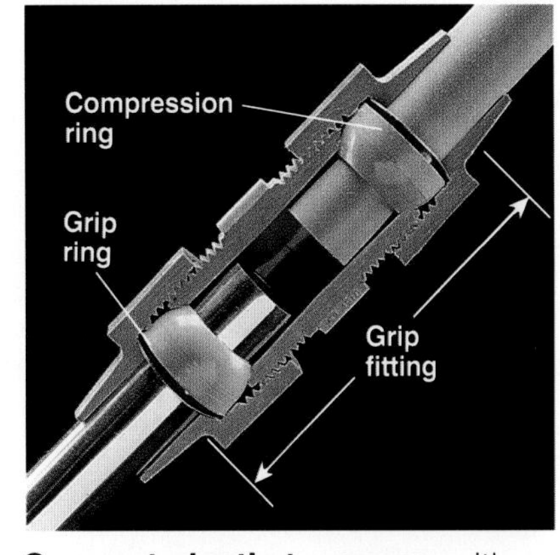

Connect plastic to copper with a grip fitting. Each side of the fitting (shown in cutaway) contains a narrow grip ring and a plastic compression ring (or rubber O-ring) that forms the seal.

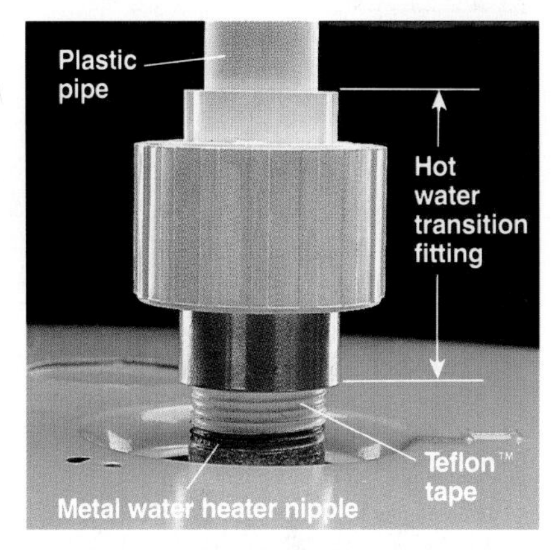

Connect metal hot water pipe to plastic with a hot water transition fitting that prevents leaks caused by different expansion rates of materials. Metal pipe threads are wrapped with Teflon tape. Plastic pipe is solvent-glued to fitting.

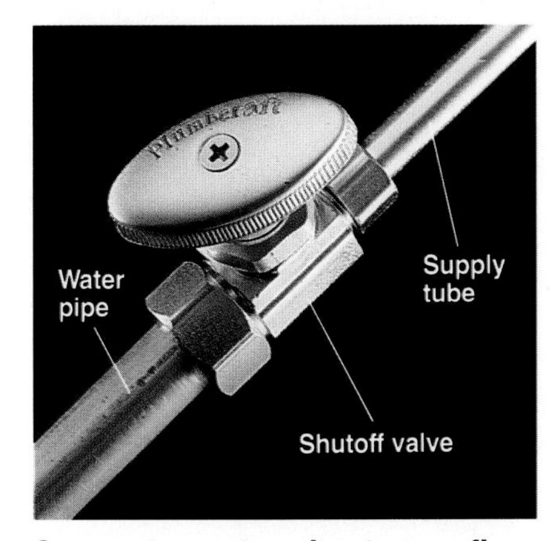

Connect a water pipe to any fixture supply tube, using a shutoff valve (pages 192 to 193).

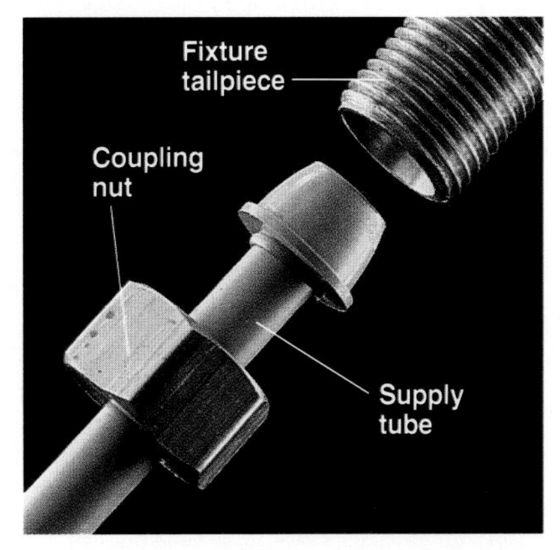

Connect any supply tube to a fixture tailpiece with a coupling nut. Coupling nut seals the bell-shaped end of supply tube against the fixture tailpiece.

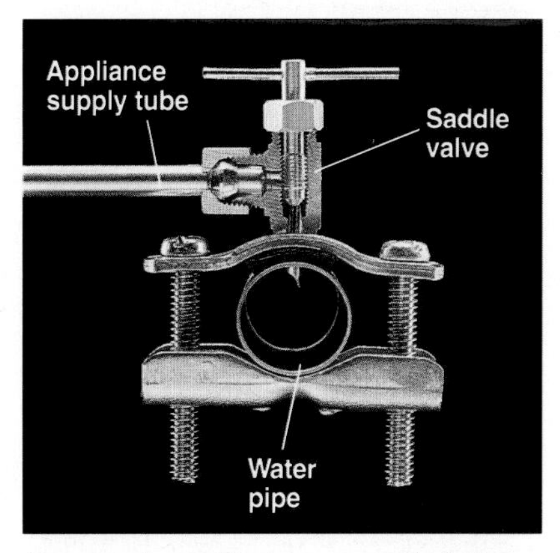

Connect appliance supply tube to copper pipe with a saddle valve (page 207). Saddle valve (shown in cutaway) often is used to connect a refrigerator icemaker.

Soldered fittings, also called sweat fittings, often are used to join copper pipes. Correctly soldered fittings (pages 46 to 50) are strong and trouble-free. Copper pipe can also be joined with compression fittings (pages 52 to 53) or flare fittings (pages 54 to 55). See chart below.

Working with Copper

Copper is the ideal material for water supply pipes. It resists corrosion and has smooth surfaces that provide good water flow. Copper pipes are available in several diameters (page 39) but most home water supply systems use 1/2" or 3/4" pipe. Copper pipe is manufactured in rigid and flexible forms.

Rigid copper, sometimes called hard copper, is approved for home water supply systems by all local codes. It comes in three wall-thickness grades: Types M, L, and K. Type M is the thinnest, the least expensive, and a good choice for do-it-yourself home plumbing.

Rigid Type L usually is required by codes for commercial plumbing systems. Because it is strong and solders easily, Type L may be preferred by some professional plumbers and do-it-yourselfers for home use. Type K has the heaviest wall thickness, and is used most often for underground water service lines.

Flexible copper, also called soft copper, comes in two wall-thickness grades: Types L and K. Both are approved for most home water supply systems, although flexible Type L copper is used primarily for gas service lines. Because it is bendable and will resist a mild frost, Type L may be installed as part of a water supply system in unheated indoor areas, like crawl spaces. Type K is used for underground water service lines.

A third form of copper, called DWV, is used for drain systems. Because most codes now allow low-cost plastic pipes for drain systems, DWV copper is seldom used.

Copper pipes are connected with soldered, compression, or flare fittings (see chart below). Always follow your local Code for the correct types of pipes and fittings allowed in your area.

Copper Pipe & Fitting Chart

	Rigid Copper			Flexible Copper		
Fitting Method	**Type M**	**Type L**	**Type K**	**Type L**	**Type K**	**General Comments**
Soldered	yes	yes	yes	yes	yes	Inexpensive, strong, and trouble-free fitting method. Requires some skill.
Compression	yes	not recommended		yes	yes	Easy to use. Allows pipes or fixtures to be repaired or replaced readily. More expensive than solder. Best used on flexible copper.
Flare	no	no	no	yes	yes	Use only with flexible copper pipes. Usually used as a gas-line fitting. Requires some skill.

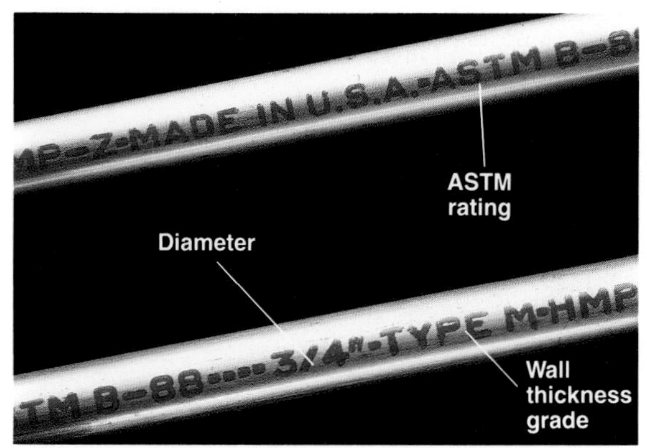

Grade stamp information includes pipe diameter, the wall-thickness grade, and a stamp of approval from the ASTM (American Society for Testing and Materials). Type M pipe is identified by red lettering, Type L by blue lettering.

Bend flexible copper pipe with a coil-spring tubing bender to avoid kinks. Select a bender that matches the outside diameter of the pipe. Slip bender over pipe using a twisting motion. Bend pipe slowly until it reaches the correct angle, but not more than 90°

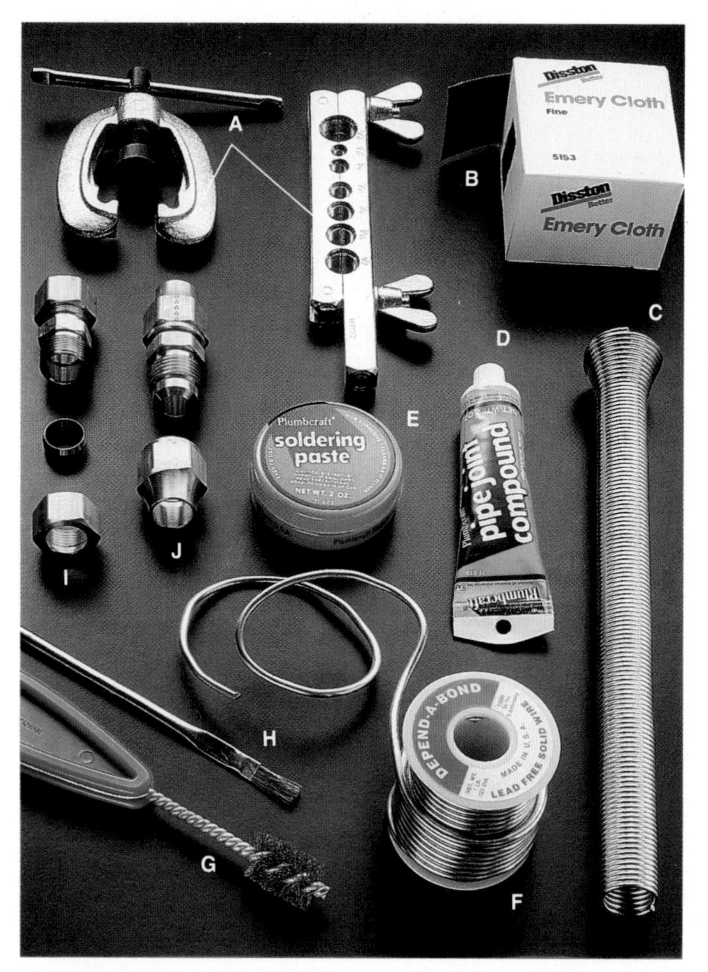

Specialty tools & materials for working with copper include: flaring tool (A), emery cloth (B), coil-spring tubing bender (C), pipe joint compound (D), self-cleaning soldering paste (flux) (E), lead-free solder (F), wire brush (G), flux brush (H), compression fitting (I), flare fitting (J).

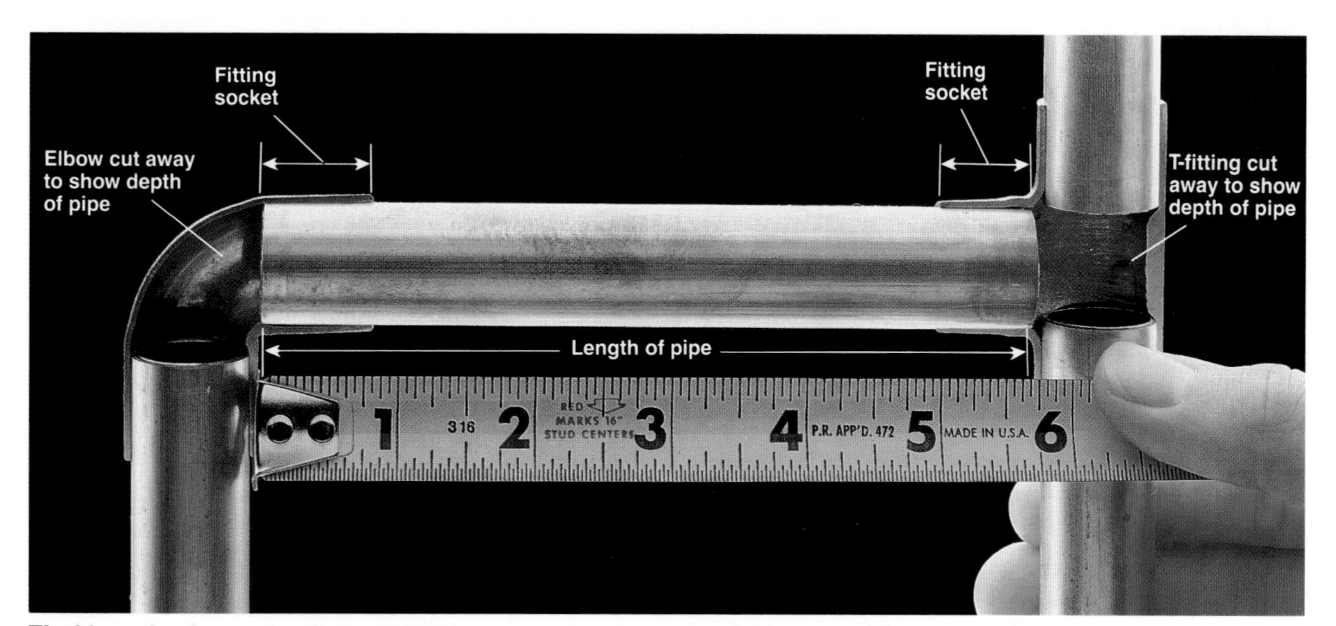

Find length of copper pipe needed by measuring between the bottom of the copper fitting sockets (fittings shown in cutaway). Mark length on the pipe with a felt-tipped pen.

Cutting & Soldering Copper

The best way to cut rigid and flexible copper pipe is with a tubing cutter. A tubing cutter makes a smooth, straight cut, an important first step toward making a watertight joint. Remove any metal burrs on the cut edges with a reaming tool or round file.

Copper can be cut with a hacksaw. A hacksaw is useful in tight areas where a tubing cutter will not fit. Take care to make a smooth, straight cut when cutting with a hacksaw.

A soldered pipe joint, also called a sweated joint, is made by heating a copper or brass fitting with a propane torch until the fitting is just hot enough to melt metal solder. The heat draws the solder into the gap between the fitting and pipe to form a watertight seal. A fitting that is overheated or unevenly heated will not draw in solder. Copper pipes and fittings must be clean and dry to form a watertight seal.

Protect wood from heat of the torch flame while soldering, using a double layer (two 18" × 18" pieces) of 26-gauge sheet metal. Buy sheet metal at hardware stores or building supply centers, and keep it to use with all soldering projects.

Everything You Need:

Tools: tubing cutter with reaming tip (or hacksaw and round file), wire brush, flux brush, propane torch, spark lighter (or matches), adjustable wrench, channel-type pliers.

Materials: copper pipe, copper fittings, emery cloth, soldering paste (flux), sheet metal, lead-free solder, rag.

Soldering Tips

Use caution when soldering copper. Pipes and fittings become very hot and must be allowed to cool before handling.

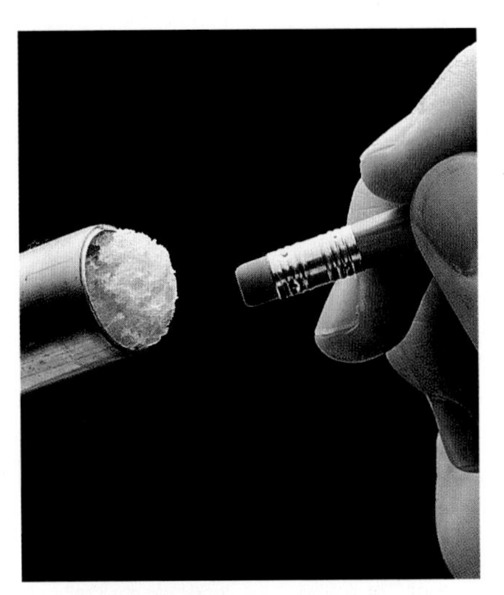

Keep joint dry when soldering existing water pipes by plugging the pipe with bread. Bread absorbs moisture that may ruin the soldering process and cause pinhole leaks. The bread dissolves when water is turned back on.

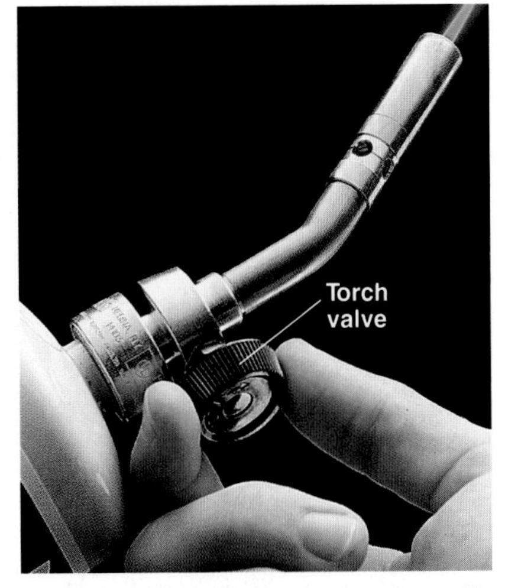

Prevent accidents by shutting off propane torch immediately after use. Make sure valve is closed completely.

How to Cut Rigid & Flexible Copper Pipe

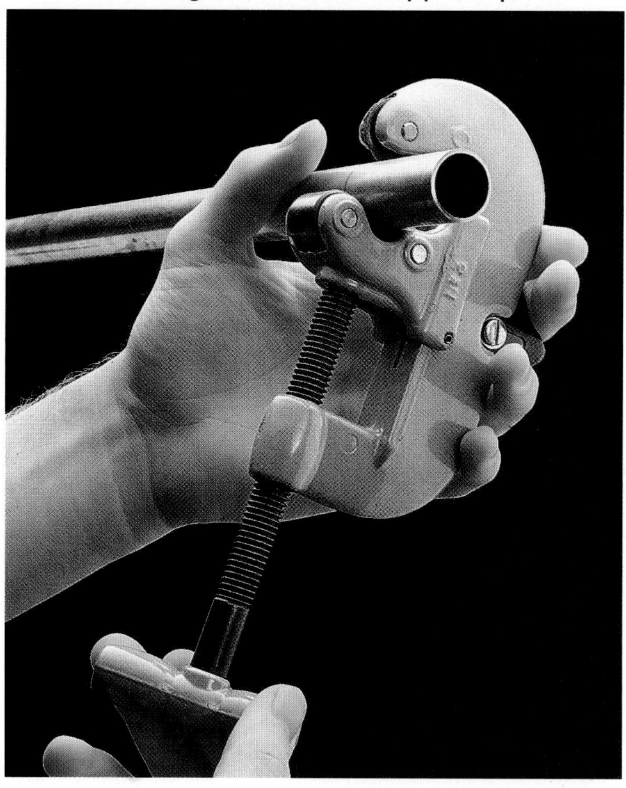

1 Place tubing cutter over the pipe and tighten the handle so that pipe rests on both rollers, and cutting wheel is on marked line.

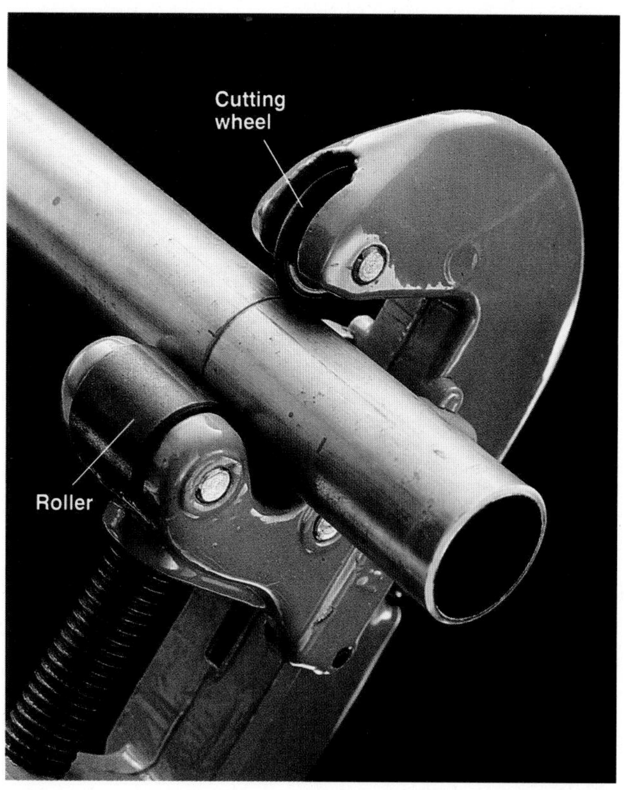

2 Turn tubing cutter one rotation so that cutting wheel scores a continuous straight line around the pipe.

3 Rotate the cutter in the opposite direction, tightening the handle slightly after every two rotations, until cut is complete.

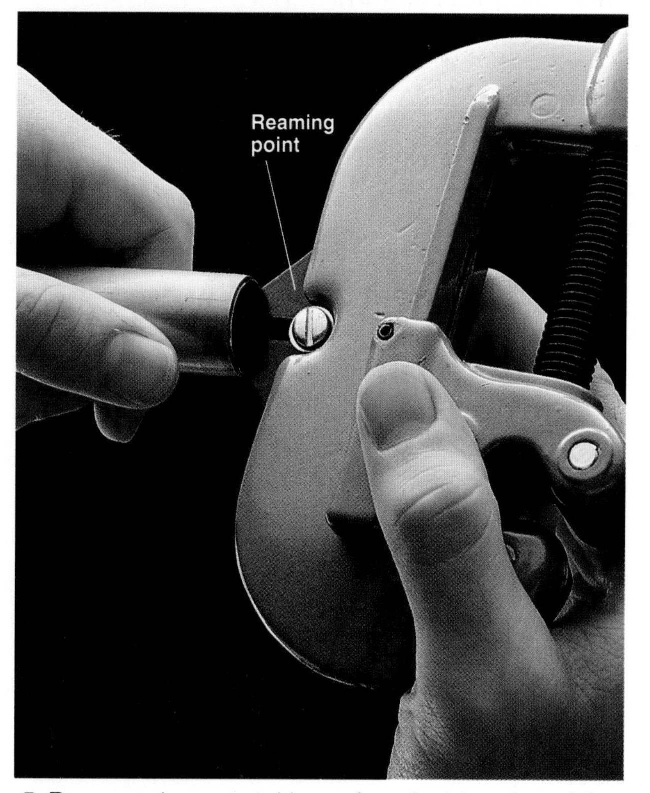

4 Remove sharp metal burrs from inside edge of the cut pipe, using the reaming point on the tubing cutter, or a round file.

How to Solder Copper Pipes & Fittings

1 Clean end of each pipe by sanding with emery cloth. Ends must be free of dirt and grease to ensure that the solder forms a good seal.

2 Clean inside of each fitting by scouring with a wire brush or emery cloth.

3 Apply a thin layer of soldering paste (flux) to end of each pipe, using a flux brush. Soldering paste should cover about 1" of pipe end.

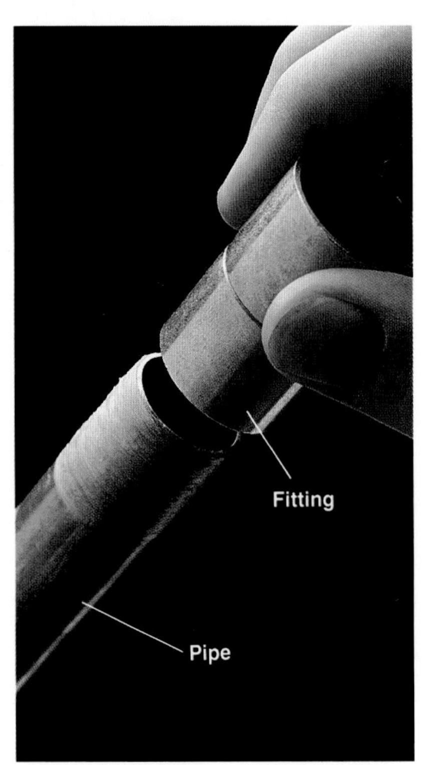

4 Assemble each joint by inserting the pipe into fitting so it is tight against the bottom of the fitting sockets. Twist each fitting slightly to spread soldering paste.

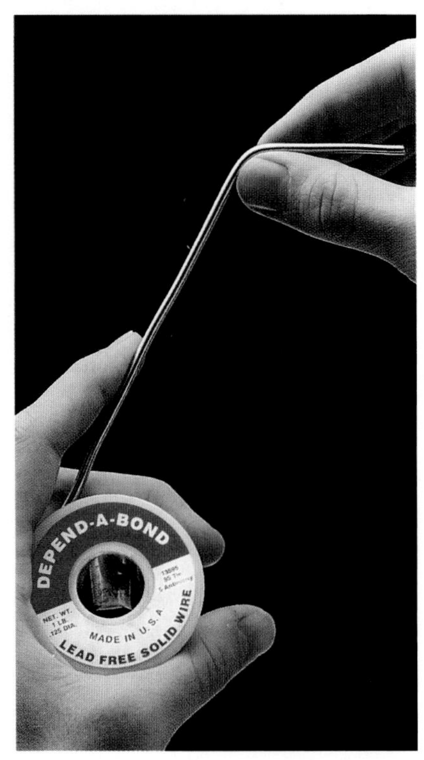

5 Prepare the wire solder by unwinding 8" to 10" of wire from spool. Bend the first 2" of the wire to a 90° angle.

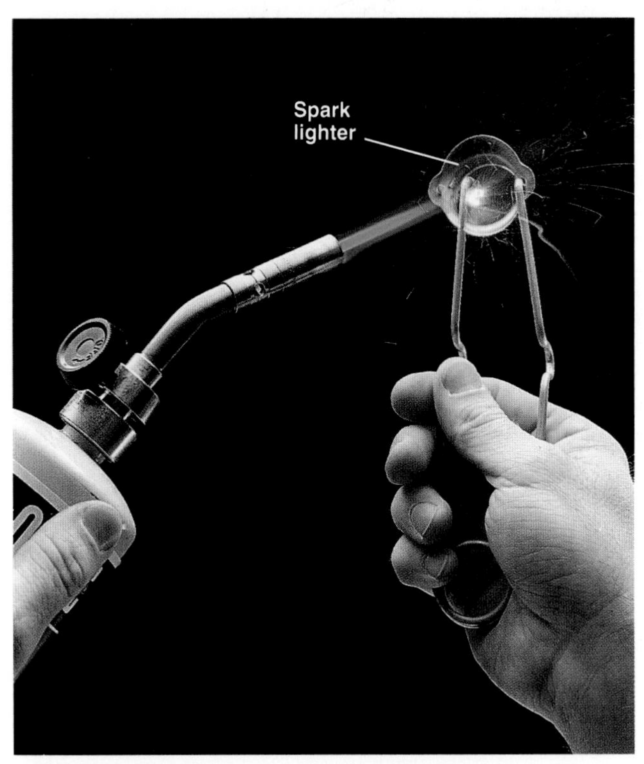

6 Light propane torch by opening valve and striking a spark lighter or a match next to the torch nozzle until the gas ignites.

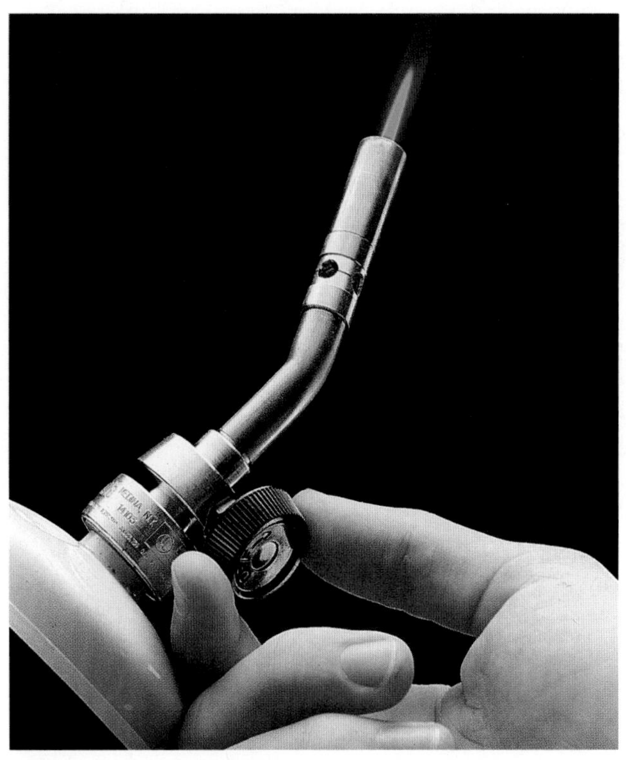

7 Adjust the torch valve until the inner portion of the flame is 1" to 2" long.

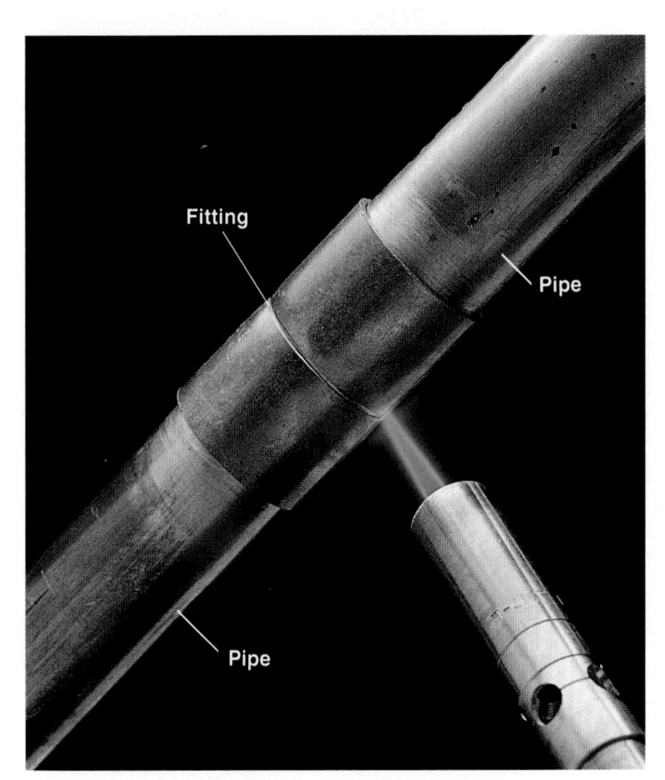

8 Hold flame tip against middle of fitting for 4 to 5 seconds, until soldering paste begins to sizzle.

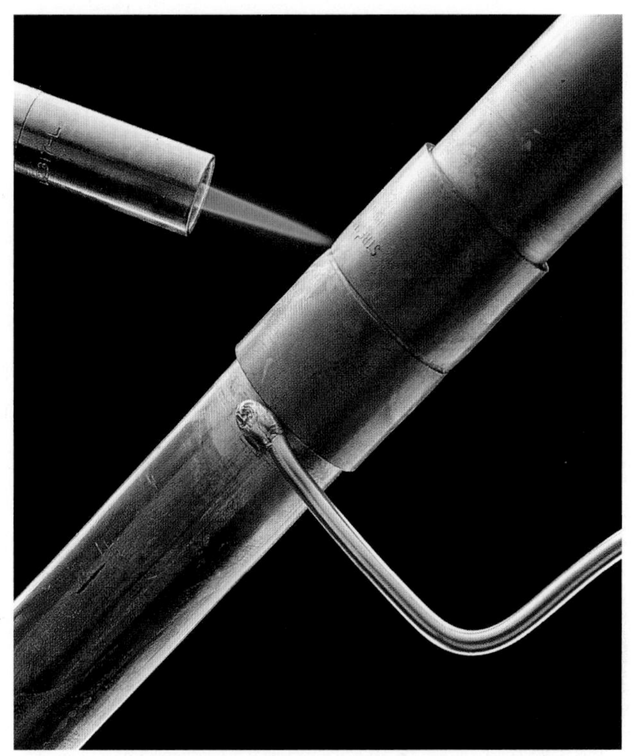

9 Heat other side of copper fitting to ensure that heat is distributed evenly. Touch solder to pipe. If solder melts, pipe is ready to be soldered.

(continued next page)

How to Solder Copper Pipes & Fittings (continued)

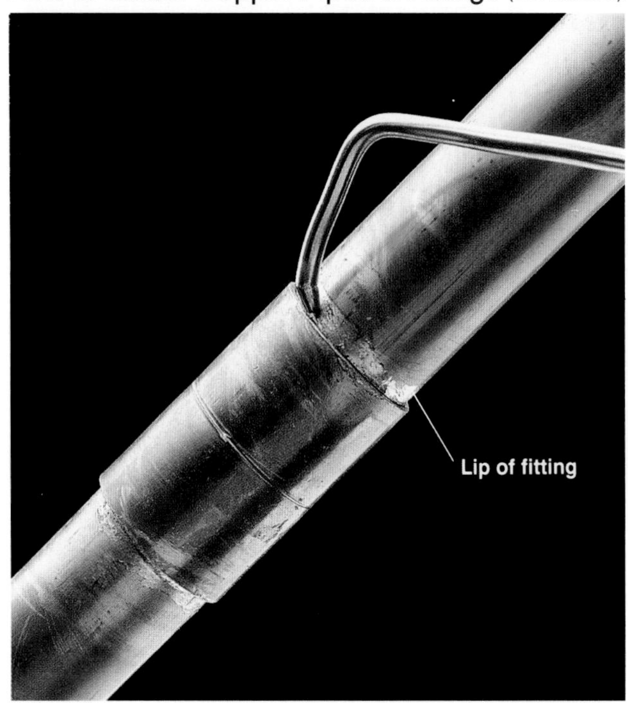

10 When pipe is hot enough to melt solder, remove torch and quickly push ½" to ¾" of solder into each joint. Capillary action fills joint with liquid solder. A correctly soldered joint should show a thin bead of solder around the lip of the fitting.

11 Allow the joint to cool briefly, then wipe away excess solder with a wet rag. **Use caution:** the pipes will be hot and may cause the damp cloth to steam. If joints leak, disassemble and resolder.

How to Solder Brass Valves

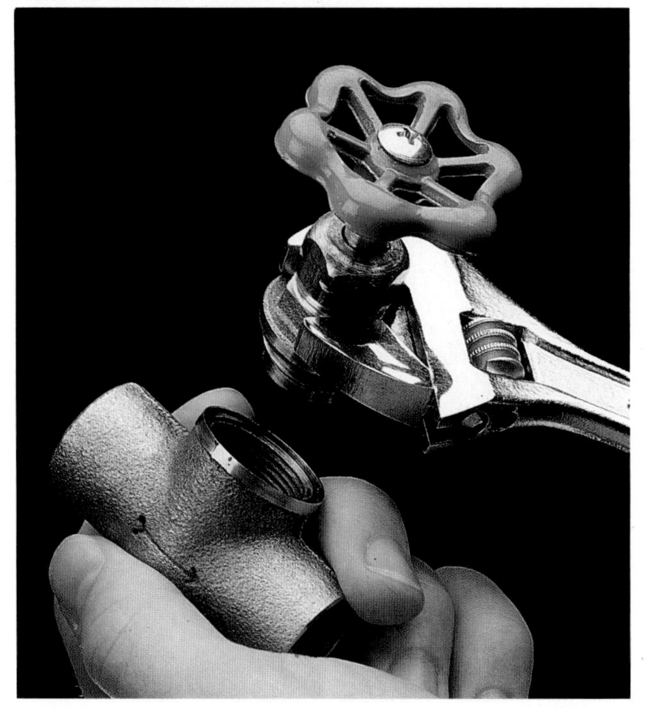

1 Remove the valve stem with an adjustable wrench. Removing the stem prevents heat damage to rubber or plastic stem parts while soldering. Prepare the copper pipes (page 48) and assemble joints.

2 Light propane torch (page 49). Heat body of valve, moving flame to distribute heat evenly. Brass is denser than copper, so it requires more heating time before joints will draw solder. Apply solder (pages 48 to 50). Let metal cool, then reassemble valve.

How to Take Apart Soldered Joints

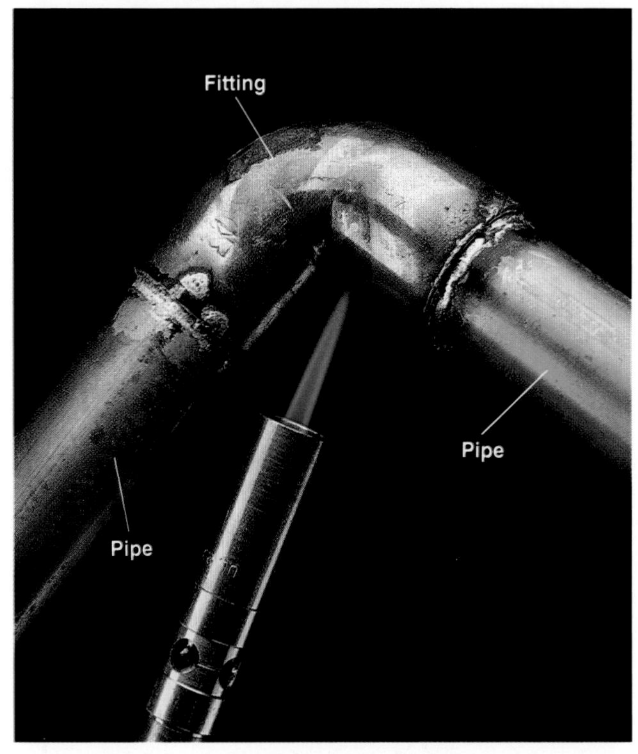

1 Turn off the water (page 12) and drain the pipes by opening the highest and lowest faucets in the house. Light propane torch (page 49). Hold flame tip to the fitting until the solder becomes shiny and begins to melt.

2 Use channel-type pliers to separate the pipes from the fitting.

3 Remove old solder by heating ends of pipe with propane torch. Use dry rag to wipe away melted solder quickly. **Caution: pipes will be hot.**

4 Use emery cloth to polish ends of pipe down to bare metal. Never reuse old fittings.

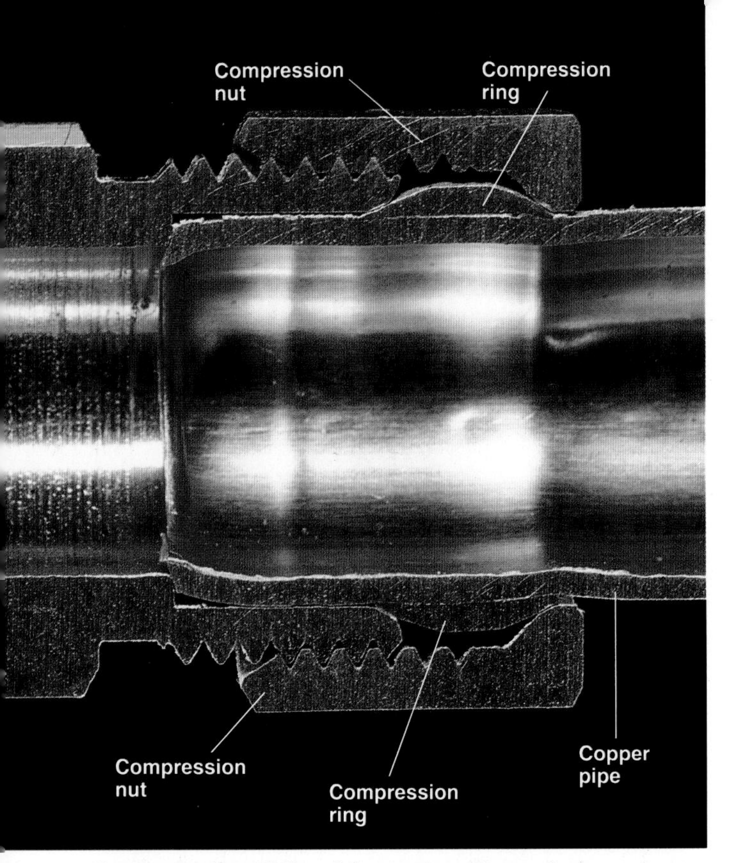

Compression fitting (shown in cutaway) shows how threaded compression nut forms seal by forcing the compression ring against the copper pipe. Compression ring is covered with pipe joint compound before assembling to ensure a perfect seal.

Using Compression Fittings

Compression fittings are used to make connections that may need to be taken apart. Compression fittings are easy to disconnect, and often are used to install supply tubes and fixture shutoff valves (pages 192 to 193, and sequence below). Use compression fittings in places where it is unsafe or difficult to solder, such as in a crawl space.

Compression fittings are used most often with flexible copper pipe. Flexible copper is soft enough to allow the compression ring to seat snugly, creating a watertight seal. Compression fittings also may be used to make connections with Type M rigid copper pipe. See the chart on page 44.

Everything You Need:

Tools: felt-tipped pen, tubing cutter or hacksaw, adjustable wrenches.

Materials: brass compression fittings, pipe joint compound.

How to Attach Supply Tubes to Fixture Shutoff Valves with Compression Fittings

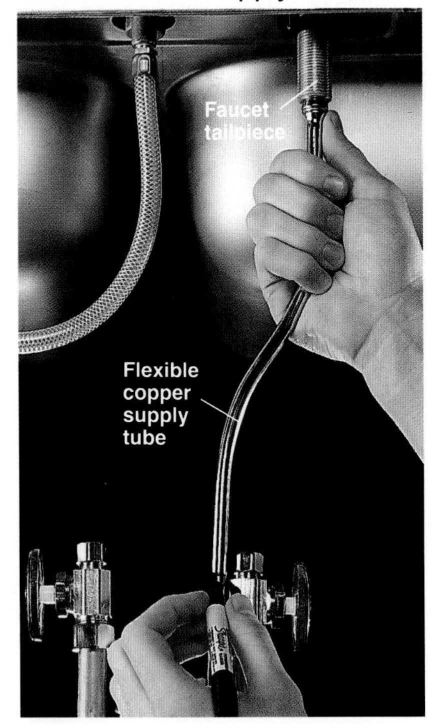

1 Bend flexible copper supply tube, and mark to length. Include 1/2" for portion that will fit inside valve. Cut tube (page 47).

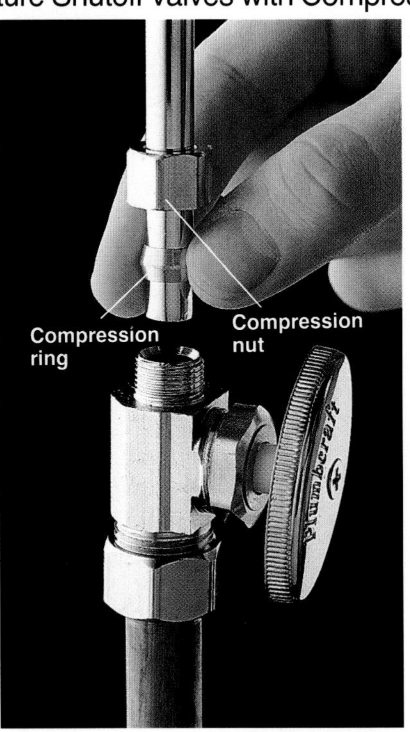

2 Slide the compression nut and then the compression ring over end of pipe. Threads of nut should face the valve.

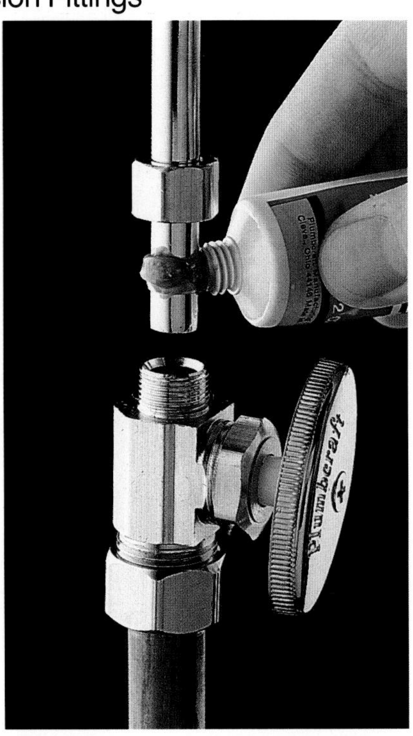

3 Apply a layer of pipe joint compound over the compression ring. Joint compound helps ensure a watertight seal.

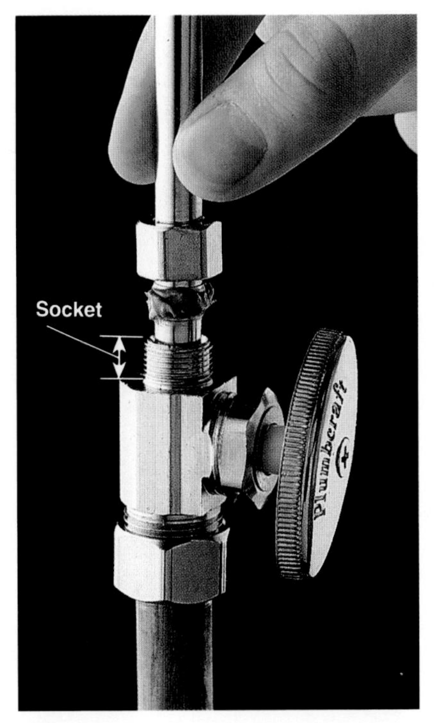

4 Insert end of pipe into fitting so it fits flush against bottom of fitting socket.

5 Slide compression ring and nut against threads of valve. Hand-tighten nut onto valve.

6 Tighten compression nut with adjustable wrenches. Do not overtighten. Turn on water and watch for leaks. If fitting leaks, tighten nut gently.

How to Join Two Copper Pipes with a Compression Union Fitting

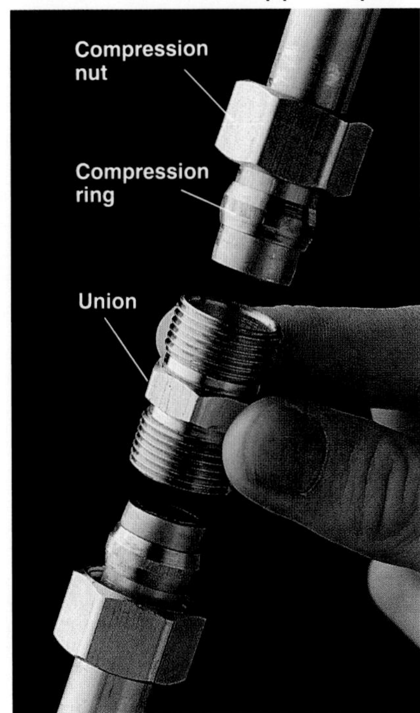

1 Slide compression nuts and rings over ends of pipes. Place threaded union between pipes.

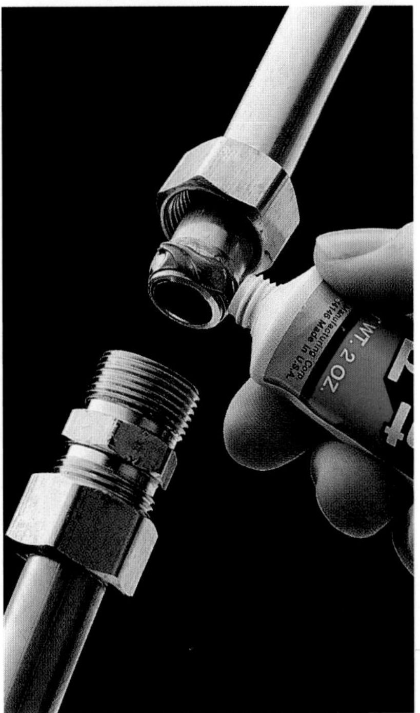

2 Apply a layer of pipe joint compound to compression rings, then screw compression nuts onto threaded union.

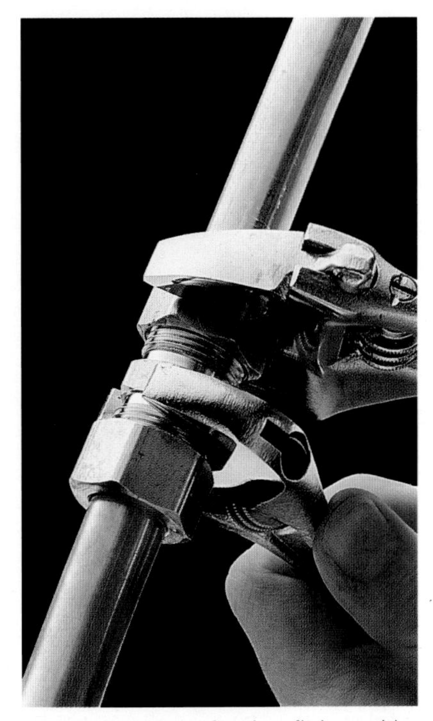

3 Hold center of union fitting with an adjustable wrench, and use another wrench to tighten each compression nut one complete turn. Turn on water. If fitting leaks, tighten nuts gently.

Using Flare Fittings

Flare nut

Flared end of pipe

Brass union fitting

Flare nut

Flexible copper pipe

Flared end of pipe

Flare fitting (shown in cutaway) shows how flared end of flexible copper pipe forms seal against the head of a brass union fitting.

Flare fittings are used more often for flexible copper gas lines. Flare fittings may be used with flexible copper water supply pipes, but they cannot be used where the connections will be concealed inside walls. Always check your local Code regarding the use of flare fittings.

Flare fittings are easy to disconnect. Use a flare fitting in places where it is unsafe or difficult to solder, such as in a crawl space.

Everything You Need:

Tools: two-piece flaring tool, adjustable wrenches.

Materials: brass flare fittings.

How to Join Two Copper Pipes with a Flare Union Fitting

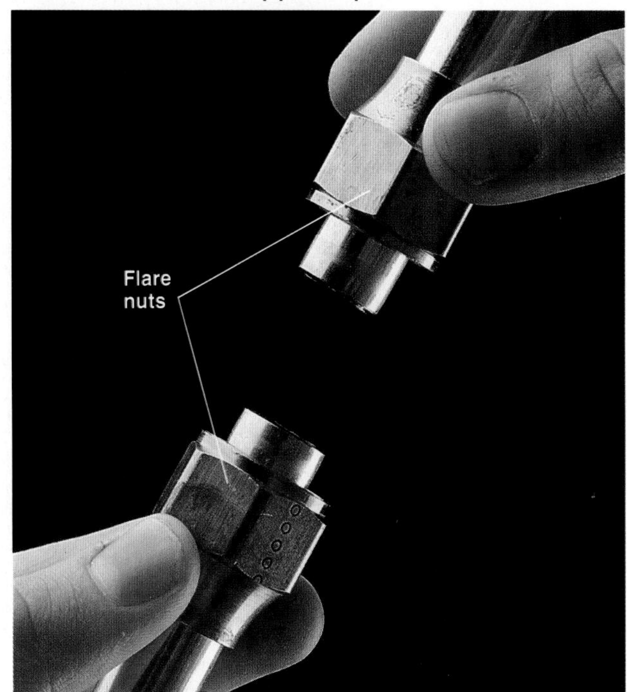

1 Slide flare nuts onto ends of pipes. Nuts must be placed on pipes before ends can be flared. Ream inside of pipe to create smooth edge.

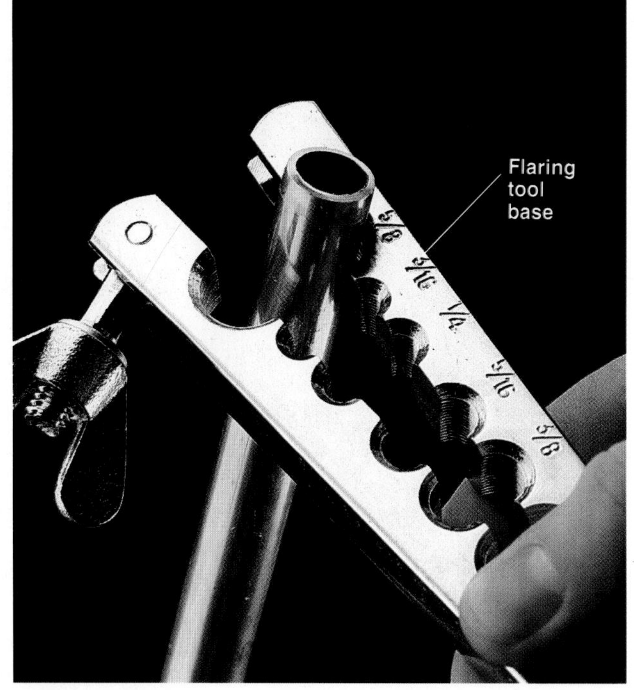

2 Select hole in flaring tool base that matches outside diameter of pipe. Open base, and place end of pipe inside hole.

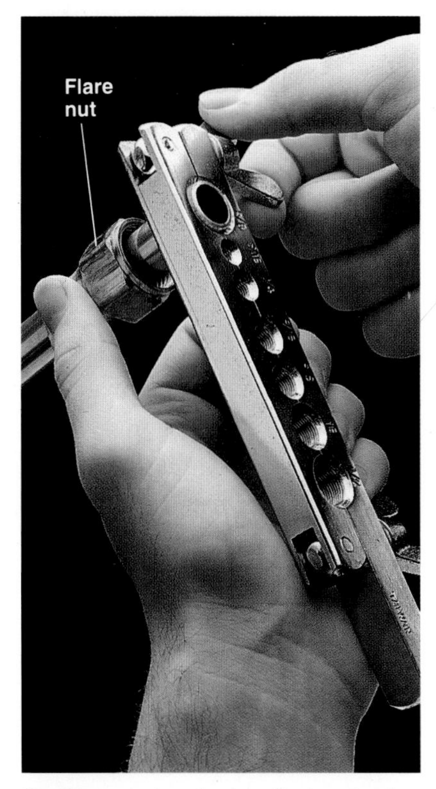

3 Clamp pipe inside flaring tool base. End of pipe must be flush with flat surface of base.

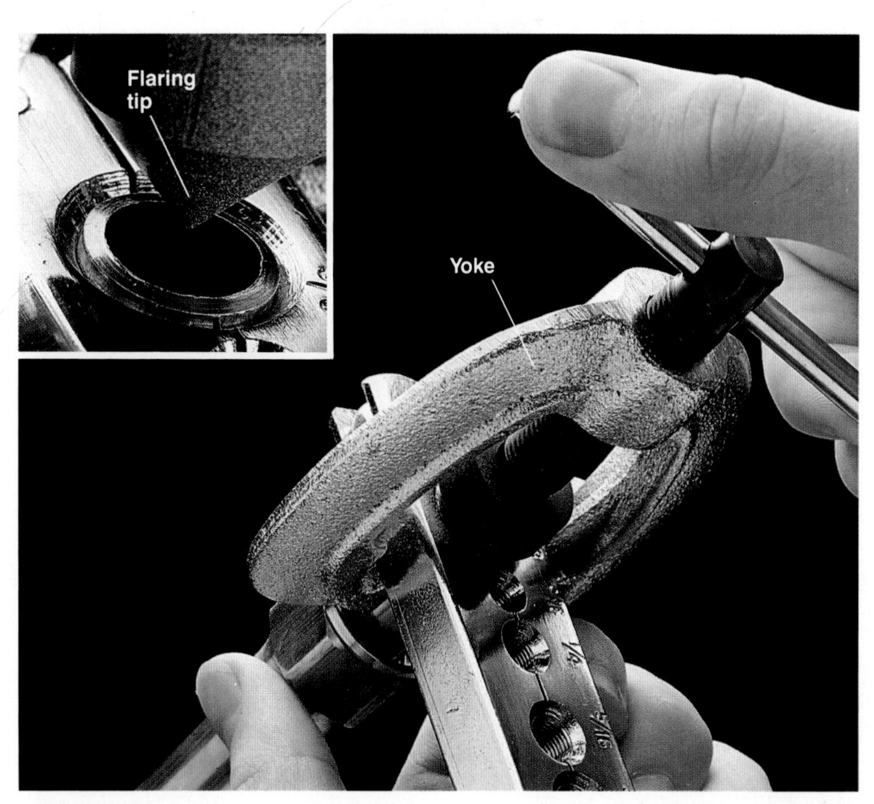

4 Slip yoke of flaring tool around base. Center flaring tip of yoke over end of pipe (inset photo above). Tighten handle of yoke to shape the end of the pipe. Flare is completed when handle cannot be turned further.

5 Remove yoke, and remove pipe from base. Repeat flaring for other pipe.

6 Place flare union between flared ends of pipe, and screw flare nuts onto union.

7 Hold center of flare union with adjustable wrench, and use another wrench to tighten flare nuts one complete turn. Turn on water. If fitting leaks, tighten nuts.

Working with Plastics

Plastic pipes and fittings are popular with do-it-yourselfers because they are lightweight, inexpensive, and easy to use. Local plumbing codes increasingly are approving the use of plastics for home plumbing.

Plastic pipes are available in rigid and flexible forms. Rigid plastics include ABS (acrylonitrile butadiene styrene), PVC (polyvinyl chloride), and CPVC (chlorinated polyvinyl chloride). The most commonly used flexible plastics are PB (polybutylene and PE (polyethylene).

ABS and PVC are used in drain systems. PVC is a newer form of plastic that resists chemical damage and heat better than ABS. It is approved for above-ground use by all plumbing codes. However, some codes still require cast-iron pipe for main drains that run under concrete slabs.

CPVC and PB are used in water supply systems. Rigid CPVC pipe and fittings are less expensive than PB, but flexible PB pipe is a good choice in cramped locations, because it bends easily and requires fewer fittings.

Plastic pipes can be joined to existing iron or copper pipes using transition fittings (page 43), but different types of plastic should not be joined. For example, if your drain pipes are ABS plastic, use only ABS pipes and fittings when making repairs and replacements.

Prolonged exposure to sunlight eventually can weaken plastic plumbing pipe, so plastics should not be installed or stored in areas that receive constant direct sunlight.

Metal pipe
Metal pipe
Jumper wire
Ground clamp
Plastic pipe

Caution: Your home electrical system could be grounded through metal water pipes. When adding plastic pipes to a metal plumbing system, make sure the electrical ground circuit is not broken. Use ground clamps and jumper wires, available at any hardware store, to bypass the plastic transition and complete the electrical ground circuit. Clamps must be firmly attached to bare metal on both sides of the plastic pipe.

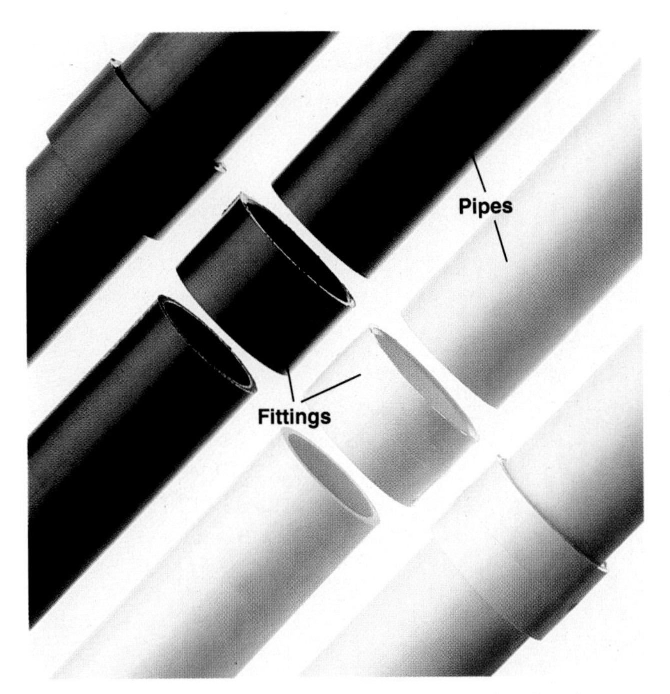

Solvent-glued fittings are used on rigid plastic pipes. Solvent dissolves a thin layer of plastic, and bonds the pipe and fitting together.

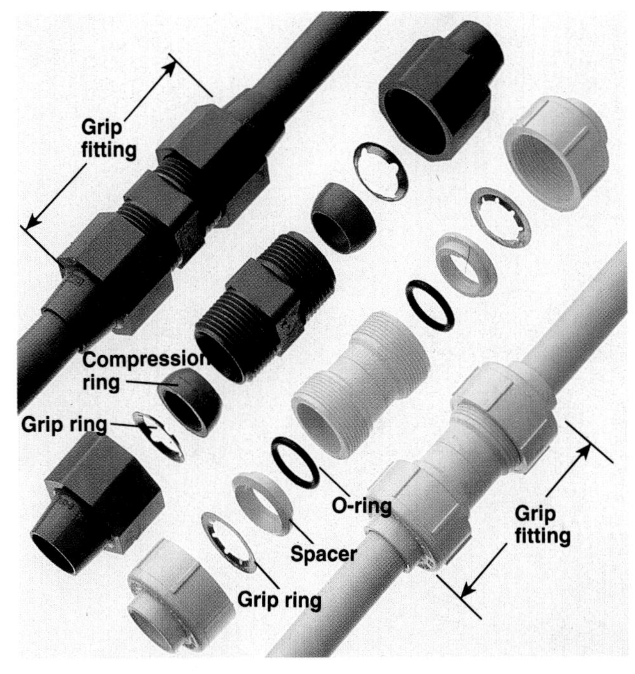

Grip fittings are used to join flexible PB pipes, and can also be used for CPVC pipes. Grip fittings come in two styles. One type (left) resembles a copper compression fitting. It has a metal grip ring and a plastic compression ring. The other type (right) has a rubber O-ring instead of a compression ring.

Plastic Pipe Grade Stamps

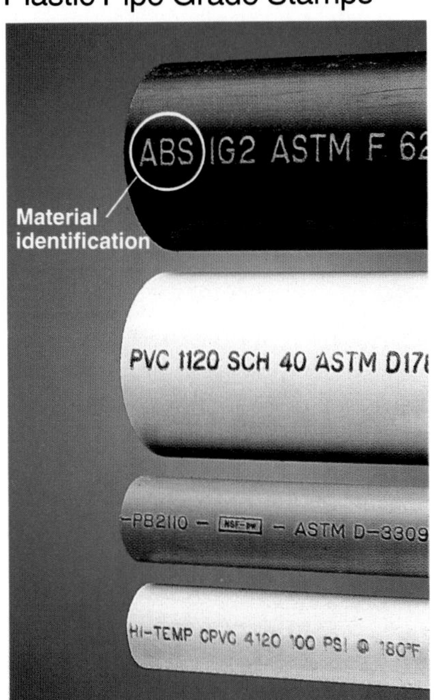

Material identification: For sink traps and drain pipes, use PVC or ABS pipe. For water supply pipes, use PB or CPVC pipe. PE is used for outdoor cold water supply.

NSF rating: For sink traps and drains, choose PVC or ABS pipe that has a DWV (drain-waste-vent) rating from the National Sanitation Foundation (NSF). For water supply pipes, choose PB or CPVC pipe that has a PW (pressurized water) rating.

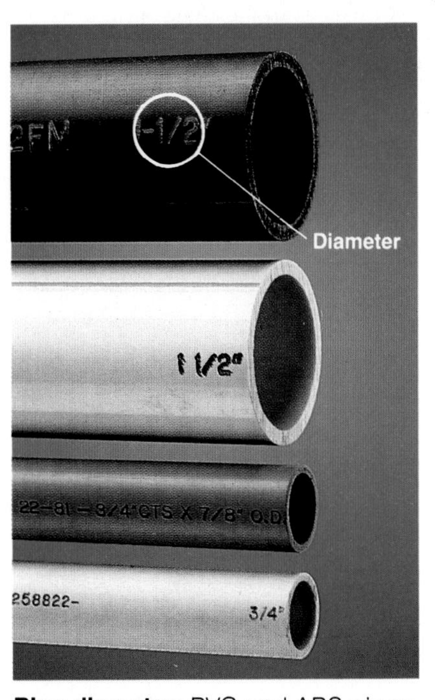

Pipe diameter: PVC and ABS pipes for drains usually have an inside diameter of 1¼" to 4". PB and CPVC pipes for water supply usually have an inside diameter of ½" or ¾".

Cutting & Fitting Plastic Pipe

Cut rigid ABS, PVC, or CPVC plastic pipes with a tubing cutter, or with any saw. Cuts must be straight to ensure watertight joints.

Rigid plastics are joined with plastic fittings and solvent glue. Use a solvent glue that is made for the type of plastic pipe you are installing. For example, do not use ABS solvent on PVC pipe. Some solvent glues, called "all-purpose" or "universal" solvents, may be used on all types of plastic pipe.

Solvent glue hardens in about 30 seconds, so test-fit all plastic pipes and fittings before gluing the first joint. For best results, the surfaces of plastic pipes and fittings should be dulled with emery cloth and liquid primer before they are joined.

Liquid solvent glues and primers are toxic and flammable. Provide adequate ventilation when fitting plastics, and store the products away from any source of heat.

Cut flexible PB and PE pipes with a plastic tubing cutter, or with a knife. Make sure cut ends of pipe are straight. Join these pipes with plastic grip fittings. Grip fittings also are used to join rigid or flexible plastic pipes to copper plumbing pipes (page 43).

Specialty materials for plastics include: solvent glues and primer (A), solvent-glue fittings (B), emery cloth (C), plastic grip fittings (D), and petroleum jelly (E).

Everything You Need:

Tools: tape measure, felt-tipped pen, tubing cutter (or miter box or hacksaw), utility knife, channel-type pliers.

Materials: plastic pipe, fittings, emery cloth, plastic pipe primer, solvent glue, rag, petroleum jelly.

Measuring Plastic Pipe

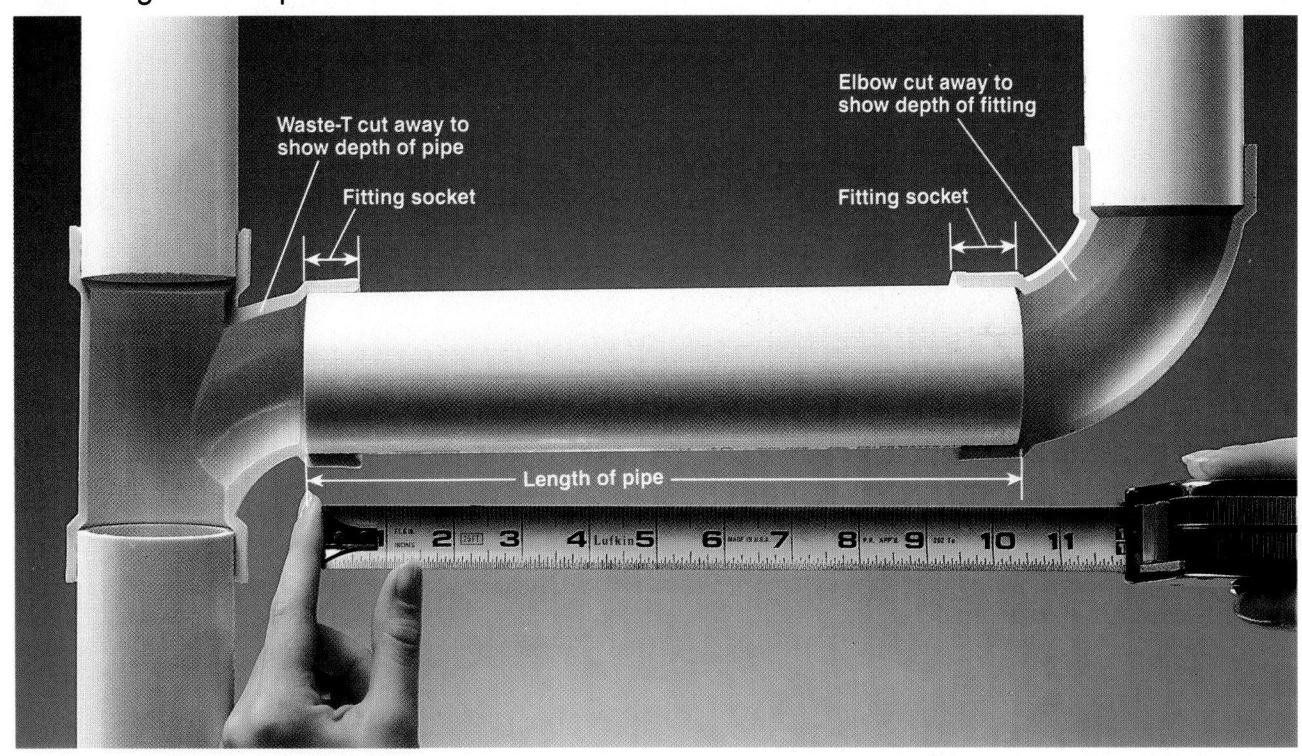

Find length of plastic pipe needed by measuring between the bottoms of the fitting sockets (fittings shown in cutaway). Mark the length on the pipe with a felt-tipped pen.

How to Cut Rigid Plastic Pipe

Tubing cutter: Tighten tool around pipe so cutting wheel is on marked line (page 47). Rotate tool around pipe, tightening screw every two rotations, until pipe snaps.

Miter box: Make straight cuts on all types of plastic pipe with a power or hand miter box.

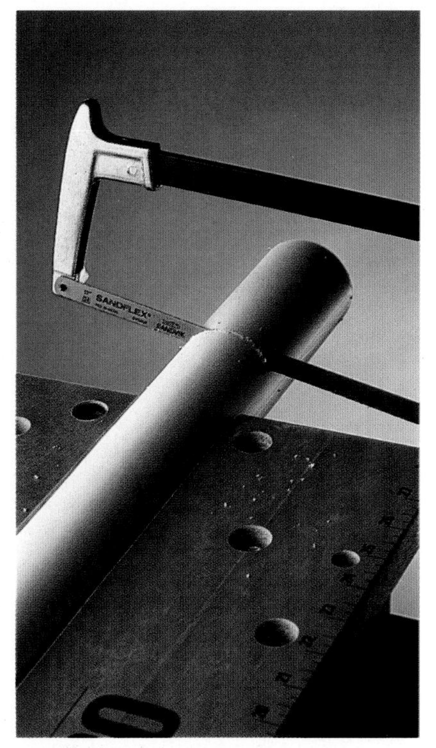

Hacksaw: Clamp plastic pipe in a portable gripping bench or a vise, and keep the hacksaw blade straight while sawing.

How to Solvent-glue Rigid Plastic Pipe

1 Remove rough burrs on cut ends of plastic pipe, using a utility knife.

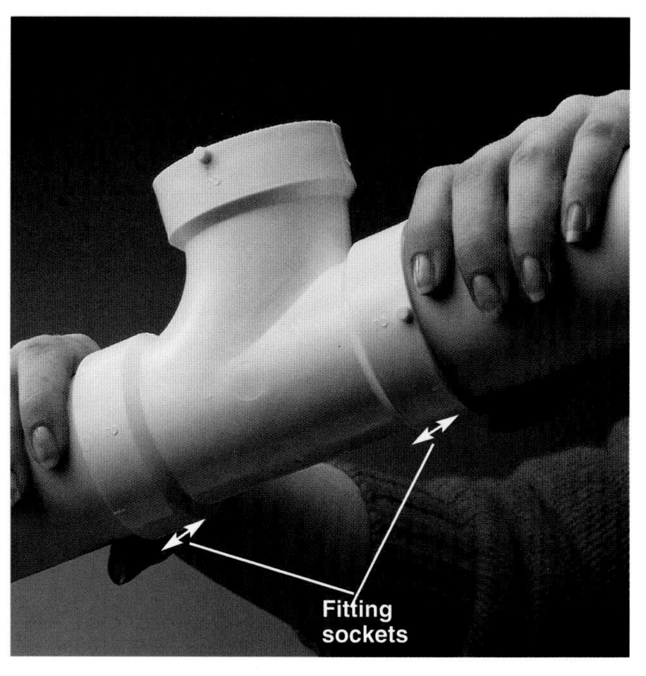

2 Test-fit all pipes and fittings. Pipes should fit tightly against the bottom of the fitting sockets.

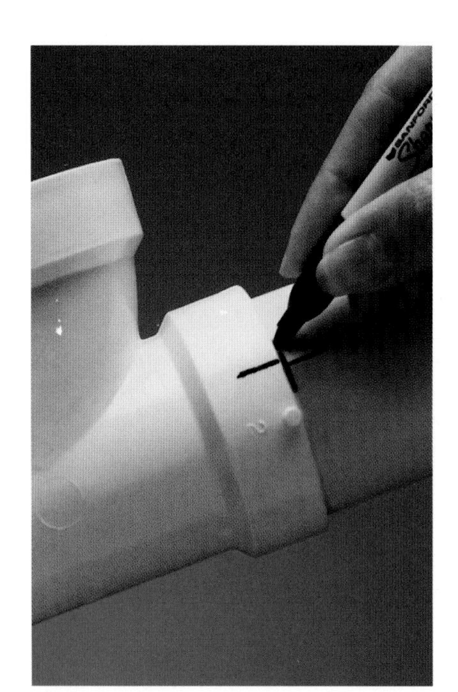

3 Mark depth of the fitting sockets on pipes. Take pipes apart. Clean ends of pipes and the fitting sockets with emery cloth.

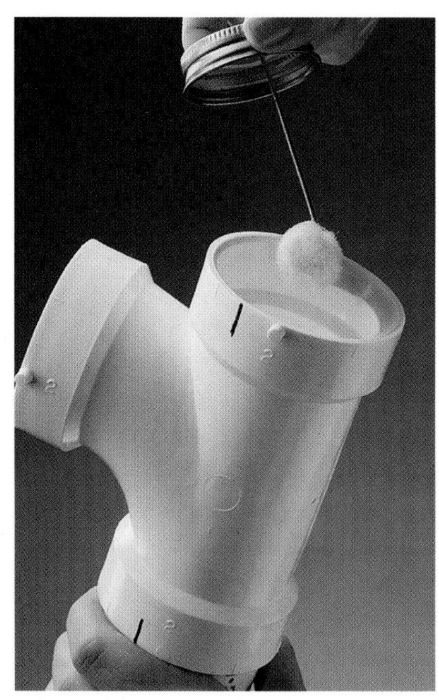

4 Apply plastic pipe primer to the ends of the pipes, and to the insides of the fitting sockets. Primer dulls glossy surfaces and ensures a good seal.

5 Solvent-glue each joint by applying a thick coat of solvent glue to end of pipe. Apply a thin coat of solvent glue to inside surface of fitting socket. Work quickly: solvent glue hardens in about 30 seconds.

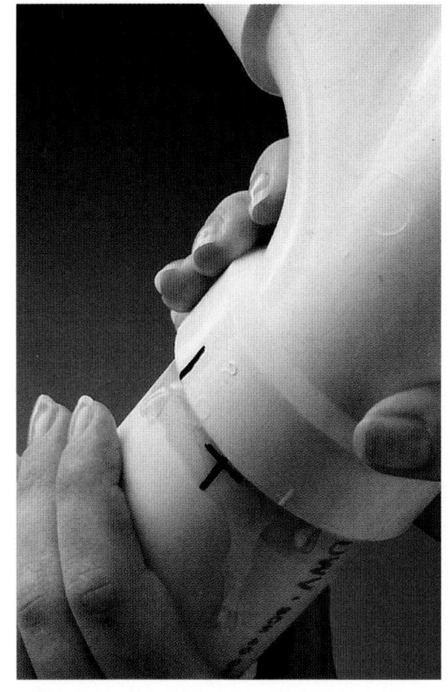

6 Quickly position pipe and fitting so that alignment marks are offset by about 2 inches. Force pipe into fitting until the end fits flush against the bottom of the socket. Twist pipe into alignment (step 7).

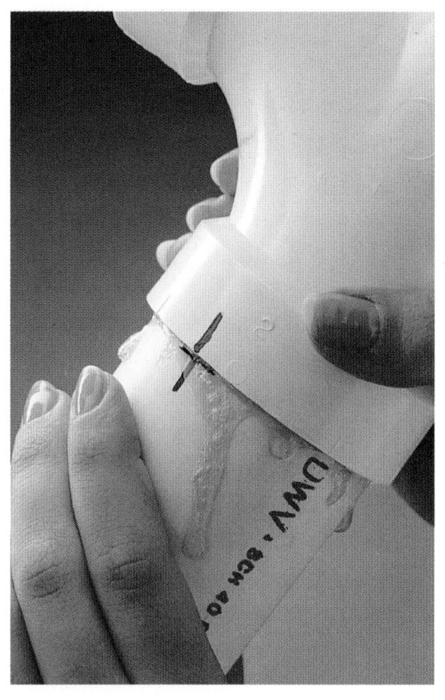

7 Spread solvent by twisting the pipe until marks are aligned. Hold pipe in place for about 20 seconds to prevent joint from slipping.

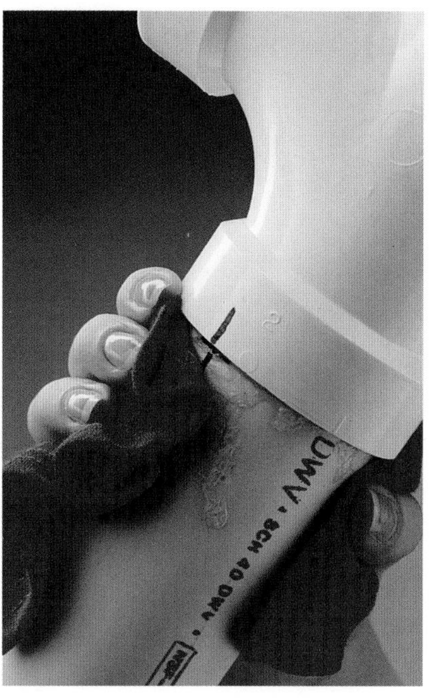

8 Wipe away excess solvent glue with a rag. Do not disturb joint for 30 minutes after gluing.

How to Cut & Fit Flexible Plastic Pipe

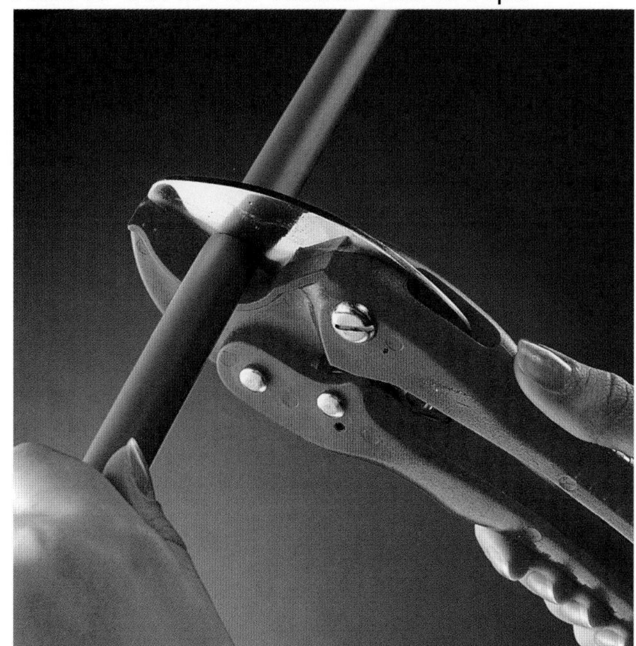

1 Cut flexible PB or PE pipe with a plastic tubing cutter, available at home centers, or with a miter box. (Flexible pipe also can be cut with a miter box or a sharp knife.) Remove any rough burrs with a utility knife.

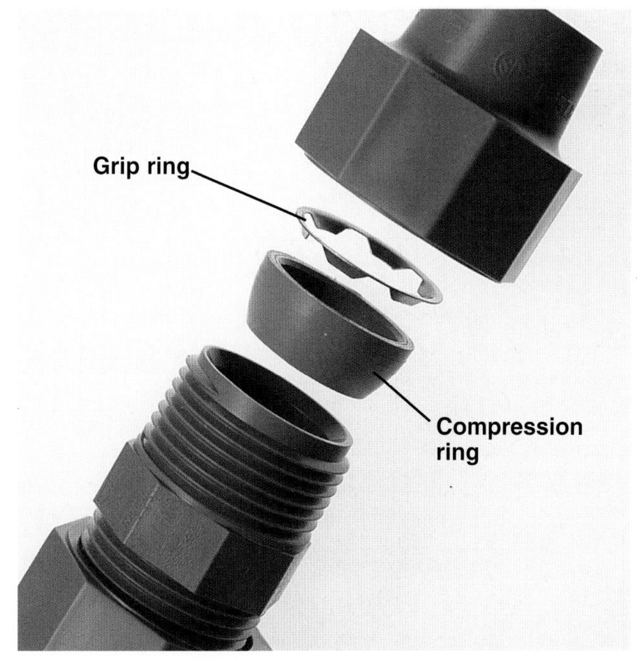

2 Take each grip fitting apart and make sure that the grip ring and the compression ring or O-ring are positioned properly (page 57). Loosely reassemble the fitting.

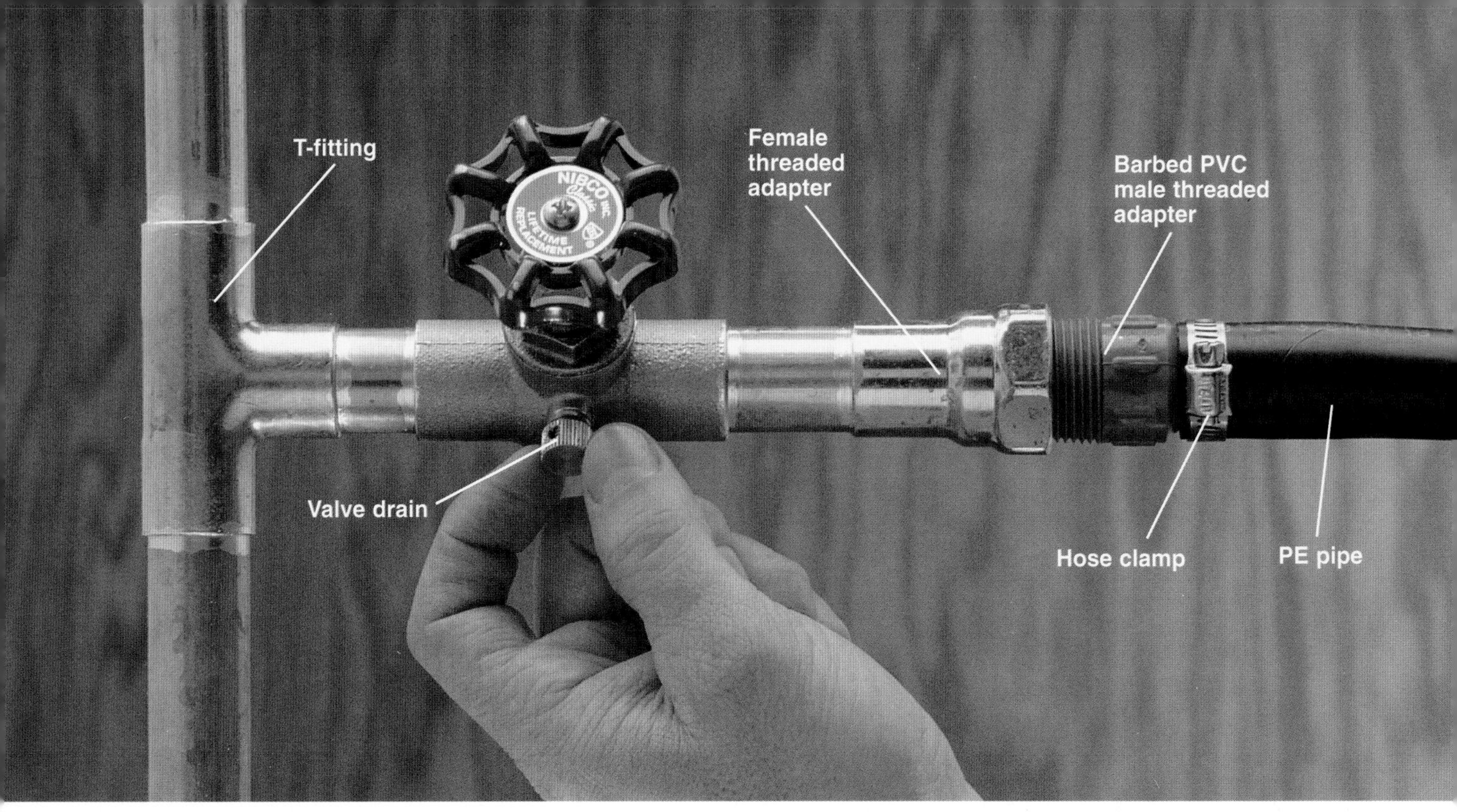

Connect PE pipe to an existing cold water supply pipe by splicing in a T-fitting to the copper pipe and attaching a drain-and-waste shutoff valve and a female threaded adapter. Screw a barbed PVC male threaded adapter into the copper fitting, then attach the PE pipe. The drain-and-waste valve allows you to blow the PE line free of water when winterizing the system.

Working with Flexible Plastic Pipe

Flexible PE (polyethylene) pipe is used for underground cold water lines. Very inexpensive, PE pipe is commonly used for automatic lawn sprinkler systems and for extending cold water supply to utility sinks in detached garages and sheds.

Unlike other plastics, PE is not solvent-glued, but is joined using "barbed" rigid PVC fittings and stainless steel hose clamps. In cold climates, outdoor plumbing lines should be shut off and drained for winter.

How to Join Flexible Plastic Pipe

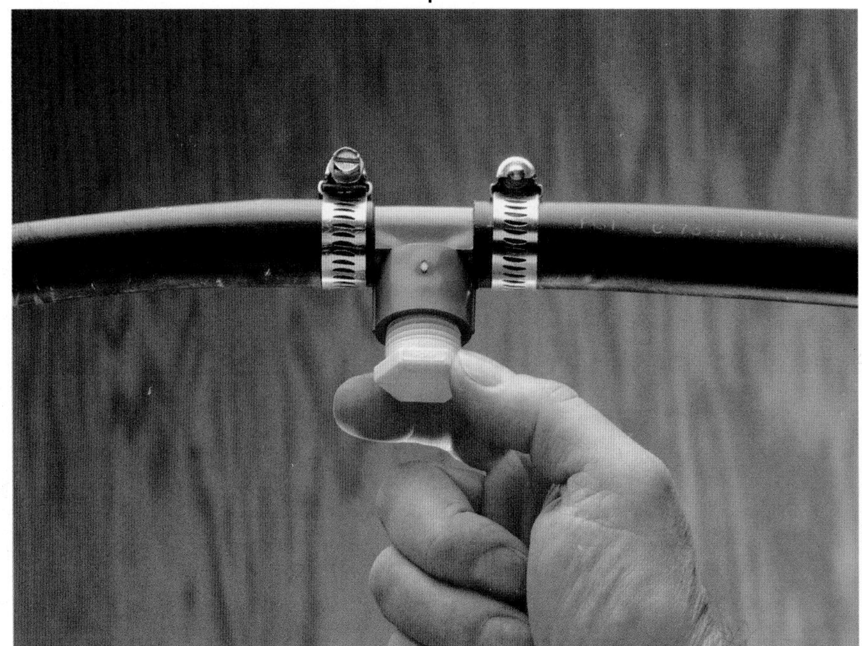

Use barbed PVC fittings to connect lengths of PE pipe. Slide stainless steel hose clamps over the pipe, then force the ends of the pipe over the barbed portion of the fitting. Slide the clamps to the ends of the pipe, then tighten securely with a wrench or screwdriver.

How to Join PB Pipe

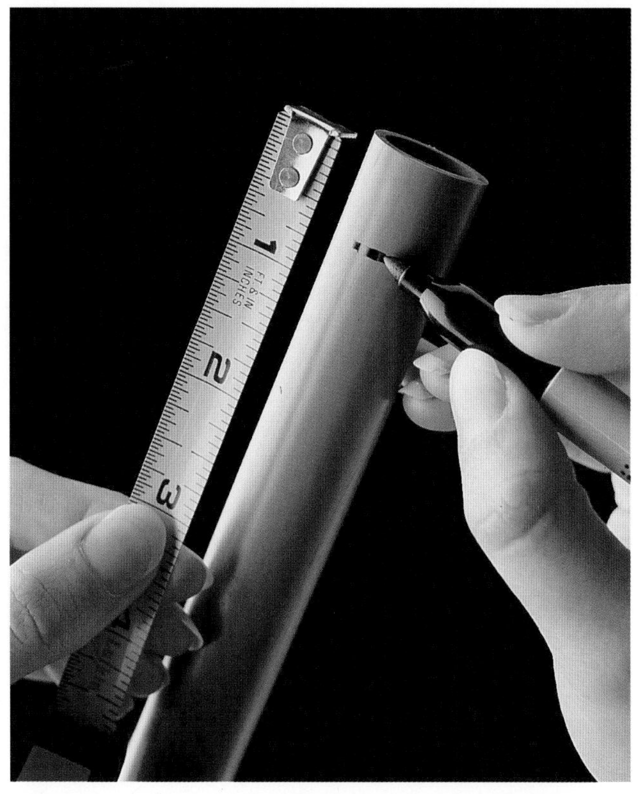

1 Make a mark on the pipe showing the depth of the fitting socket, using a felt-tipped pen. Round off the edges of the pipe with emery cloth.

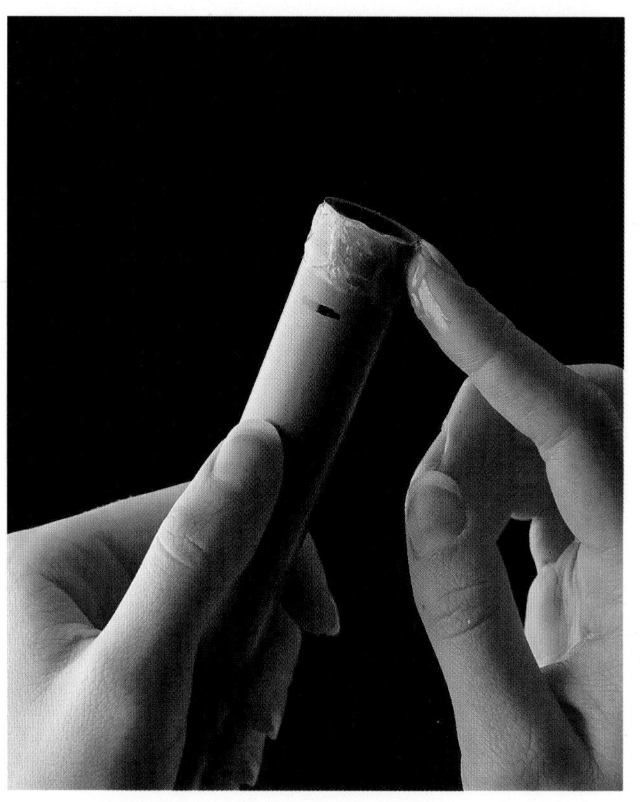

2 Lubricate the end of the pipe with petroleum jelly. Lubricated tip makes it easier to insert pipes into grip fitting.

3 Force end of pipe into fitting up to the mark on the pipe. Hand-tighten coupling nut.

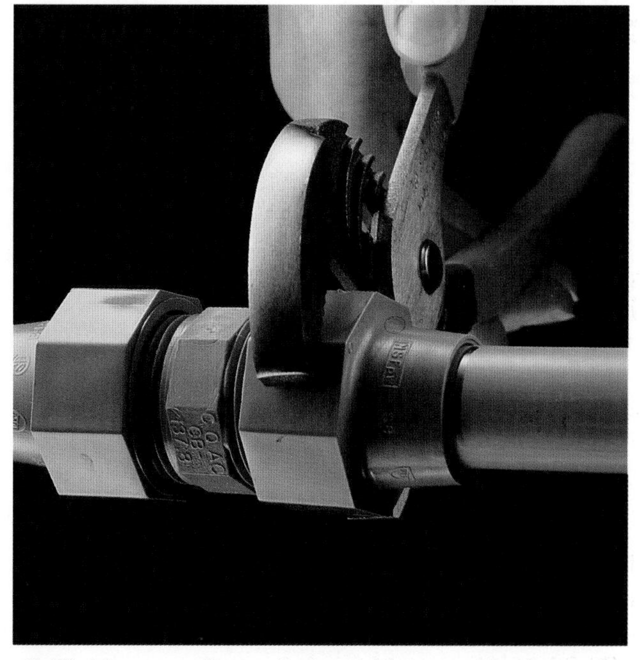

4 Tighten coupling nut about 1/2 turn with channel-type pliers. Turn on water and test the fitting. If the fitting leaks, tighten coupling nut slightly.

Working with Galvanized Iron

Galvanized iron pipe often is found in older homes, where it is used for water supply and small drain lines. It can be identified by the zinc coating that gives it a silver color, and by the threaded fittings used to connect pipes.

Galvanized iron pipes and fittings will corrode with age and eventually must be replaced. Low water pressure may be a sign that the insides of galvanized pipes have a buildup of rust. Blockage usually occurs in elbow fittings. Never try to clean the insides of galvanized iron pipes. Instead, remove and replace them as soon as possible.

Galvanized iron pipe and fittings are available at hardware stores and home improvement centers. Always specify the interior diameter (I.D.) when purchasing galvanized pipes and fittings. Pre-threaded pipes, called *nipples,* are available in lengths from 1" to 1 foot. If you need a longer length, have the store cut and thread the pipe to your dimensions.

Old galvanized iron can be difficult to repair. Fittings often are rusted in place, and what seems like a small job may become a large project. For example, cutting apart a section of pipe to replace a leaky fitting may reveal that adjacent pipes are also in need of replacement. If your job takes an unexpected amount of time, you can cap off any open lines and restore water to the rest of your house. Before you begin a repair, have on hand nipples and end caps that match your pipes.

Taking apart a system of galvanized iron pipes and fittings is time-consuming. Disassembly must start at the end of a pipe run, and each piece must be unscrewed before the next piece can be removed. Reaching the middle of a run to replace a section of pipe can be a long and tedious job. Instead, use a special three-piece fitting called a union. A union makes it possible to remove a section of pipe or a fitting without having to take the entire system apart.

Note: Galvanized iron is sometimes confused with "black iron." Both types have similar sizes and fittings. Black iron is used only for gas lines.

Measure old pipe. Include ½" at each end for the threaded portion of the pipe inside fitting. Bring overall measurement to the store when shopping for replacement parts.

Everything You Need:

Tools: tape measure, reciprocating saw with metal-cutting blade or a hacksaw, pipe wrenches, propane torch, wire brush.

Materials: nipples, end caps, union fitting, pipe joint compound, replacement fittings (if needed).

How to Remove & Replace a Galvanized Iron Pipe

1 Cut through galvanized iron pipe with a reciprocating saw and a metal-cutting blade, or with a hacksaw.

2 Hold fitting with one pipe wrench, and use another wrench to remove old pipe. Jaws of wrenches should face opposite directions. Always move wrench handle toward jaw opening.

3 Remove any corroded fittings using two pipe wrenches. With jaws facing in opposite directions, use one wrench to turn fitting and the other to hold the pipe. Clean pipe threads with a wire brush.

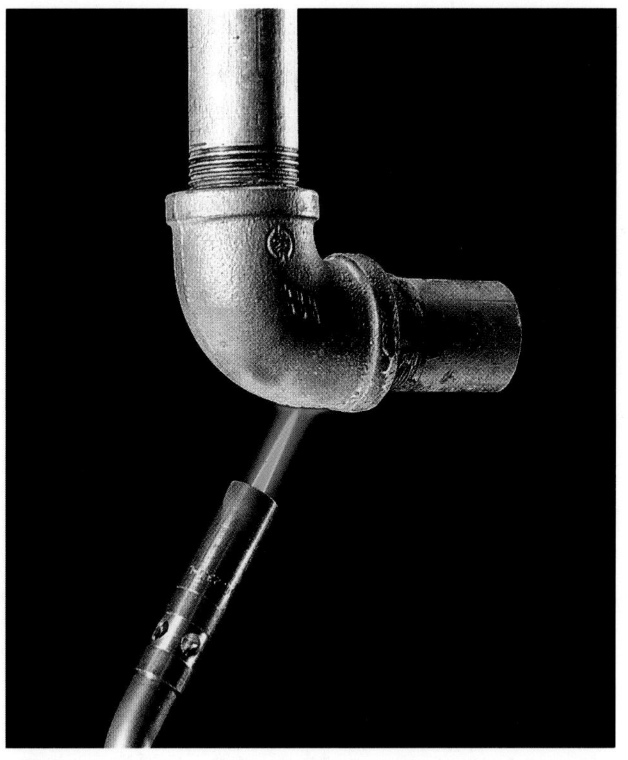

4 Heat stubborn fittings with a propane torch to make them easier to remove. Apply flame for 5 to 10 seconds. Protect wood or other flammable materials from heat, using a double layer of sheet metal (page 46).

(continued next page)

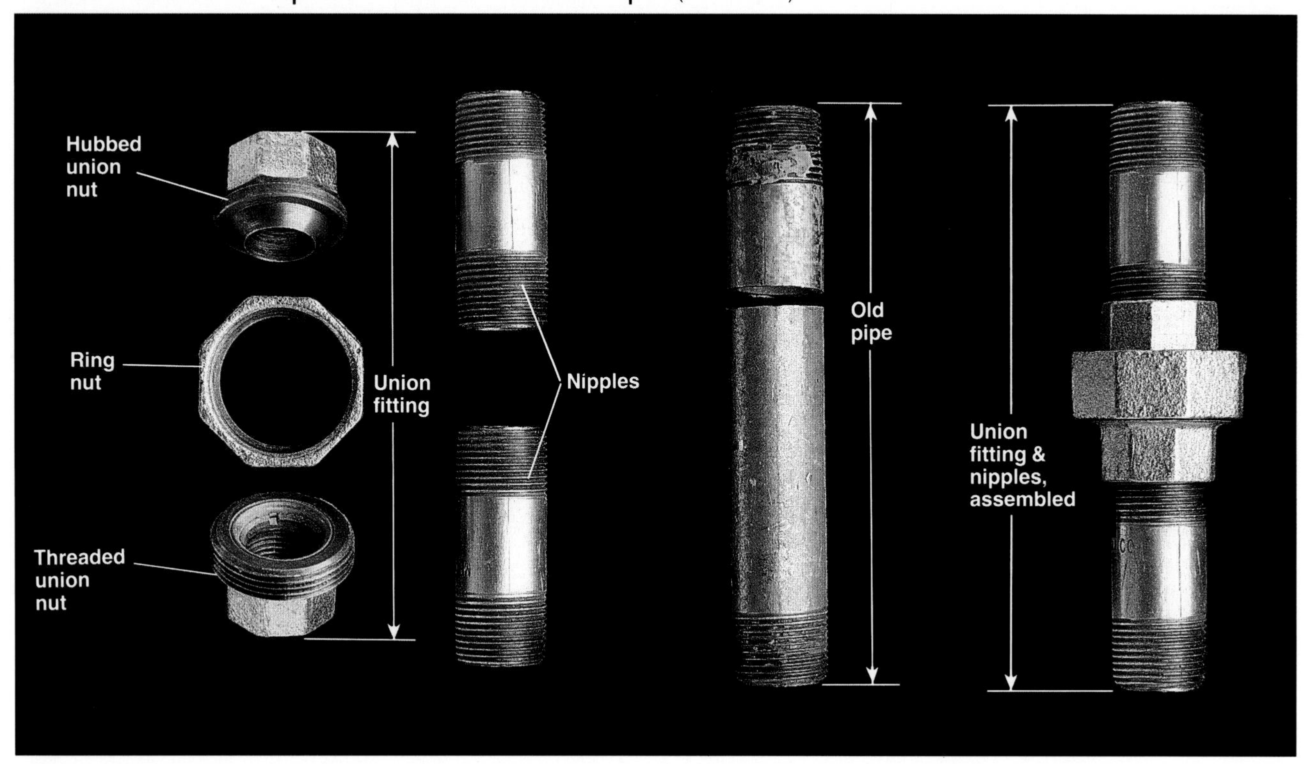

5 Replace a section of galvanized iron pipe with a union fitting and two threaded pipes (nipples). When assembled, the union and nipples must equal the length of the pipe that is being replaced.

6 Apply a bead of pipe joint compound around threaded ends of all pipes and nipples. Spread compound evenly over threads with fingertip.

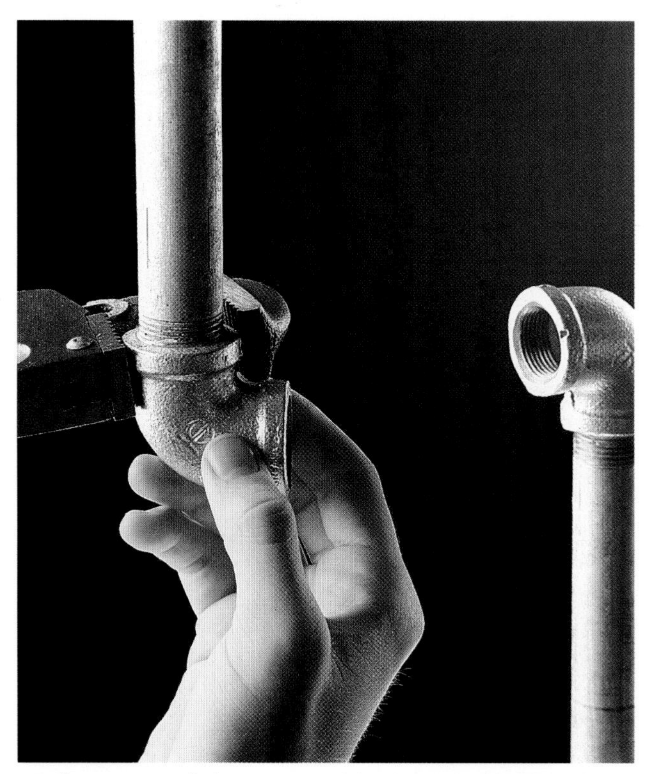

7 Screw new fittings onto pipe threads. Tighten fittings with two pipe wrenches, leaving them about ⅛ turn out of alignment to allow assembly of union.

8 Screw first nipple into fitting, and tighten with pipe wrench.

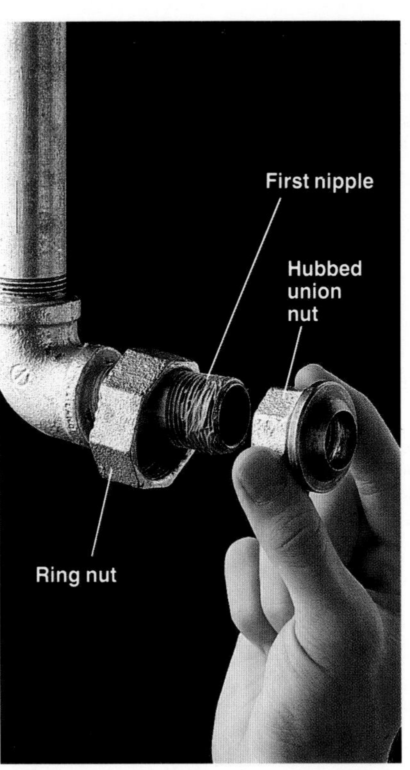

9 Slide ring nut onto the installed nipple, then screw the hubbed union nut onto the nipple and tighten with a pipe wrench.

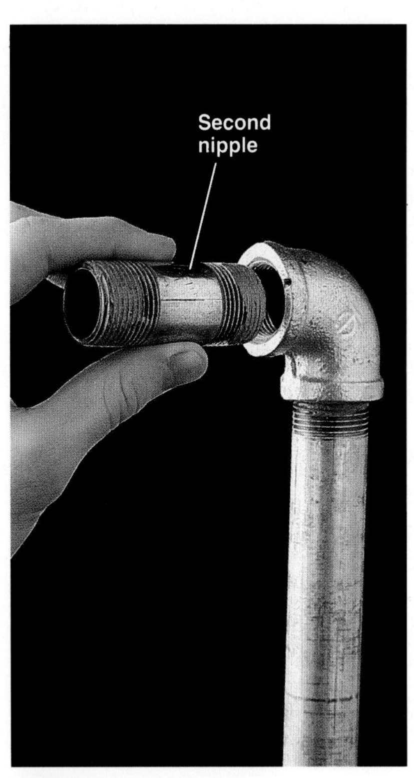

10 Screw second nipple onto other fitting. Tighten with a pipe wrench.

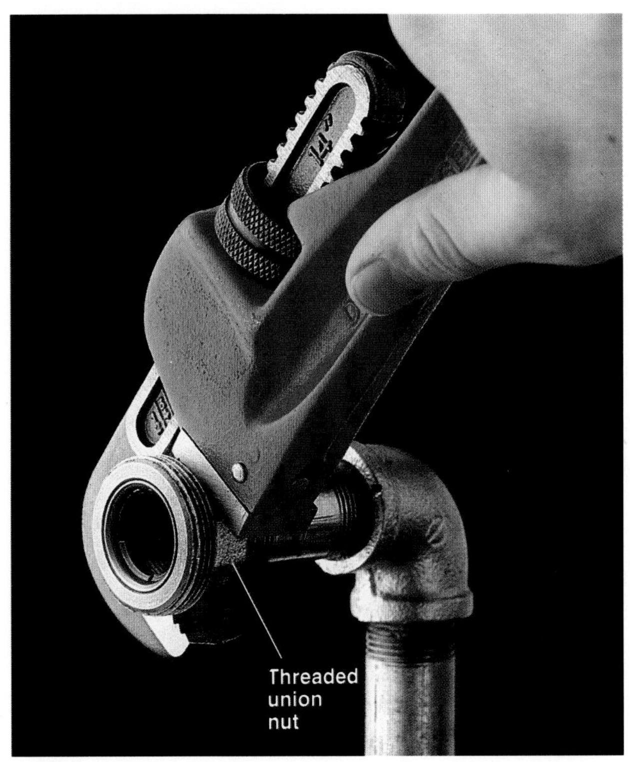

11 Screw threaded union nut onto second nipple. Tighten with a pipe wrench. Turn pipes into alignment, so that lip of hubbed union nut fits inside threaded union nut.

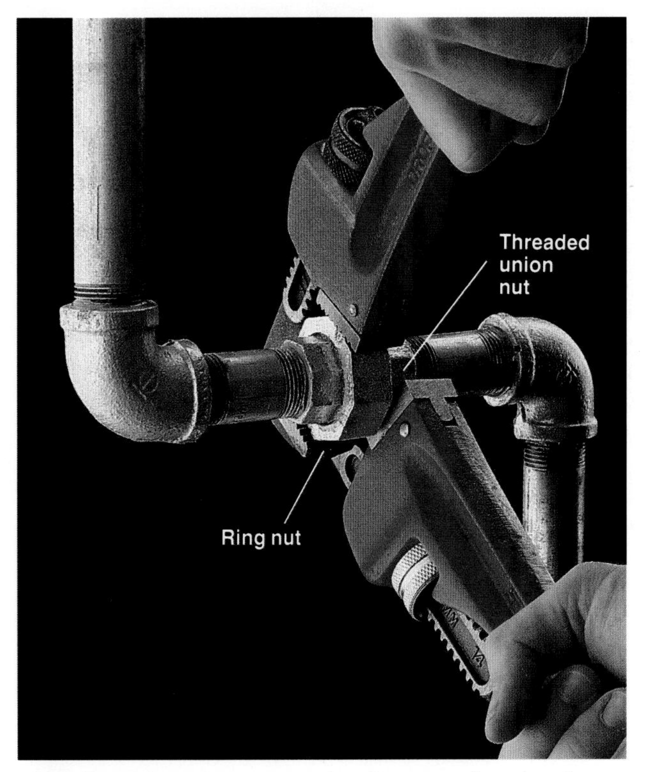

12 Complete the connection by screwing the ring nut onto the threaded union nut. Tighten ring nut with pipe wrenches.

Working with Cast Iron

Cast-iron pipe often is found in older homes, where it is used for large drain-waste-vent pipes, especially the main stack and sewer service lines. It can be identified by its dark color, rough surface, and large size. Cast-iron pipes in home drains usually are 3" or more in diameter.

Cast-iron pipes may rust through or hubbed fittings (below) may leak. If your house is more than 30 years old, you may find it necessary to replace a cast-iron pipe or joint.

Cast iron is heavy and difficult to cut and fit. One 5-ft. section of 4" pipe weighs 60 pounds. For this reason, leaky cast-iron pipe usually is replaced with a new plastic pipe of the same diameter. Plastic pipe can be joined to cast iron easily, using a banded coupling (below).

Cast iron is best cut with a rented tool called a *snap cutter.* Snap cutter designs vary, so follow the rental dealer's instructions for using the tool.

Everything You Need:

Tools: tape measure, chalk, adjustable wrenches, rented cast iron snap cutter (or hacksaw), ratchet wrench, screwdriver.

Materials: riser clamps or strap hangers, two wood blocks, 2½" wallboard screws, banded couplings, plastic replacement pipe.

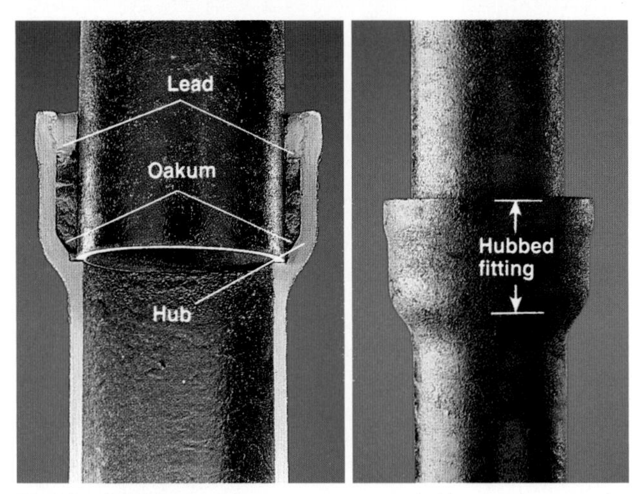

Hubbed fittings (shown cut away, left) may be used to join old cast-iron pipe. Hubbed pipe has a straight end and a flared end. The straight end of one pipe fits inside the hub of the next pipe. Joints are sealed with packing material (oakum) and lead. Repair leaky joints by cutting out the entire hubbed fitting and replacing with plastic pipe.

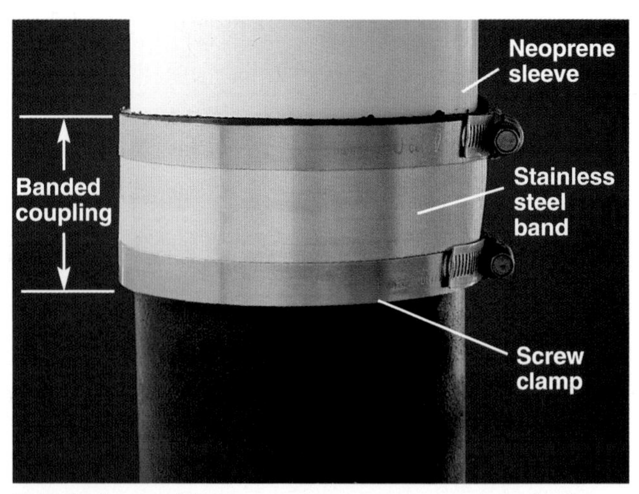

Banded couplings may be used to replace leaky cast-iron pipe with a PVC or ABS plastic pipe. The new plastic pipe is connected to the remaining cast-iron pipe with banded coupling. Banded coupling has a neoprene sleeve that seals the joint. Pipes are held together with stainless steel bands and screw clamps.

Before cutting a horizontal run of cast-iron drain pipe, make sure it is supported with strap hangers every 5 feet and at every joint connection.

Before cutting a vertical run of cast-iron pipe, make sure it is supported at every floor level with a riser clamp. Never cut apart pipe that is not supported.

How to Repair & Replace a Section of Cast-iron Pipe

1 Use chalk to mark cut lines on the cast-iron pipe. If replacing a leaky hub, mark at least 6" on each side of hub.

2 Support lower section of pipe by installing a riser clamp flush against bottom plate or floor.

3 Support upper section of pipe by installing a riser clamp 6" above pipe section to be replaced. Attach wood blocks to the studs with 2½" wallboard screws, so that the riser clamp rests on tops of blocks.

(continued next page)

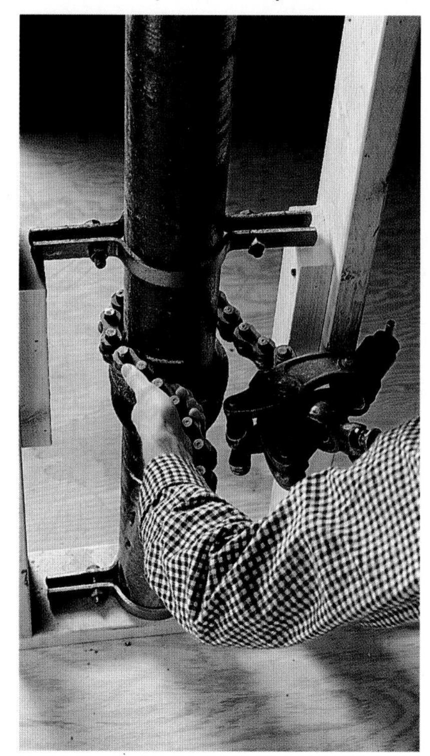

4 Wrap chain of the snap cutter around the pipe, so that the cutting wheels are against the chalk line.

5 Tighten the chain and snap the pipe according to the tool manufacturer's directions.

6 Repeat cutting at the other chalk line. Remove cut section of pipe.

7 Cut a length of PVC or ABS plastic pipe to be ½" shorter than the section of cast-iron pipe that has been cut away.

8 Slip a banded coupling and a neoprene sleeve onto each end of the cast-iron pipe.

9 Make sure the cast-iron pipe is seated snugly against the rubber separator ring molded into the interior of the sleeve.

10 Fold back the end of each neoprene sleeve, until the molded separator ring on the inside of the sleeve is visible.

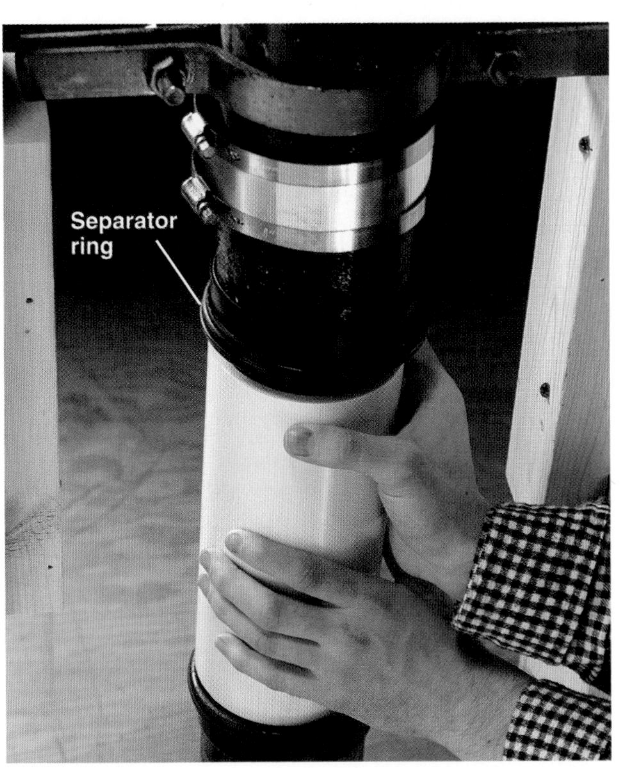

11 Position the new plastic pipe so it is aligned with the cast-iron pipes.

12 Roll the ends of the neoprene sleeves over the ends of the new plastic pipe.

13 Slide stainless steel bands and clamps over the neoprene sleeves.

14 Tighten the screw clamps with a ratchet wrench or screwdriver.

New Installation

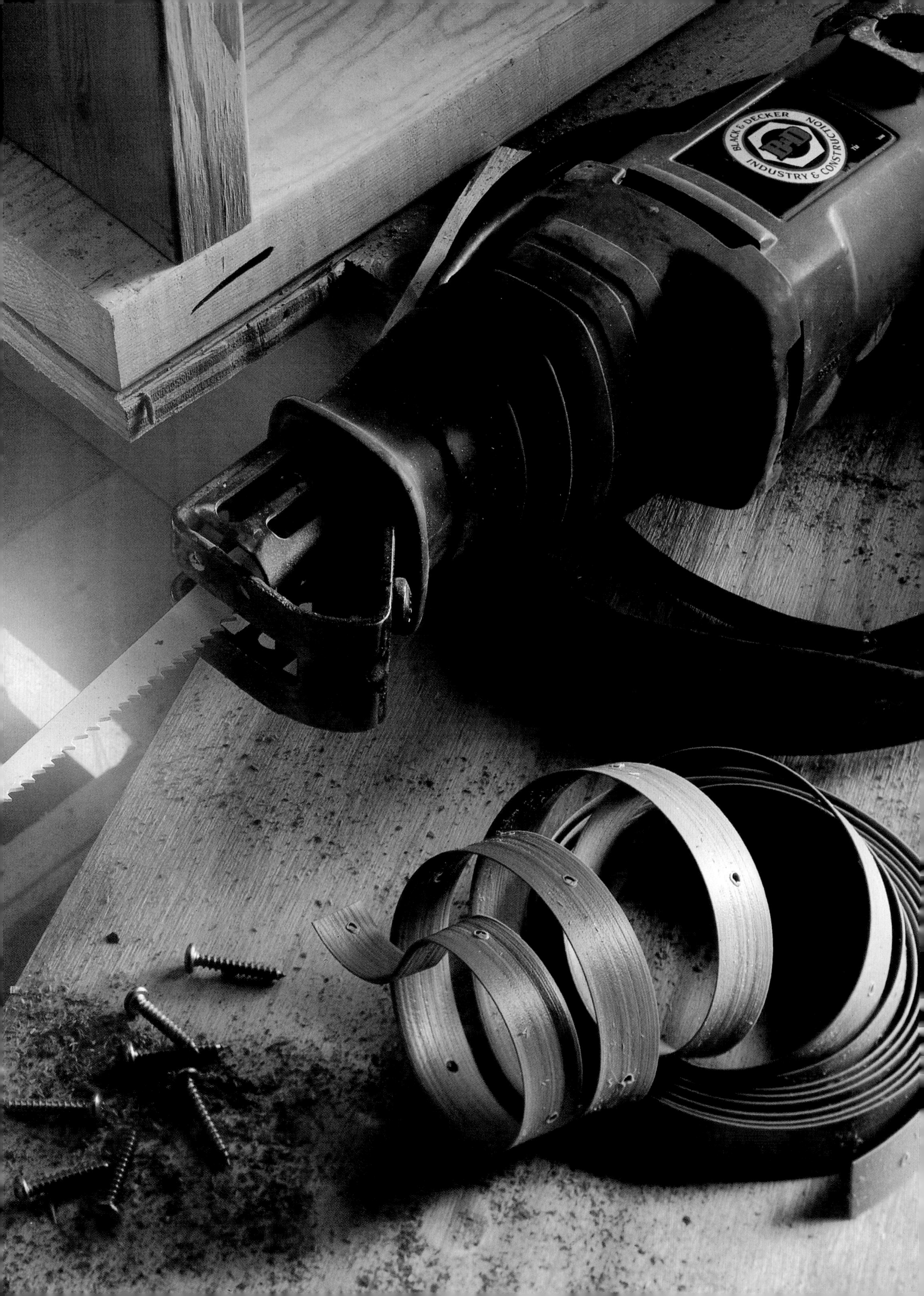
BLACK & DECKER
INDUSTRY & CONSTRUCTION

Use 2 x 6 studs to frame "wet walls" when constructing a new bathroom or kitchen. Thicker walls provide more room to run drain pipes and main waste-vent stacks, making installation much easier.

Installing New Plumbing

A major plumbing project is a complicated affair that often requires demolition and carpentry skills. Bathroom or kitchen plumbing may be unusable for several days while completing the work, so make sure you have a backup bathroom or kitchen space to use during this time.

To ensure that your project goes quickly, always buy plenty of pipe and fittings—at least 25% more than you think you need. Making several extra trips to the building center for last-minute fittings is a nuisance, and it can add many hours of time to your project. Always purchase from a reputable retailer that will allow you to return leftover fittings for credit.

The how-to projects on the following pages demonstrate standard plumbing techniques, but should not be used as a literal blueprint for your own work. Pipe and fitting sizes, fixture layout, and pipe routing will always vary according to individual circumstances. When planning your project, carefully read all the information in the Planning section, especially the material on Understanding Plumbing Codes (pages 24 to 29). Before you begin work, create a detailed plumbing plan to guide your work and help you obtain the required permits. This section includes information on:

- Plumbing Bathrooms (pages 78 to 127)
- Plumbing a Kitchen (pages 122 to 141)
- Installing Outdoor Plumbing (pages 142 to 147)

Tips for Installing New Plumbing

Use masking tape to mark the locations of fixtures and pipes on the walls and floors. Read the layout specifications that come with each sink, tub, or toilet, then mark the drain and supply lines accordingly. Position the fixtures on the floor, and outline them with tape. Measure and adjust until the arrangement is comfortable to you and meets minimum clearance specifications. If you are working in a finished room, prevent damage to wallpaper or paint by using self-adhesive notes to mark the walls.

Consider the location of cabinets when roughing in the water supply and drain stub-outs. You may want to temporarily position the cabinets in their final locations before completing the drain and water supply runs.

Install control valves at the points where the new branch supply lines meet the main distribution pipes. By installing valves, you can continue to supply the rest of the house with water while you are working on the new branches.

(continued next page)

Tips for Installing New Plumbing (continued)

Framing Member	Maximum Hole Size	Maximum Notch Size
2 × 4 loadbearing stud	$1\frac{7}{16}$" diameter	$\frac{7}{8}$" deep
2 × 4 non-loadbearing stud	$2\frac{1}{2}$" diameter	$1\frac{7}{16}$" deep
2 × 6 loadbearing stud	$2\frac{1}{4}$" diameter	$1\frac{3}{8}$" deep
2 × 6 non-loadbearing stud	$3\frac{5}{16}$" diameter	$2\frac{3}{16}$" deep
2 × 6 joists	$1\frac{1}{2}$" diameter	$\frac{7}{8}$" deep
2 × 8 joists	$2\frac{3}{8}$" diameter	$1\frac{1}{4}$" deep
2 × 10 joists	$3\frac{1}{16}$" diameter	$1\frac{1}{2}$" deep
2 × 12 joists	$3\frac{3}{4}$" diameter	$1\frac{7}{8}$" deep

Framing member chart shows the maximum sizes for holes and notches that can be cut into studs and joists when running pipes. Where possible, use notches rather than bored holes, because pipe installation is usually easer. When boring holes, there must be at least $\frac{5}{8}$" of wood between the edge of a stud and the hole, and at least 2" between the edge of a joist and the hole. Joists can be notched only in the end one-third of the overall span; never in the middle one-third of the joist. When two pipes are run through a stud, the pipes should be stacked one over the other, never side by side.

Create access panels so that in the future you will be able to service fixture fittings and shutoff valves located inside the walls. Frame an opening between studs, then trim the opening with wood moldings. Cover the opening with a removable plywood panel the same thickness as the wall surface, then finish it to match the surrounding walls.

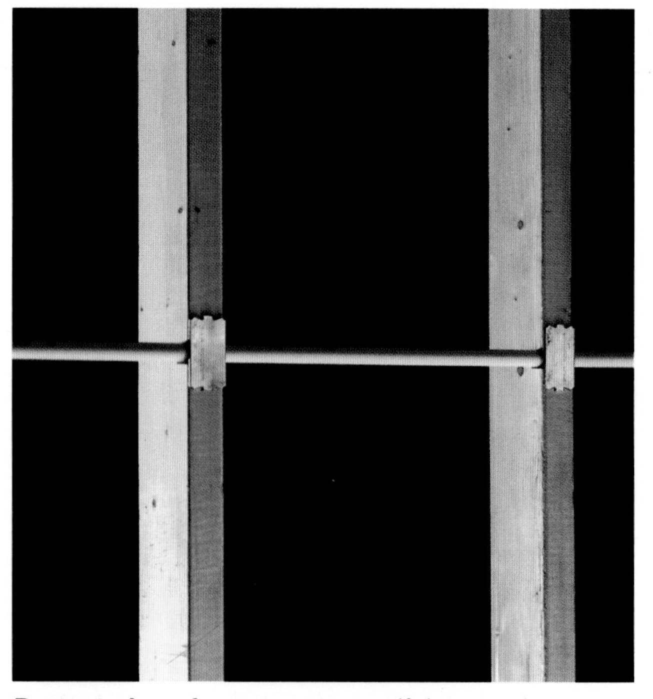

Protect pipes from punctures, if they are less than $1\frac{1}{4}$" from the front face of wall studs or joists, by attaching metal protector plates to the framing members.

Test-fit materials before solvent-gluing or soldering joints. Test-fitting ensures that you have the correct fittings and enough pipe to do the job, and can help you avoid lengthy delays during installation.

Support pipes adequately. Horizontal and vertical runs of DWV and water supply pipe must be supported at minimum intervals, which are specified by your local Plumbing Code (page 28). A variety of metal and plastic materials are available for supporting plumbing pipes (page 40).

Use plastic bushings to help hold plumbing pipes securely in holes bored through wall plates, studs, and joists. Bushings can help to cushion the pipes, preventing wear and reducing rattling.

Install extra T-fittings on new drain and vent lines so that you can pressure-test the system when the building inspector reviews your installation (pages 30 to 37). A new DWV line should have these extra T-fittings near the points where the new branch drains and vent pipes reach the main wast-vent stack.

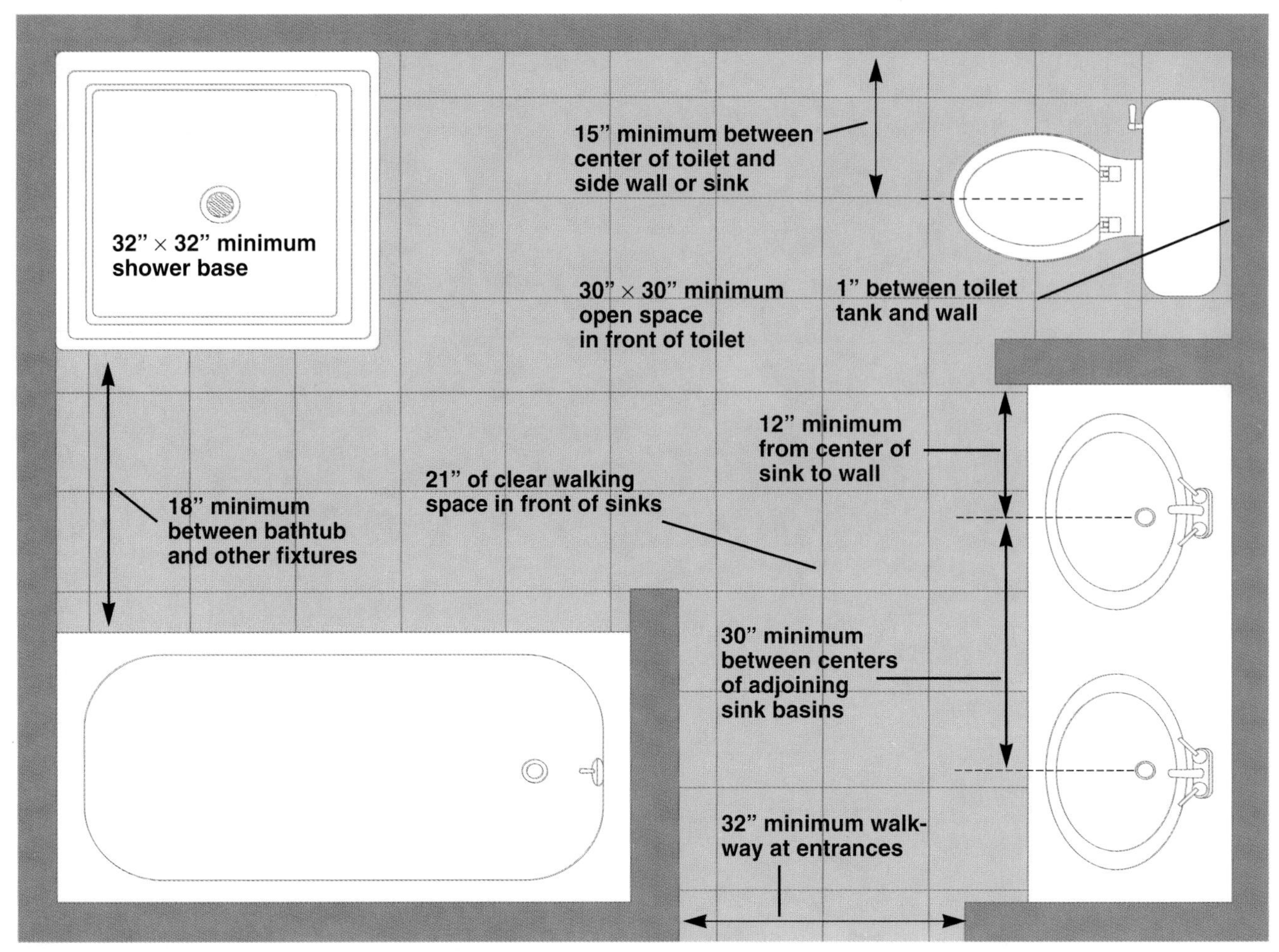

Follow minimum clearance guidelines when planning the locations of bathroom fixtures. Unobstructed access to the fixtures is fundamental to creating a comfortable, safe, and efficient bathroom.

Plumbing Bathrooms

Adding a new bathroom or updating an old one is a sure way to add value to your home. The personal comfort gained from a custom master bathroom can add a new dimension to your relaxation time. And adding a full bath in the basement or a half bath next to the kitchen offers convenience for both family members and guests. When planned and built correctly, a new or remodeled bathroom also improves the resale value of your home.

The first step when planning a bathroom is determining the type of bathroom you want. Do you have your heart set on expanding into a spare bedroom to create the ultimate master bathroom, or do you simply need a functional half bath for convenience? In this section, you'll see three demonstration projects that represent the full range of bathrooms, from a spacious master bathroom to a simple half bath (opposite page).

Next, you'll need to decide on the type of fixtures you need. Visit your local building centers to determine the range of fixtures available and their prices, and visit model homes and remodeling exhibitions to see how professionals arrange and install the fixtures you plan to use.

Once you have determined the scope of your project and settled on a budget, you can develop working plans for your bathroom. When creating a floor plan, always follow minimum clearance guidelines (above), and think about where the drain and water supply pipes will run. You can save yourself many hours of work by positioning fixtures so the pipes can be routed with simple, straight runs rather than with many complicated bends. Make sure that your project complies with local Plumbing Code regulations for bathroom plumbing.

Master bathroom can include luxury features, such as a large whirlpool tub or multi-jet shower. Our project contains both these features, as well as a pedestal vanity sink and toilet. A spacious bathroom may require substantial construction work if you intend to expand into an adjoining room. See pages 80 to 87.

Basement bathroom is ideal if you have bedrooms or finished recreation areas in your basement. Our project includes a shower, toilet, and vanity sink. Plumbing a basement bathroom may require that you break into the concrete floor to connect drain pipes. See pages 88 to 93.

Half bath can be easily added to a room that shares a "wet wall" with a kitchen or other bathroom. Our project includes a toilet and vanity sink. See pages 94 to 95.

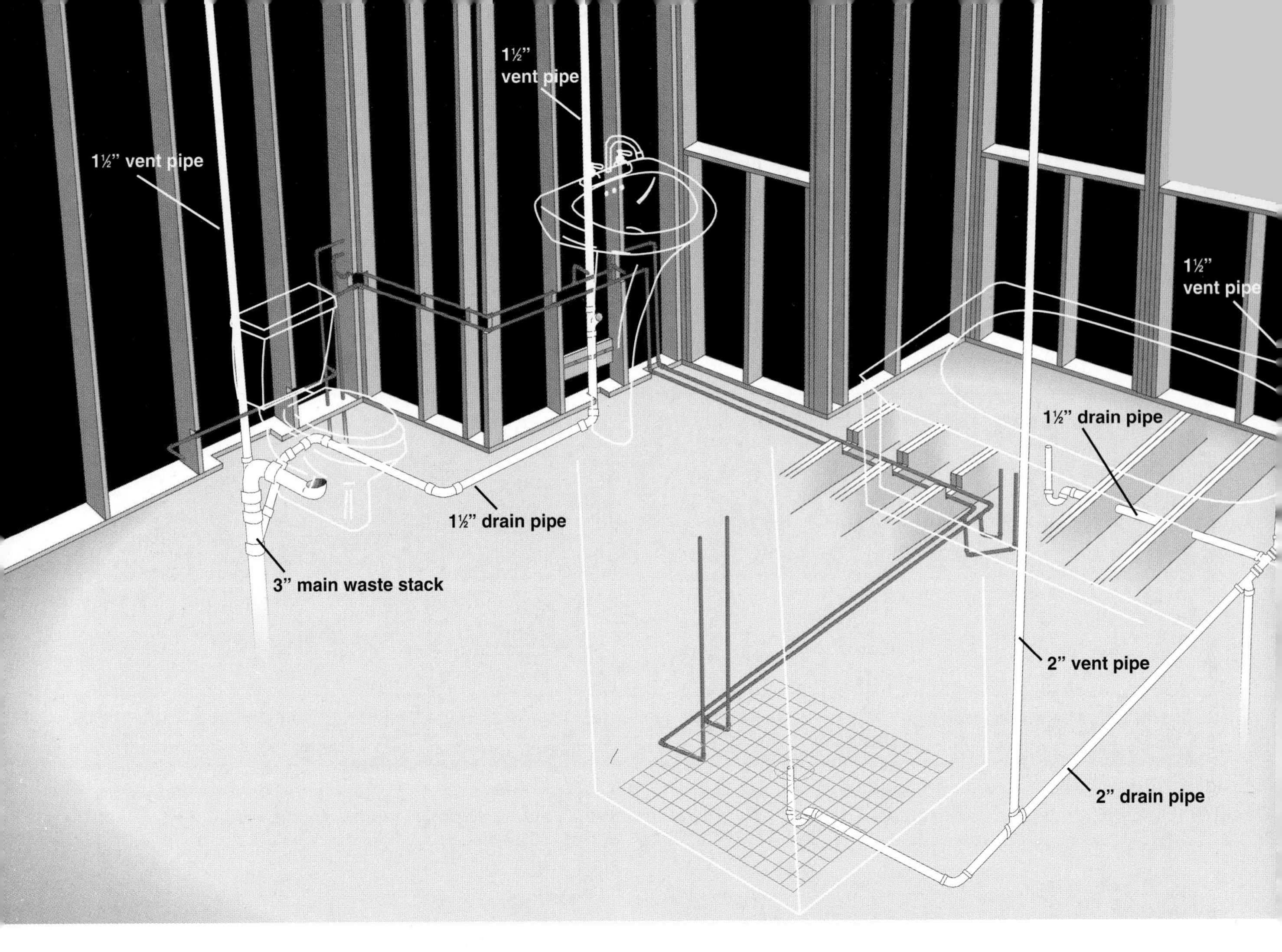

Plumbing a Master Bathroom

A large bathroom has more plumbing fixtures and consumes more water than any other room in your house. For this reason, a master bath has special plumbing needs.

Frame bathroom "wet walls" with 2 × 6 studs, to provide plenty of room for running 3" pipes and fittings. If your bathroom includes a heavy whirlpool tub, you will need to strengthen the floor by installing "sister" joists alongside the existing floor joists underneath the tub.

For convenience, our project is divided into the following sequences:

- How to Install DWV Pipes for the Toilet & Sink (pages 81 to 83)
- How to Install DWV Pipes for the Tub & Shower (pages 84 to 85)
- How to Connect the Drain Pipes & Vent Pipes to the Main Waste-Vent Stack (page 86)
- How to Install the Water Supply Pipes (page 87)

Our demonstration bathroom is a second-story master bath. We are installing a 3" vertical drain pipe to service the toilet and the vanity sink, and a 2" vertical pipe to handle the tub and shower drains. The branch drains for the sink and bathtub are 1½" pipes; for the shower, 2" pipe. Each fixture has its own vent pipe extending up into the attic, where they are joined together and connected to the main stack.

How to Install DWV Pipes for the Toilet & Sink

1 Use masking tape to outline the locations of the fixtures and pipe runs on the subfloor and walls. Mark the location for a 3" vertical drain pipe on the sole plate in the wall behind the toilet. Mark a 4½"-diameter circle for the toilet drain on the subfloor.

2 Cut out the drain opening for the toilet, using a jig saw. Mark and remove a section of flooring around the toilet area, large enough to provide access for installing the toilet drain and for running drain pipe from the sink. Use a circular saw with blade set to the thickness of the flooring to cut through the subfloor.

3 If a floor joist interferes with the toilet drain, cut away a short section of the joist and box-frame the area with double headers.The framed opening should be just large enough to install the toilet and sink drains.

4 To create a path for the vertical 3" drain pipe, cut a 4½" × 12" notch in the sole plate of the wall behind the toilet. Make a similar cutout in the double wall plate at the bottom of the joist cavity. From the basement, locate the point directly below the cutout by measuring from a reference point, such as the main waste-vent stack.

5 Mark the location for the 3" drain pipe on the basement ceiling, then drill a 1"-diameter hole up through the center of the marked area. Direct the beam of a bright flashlight up into the hole, then return to the bathroom and look down into the wall cavity. If you can see light, return to the basement and cut a 4½"-diameter hole centered over the test hole.

(continued next page)

6 Measure and cut a length of 3" drain pipe to reach from the bathroom floor cavity to a point flush with the bottom of the ceiling joists in the basement. Solvent-glue a 3" × 3" × 1½" Y-fitting to the top of the pipe, and a low-heel vent 90° fitting above the Y. The branch inlet on the Y should face toward the sink location; the front inlet on the low-heel should face forward. Carefully lower the pipe into the wall cavity.

7 Lower the pipe so the bottom end slides through the opening in the basement ceiling. Support the pipe with vinyl pipe strap wrapped around the low-heel vent 90° fitting and screwed to framing members.

8 Use a length of 3" pipe and a 4" × 3" reducing elbow to extend the drain out to the toilet location. Make sure the drain slopes at least ⅛" per foot toward the wall, then support it with pipe strap attached to the joists. Insert a short length of pipe into the elbow, so it extends at least 2" above the subfloor. After the new drains are pressure tested, this stub-out will be cut flush with the subfloor and fitted with a toilet flange.

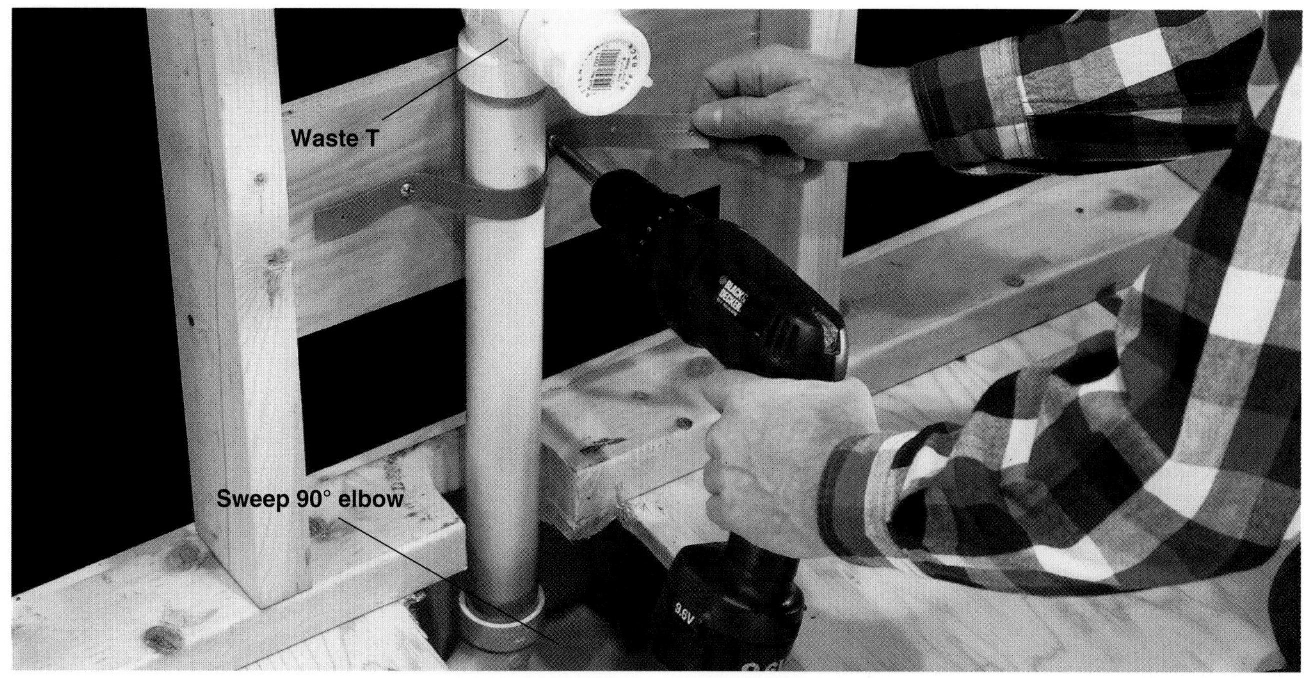

9 Notch out the sole plate and subfloor below the sink location. Cut a length of 1½" plastic drain pipe, then solvent-glue a waste T to the top of the pipe and a sweep 90° elbow to the bottom. NOTE: the distance from the subfloor to the center of the waste T should be 14" to 18". The branch of the T should face out, and the discharge on the elbow should face toward the toilet location. Adjust the pipe so the top edge of the elbow nearly touches the bottom of the sole plate. Anchor it with a ¾"-thick backing board nailed between the studs.

10 Dry-fit lengths of 1½" drain pipe and elbows to extend the sink drain to the 3" drain pipe behind the toilet. Use a right-angle drill to bore holes in joists, if needed. Make sure the horizontal drain pipe slopes at least ¼" per foot toward the vertical drain. When satisfied with the layout, solvent-glue the pieces together and support the drain pipe with vinyl pipe straps attached to the joists.

11 In the top plates of the walls behind the sink and toilet, bore ½"-diameter holes up into the attic. Insert pencils or dowels into the holes, and tape them in place. Enter the attic and locate the pencils, then clear away insulation and cut 2"-diameter holes for the vertical vent pipes. Cut and install 1½" vent pipes running from the toilet and sink drain at least 1 ft. up into the attic.

How to Install DWV Pipes for the Tub & Shower

1 On the subfloor, use masking tape to mark the locations of the tub and shower, the water supply pipes, and the tub and shower drains, according to your plumbing plan. Use a jig saw to cut out a 12"-square opening for each drain, and drill 1"-diameter holes in the subfloor for each water supply riser.

2 When installing a large whirlpool tub, cut away the subfloor to expose the full length of the joists under the tub, then screw or bolt a second joist, called a *sister,* against each existing joist. Make sure both ends of each joist are supported by loadbearing walls.

3 In a wall adjacent to the tub, establish a route for a 2" vertical waste-vent pipe running from basement to attic.This pipe should be no more than 3½ ft. from the bathtub trap. Then, mark a route for the horizontal drain pipe running from the bathtub drain to the waste-vent pipe location. Cut 3"-diameter holes through the centers of the joists for the bathtub drain.

4 Cut and install a vertical 2" drain pipe running from basement to the joist cavity adjoining the tub location, using the same technique as for the toilet drain (steps 4 to 6, pages 81 to 82). At the top of the drain pipe, use assorted fittings to create three inlets: branch inlets for the bathtub and shower drains, and a 1½" top inlet for a vent pipe running to the attic.

5 Dry-fit a 1½" drain pipe running from the bathtub drain location to the vertical waste-vent pipe in the wall. Make sure the pipe slopes ¼" per foot toward the wall. When satisfied with the layout, solvent-glue the pieces together and support the pipe with vinyl pipe straps attached to the joists.

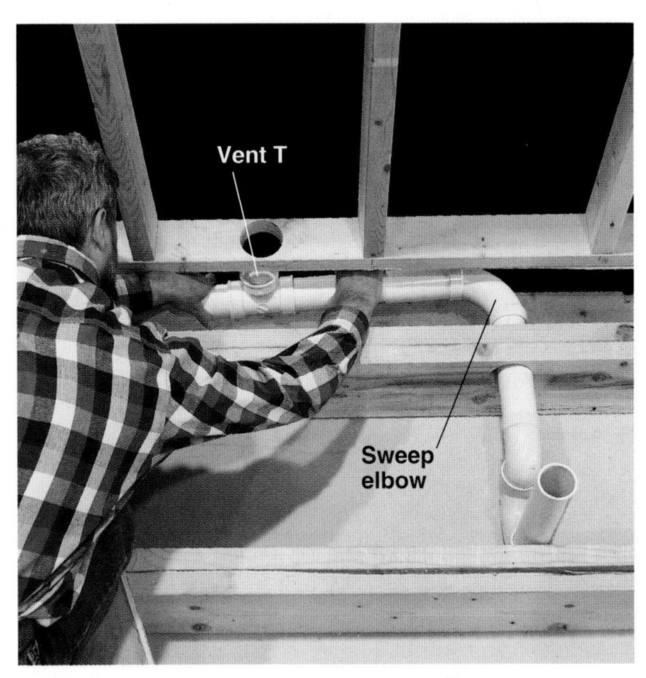

6 Dry-fit a 2" drain pipe from the shower drain to the vertical waste-vent pipe near the tub. Install a solvent-glued trap at the drain location, and cut a hole in the sole plate and insert a 2" × 2" × 1½" vent T within 5 ft. of the trap. Make sure the drain is sloped ¼" per foot downward away from the shower drain. When satisfied with the layout, solvent-glue the pipes together.

7 Cut and install vertical vent pipes for the bathtub and shower, extending up through the wall plates and at least 1 ft. into the attic. These vent pipes will be connected in the attic to the main waste-vent stack. In our project, the shower vent is a 2" pipe, while the bathtub vent is a 1½" pipe.

How to Connect the Drain Pipes to the Main Waste-Vent Stack

1 In the basement, cut into the main waste-vent stack and install the fittings necessary to connect the 3" toilet-sink drain and the 2" bathtub-shower drain. In our project, we created an assembly made of a waste T-fitting with an extra side inlet and two short lengths of pipe, then inserted it into the existing waste-vent stack using branded couplings (pages 68 to 71). Make sure the T-fittings are positioned so the drain pipes will have the proper downward slope toward the stack.

2 Dry-fit Y-fittings with 45° elbows onto the vertical 3" and 2" drain pipes. Position the horizontal drain pipes against the fittings, and mark them for cutting. When satisfied with the layout, solvent-glue the pipes together, then support the pipes every 4 ft. with vinyl pipe straps. Solvent-glue cleanout plugs on the open inlets on the Y-fittings.

How to Connect the Vent Pipes to the Main Waste-Vent Stack

1 In the attic, cut into the main waste-vent stack and install a vent T-fitting, using banded couplings. The side outlet on the vent T should face the new 2" vent pipe running down to the bathroom. Attach a test T-fitting to the vent T. NOTE: If your stack is cast iron, make sure to adequately support it before cutting into it (page 69).

2 Use elbows, vent T-fittings, reducers, and lengths of pipe as needed to link the new vent pipes to the test T-fitting on the main waste-vent stack. Vent pipes can be routed in many ways, but you should make sure the pipes have a slight downward angle to prevent moisture from collecting in the pipes. Support the pipes every 4 ft.

How to Install the Water Supply Pipes

1 After shutting off the water, cut into existing supply pipes and install T-fittings for new branch lines. Notch out studs and run copper pipes to the toilet and sink locations. Use an elbow and threaded female fitting to form the toilet stub-out. Once satisfied with the layout, solder the pipes in place.

2 Cut 1" × 4"-high notches around the wall, and extend the supply pipes to the sink location. Install reducing T-fittings and female threaded fittings for the sink faucet stub-outs. The stub-outs should be positioned about 18" above the floor, spaced 8" apart. Once satisfied with the layout, solder the joints, then insert ¾" blocking behind the stub-outs and strap them in place.

3 Extend the water supply pipes to the bathtub and shower. In our project, we removed the subfloor and notched the joists to run ¾" supply pipes from the sink to a whirlpool bathtub, then to the shower. At the bathtub, we used reducing T-fittings and elbows to create ½" risers for the tub faucet. Solder caps onto the risers; after the subfloor is replaced, the caps will be removed and replaced with shutoff valves.

4 At the shower location, use elbows to create vertical risers where the shower wet wall will be constructed. The risers should extend at least 6" above floor level. Support the risers with a ¾" backer board attached between joists. Solder caps onto the risers. After the shower stall is constructed, the caps will be removed and replaced with shutoff valves.

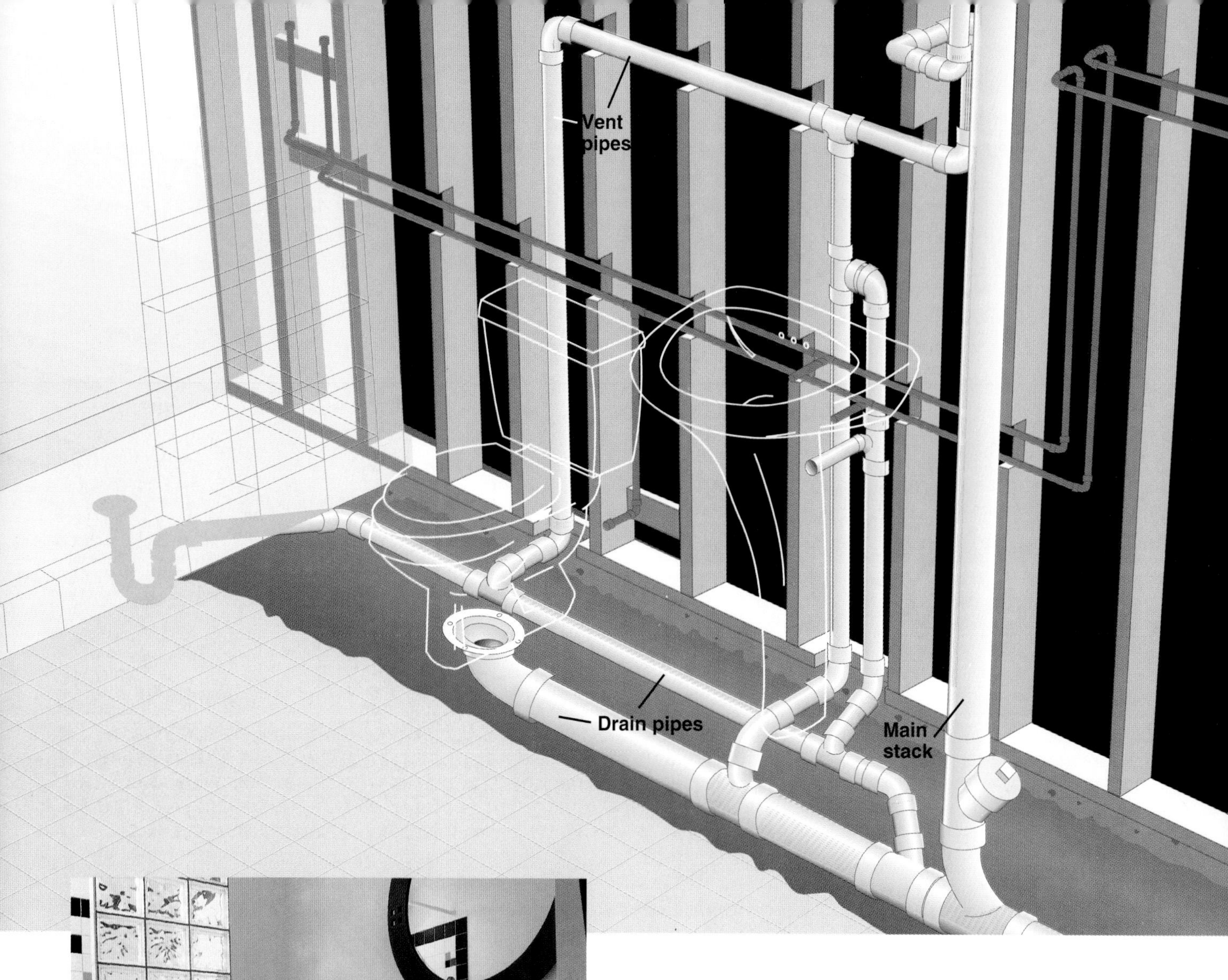

Our demonstration bathroom includes a shower, toilet, and vanity sink arranged in a line to simplify trenching. A 2" drain pipe services the new shower and sink; a 3" pipe services the new toilet. The drain pipes converge at a Y-fitting joined to the existing main drain. The shower, toilet, and sink have individual vent pipes that meet inside the wet wall before extending up into the attic, where they join the main waste-vent stack.

Plumbing a Basement Bath

When installing a basement bath, make sure you allow extra time for tearing out the concrete floor to accommodate the drains, and for construction of a wet wall to enclose supply and vent pipes. Constructing your wet wall with 2 × 6 studs and plates will provide ample room for running pipes. Be sure to schedule an inspection by a building official before you replace the concrete and cover the walls with wallboard.

Whenever possible, try to hold down costs by locating your basement bath close to existing drains and supply pipes.

How to Plumb a Basement Bath

1 Outline a 24"-wide trench on the concrete where new branch drains will run to the main drain. In our project, we ran the trench parallel to an outside wall, leaving a 6" ledge for framing a wet wall. Use a masonry chisel and hand maul to break up concrete near the stack.

2 Use a circular saw and masonry blade to cut along the outline, then break the rest of the trench into convenient chunks with a jackhammer. Remove any remaining concrete with a chisel. Excavate the trench to a depth about 2" deeper than the main drain. At vent locations for the shower and toilet, cut 3" notches in the concrete all the way to the wall.

3 Cut the 2 × 6 framing for the wet wall that will hold the pipes. Cut 3" notches in the bottom plate for the pipes, then secure the plate to the floor with construction adhesive and masonry nails. Install the top plate, then attach studs.

4 Assemble a 2" horizontal drain pipe for the sink and shower, and a 3" drain pipe for the toilet. The 2" drain pipe includes a solvent-glued trap for the shower, a vent T, and a waste T for the sink drain. The toilet drain includes a toilet bend and a vent T. Use elbows and straight lengths of pipe to extend the vent and drain pipes to the wet wall. Make sure the vent fittings angle upward from the drain pipe at least 45°.

(continued next page)

5 Use pairs of stakes with vinyl support straps slung between them to cradle drain pipes in the proper position (inset). The drain pipes should be positioned so they slope ¼" per foot down toward the main drain.

6 Assemble the fittings required to tie the new branch drains into the main drain. In our project, we will be cutting out the cleanout and sweep on the main waste-vent stack in order to install a new assembly that includes a Y-fitting to accept the two new drain pipes.

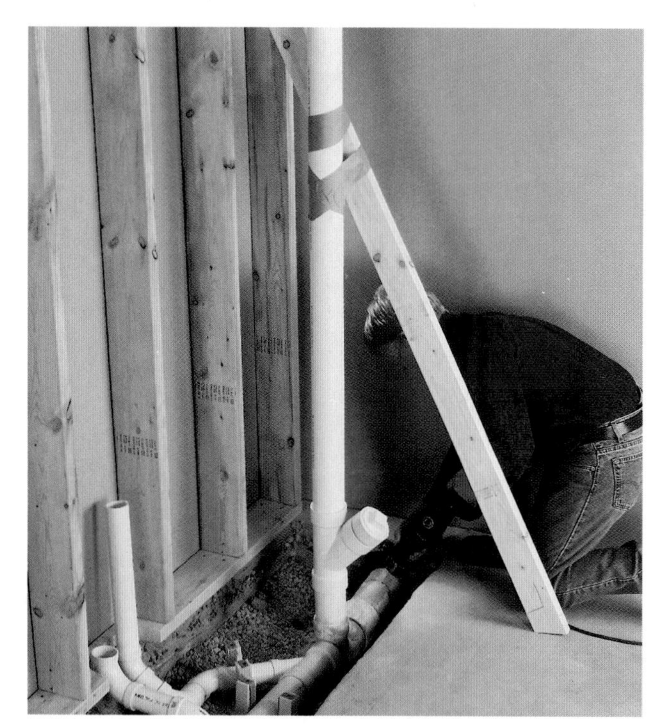

7 Support the main waste-vent stack before cutting. Use a 2 × 4 for a plastic stack, or riser clamps (page 161) for a cast-iron stack. Using a reciprocating saw (or snap cutter), cut into the main drain as close as possible to the stack.

8 Cut into the stack above the cleanout and remove the pipe and fittings. Wear rubber gloves, and have a bucket and plastic bags ready, as old pipes and fittings may be coated with messy sludge.

9 Test-fit, then solvent-glue the new cleanout and reducing Y assembly into the main drain. Support the weight of the stack by adding sand underneath the Y, but leave plenty of space around the end for connecting the new branch pipes.

10 Working from the reducing Y, solvent-glue the new drain pipes together. Be careful to maintain proper slope of the drain pipes when gluing. Be sure the toilet and shower drains extend at least 2" above the floor level.

11 Check for leaks by pouring fresh water into each new drain pipe. If no leaks appear, cap or plug the drains with rags to prevent sewer gas from leaking into the work area as you complete the installation.

(continued next page)

12 Run 2" vent pipes from the drains up the inside of the wet wall. Notch the studs and insert a horizontal vent pipe, then attach the vertical vent pipes with an elbow and vent T-fitting. Test-fit all pipes, then solvent-glue them in place.

13 Route the vent pipe from the wet wall to a point below a wall cavity running from the basement to the attic. NOTE: If there is an existing vent pipe in the basement, you can tie into this pipe rather than run the vent to the attic.

14 If you are running vent pipes in a two-story home, remove sections of wall surface as needed to bore holes for running the vent pipe through wall plates. Feed the vent pipe up into the wall cavity from the basement.

15 Wedge the vent pipe in place while you solvent-glue the fittings. Support the vent pipe at each floor with vinyl pipe straps. Do not patch the walls until your work has been inspected by a building official.

16 Cut into the main stack in the attic, and install a vent T-fitting, using banded couplings. (If the stack is cast iron, make sure to support it adequately above and below the cuts.) Attach a test T-fitting to the vent T, then join the new vent pipe to the stack, using elbows and lengths of straight pipe as needed.

17 Shut off the main water supply, cut into the water supply pipes as near as possible to the new bathroom, and install T-fittings. Install full-bore control valves on each line, then run ¾" branch supply pipes down into the wet wall by notching the top wall plate. Extend the pipes across the wall by notching the studs.

18 Use reducing T-fittings to run ½" supplies to each fixture, ending with female threaded adapters. Install backing boards, and strap the pipes in place. Attach metal protector plates over notched studs to protect pipes. After having your work approved by a building official, fill in around the pipes with dirt or sand, then mix and pour new concrete to cover the trench. Trowel the surface smooth, and let the cement cure for 3 days before installing fixtures.

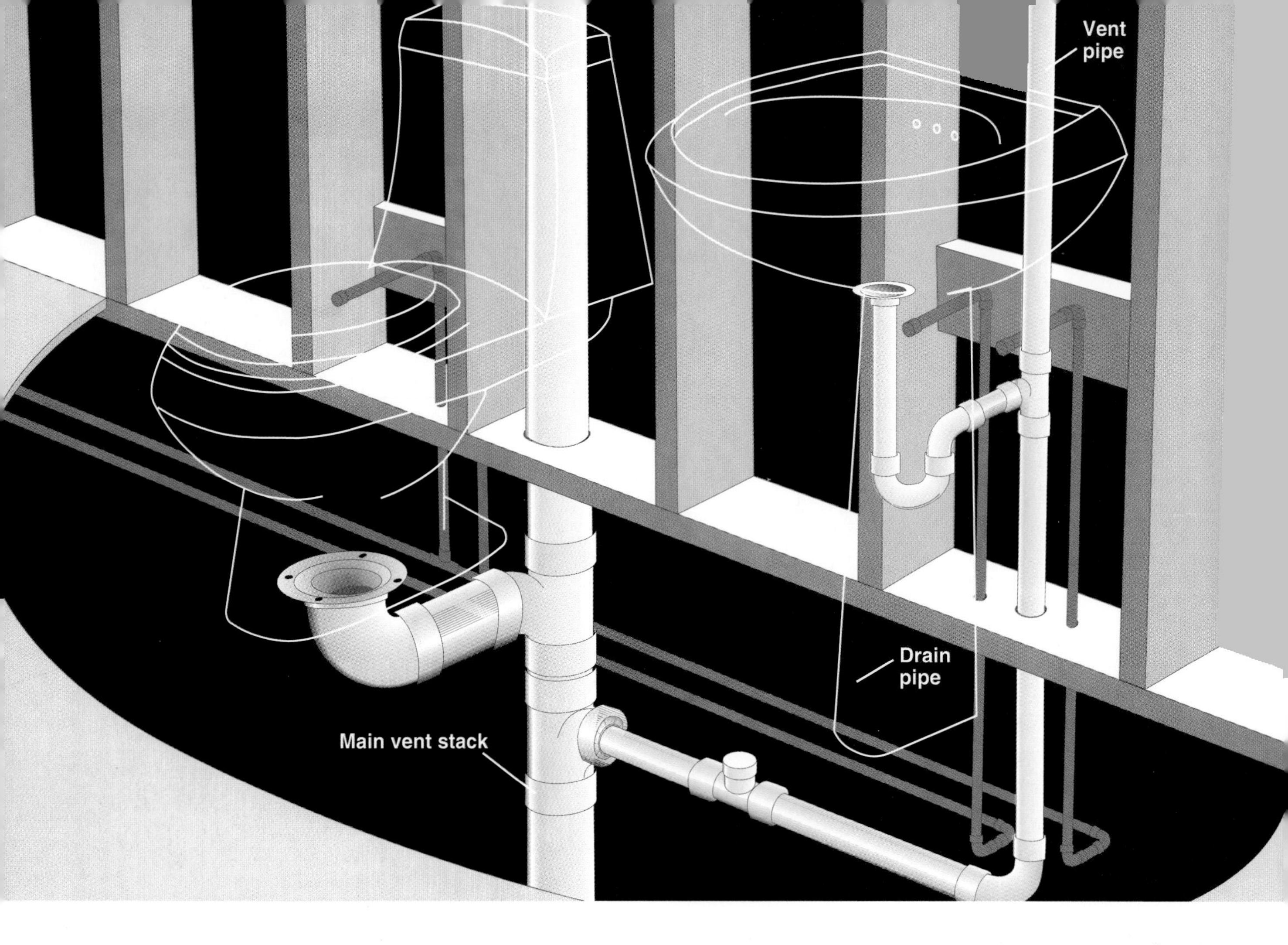

Plumbing a Half Bath

A first-story half bath is easy to install when located behind a kitchen or existing bathroom, because you can take advantage of accessible supply and DWV lines. It is possible to add a half bath on an upper story or in a location distant from existing plumbing, but the complexity and cost of the project may be increased considerably.

Be sure that the new fixtures are adequately vented. We vented the pedestal sink with a pipe that runs up the wall a few feet before turning to join the main stack. However, if there are higher fixtures draining into the main stack, you would be required to run the vent up to a point at least 6" above the highest fixture before splicing it into the main stack or an existing vent pipe. When the toilet is located within 6 ft. of the stack, as in our design, it requires no additional vent pipe.

The techniques for plumbing a half bath are similar to those used for a master bathroom. Refer to pages 80 to 87 for more detailed information.

In our half bath, the toilet and sink are close to the main stack for ease of installation, but are spaced far enough apart to meet minimum allowed distances between fixtures. Check your local Code for any restrictions in your area. Generally, there should be at least 15" from the center of the toilet drain to a side wall or fixture, and a minimum of 21" of space between the front edge of the toilet and the wall.

How to Plumb a Half Bath

1 Locate the main waste-vent stack in the wet wall, and remove the wall surface behind the planned location for the toilet and sink. Cut a 4½"-diameter hole for the toilet flange (centered 12" from the wall, for most toilets). Drill two ¾" holes through the sole plate for sink supply lines and one hole for the toilet supply line. Drill a 2" hole for the sink drain.

2 In the basement, cut away a section of the stack and insert two waste T-fittings. The top fitting should have a 3" side inlet for the toilet drain; the bottom fitting requires a 1½" reducing bushing for the sink drain. Install a toilet bend and 3" drain pipe for the toilet, and install a 1½" drain pipe with a sweep elbow for the sink.

3 Tap into water distribution pipes with ¾" × ½" reducing T-fittings, then run ½" copper supply pipes through the holes in the sole plate to the sink and toilet. Support all pipes at 4-ft. intervals with strapping attached to joists.

4 Attach drop ear elbows to the ends of the supply pipes, and anchor them to blocking installed between studs. Anchor the drain pipe to the blocking, then run a vertical vent pipe from the waste T-fitting up the wall to a point at least 6" above the highest fixture on the main stack. Then, route the vent pipe horizontally and splice it into the vent stack with a vent T.

Select the shower, bathtub, or whirlpool first when planning your new bathroom. These fixtures, available in a limited range of colors and styles, set the tone for an entire room. Sinks, cabinets, tile, and accessories are available in many styles and colors, and can be matched easily to a new tub or whirlpool.

Installing Showers, Bathtubs & Whirlpools

Installing and hooking up plumbing for bathtubs and showers is a fairly simple job. Whirlpools are more complicated because they also require electrical hookups.

The most difficult task when installing tubs, showers, and whirlpools is moving these bulky fixtures up stairways and through narrow doorways. With a two-wheel dolly and a little help, however, the job is much easier. Measure your doorways and hallways to make sure that any large fixture you buy will fit through them.

If you do not plan to remove and replace your wall surfaces, you still should cut away at least 6" of wall surface above the tub or whirlpool, allowing easier access for installing fixtures.

This section shows:

- Installing showers (pages 98 to 103)
- Installing bathtubs (pages 104 to 106)
- How to install a tub surround (page 109)
- Installing whirlpools (pages 110 to 115)

Tips for Installing Showers, Bathtubs & Whirlpools

Choose the correct tub for your plumbing set-up. Alcove-installed tubs with only one side apron are sold as either "left-hand" or "right-hand" models, depending on the location of the predrilled drain and overflow holes in the tub. To determine which type you need, face into the alcove and look to see if the tub drain will be on your right or your left.

Install extra floor support if the floor joists below the planned bathtub or whirlpool location are too small, or spaced too far apart. Generally, you should attach additional "sister" joists under the tub area (page 75) if your current joists are 2 × 10 or smaller, or if they are more than 16" apart. If you are unsure about structural support issues, contact a building inspector or a professional contractor.

Add fiberglass insulation around the body of a bathtub to reduce noise and conserve heat. Before setting the tub in position, wrap unfaced batting around the tub, and secure it with string or twine. For showers and deck-mounted whirlpools, insulate between framing members.

Chisel out mortar around plumbing stub-outs set into basement floors, to make room for drain fittings that slip over the drain pipe. Use a hand maul and masonry chisel, directing the blows away from the pipe, until you have exposed the pipe about 1" below floor level. Wear eye protection.

Installing Showers

Use prefabricated shower panels and a plastic shower base to build an inexpensive, easy-to-install shower stall. For an elegant look and greater durability, you may prefer a custom-tiled shower stall.

Building Code requires that each home have a tub in at least one bathroom, but in spare bathrooms or guest baths, you can replace the tub with a shower stall to create space for storage or a second sink.

Everything You Need:

Tools: marker, level, channel-type pliers, hacksaw, hole saw, drill, caulk gun.

Materials: 2 × 4, 1 × 4 lumber, 10d nails, pipe straps, shower pipes & fittings, dry-set mortar, soap, wood screws, wallboard screws, panel adhesive, carpet scrap, tub & tile caulk.

Ceramic tile for custom showers is installed the same way as ceramic wall tile. Ceramic shower accessories are mortared in place during the tile installation.

Optional Shower Features

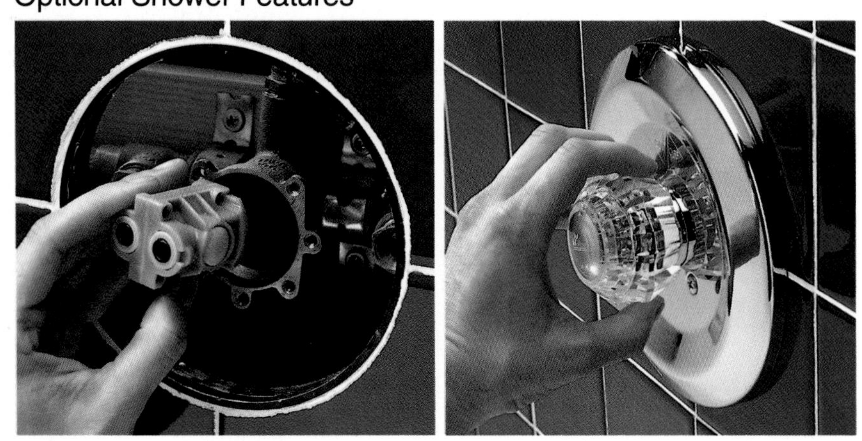

Anti-scald valves protect against sudden water temperature changes. Most types use a water-pressure regulator (left). Once installed, faucets with anti-scald valves look like standard faucets (right).

Storage cabinet with sliding doors holds bathing accessories. Some shower-surround panels have cabinets already built in.

Anatomy of a Shower

Shower stalls are available in many different sizes and styles, but the basic eiements are the same. Most shower stalls have a supply system, a drain system, and a framed shower alcove.

The supply system: The shower head and shower arm extend out from the wall, where they are connected to the shower pipe with an elbow fitting. The shower pipe rises up from the faucet body, which is supplied by hot and cold supply pipes, and is controlled by a faucet handle and shutoff valves.

The drain system: The drain cover attaches to the drain tailpiece. A rubber gasket on the tailpiece slips over the drain pipe, leading to the P-trap and the branch drain.

Shower stall: A layer of mortar is applied to the subfloor to create a bed for the shower base. Water-resistant wallboard or cementboard is attached to the stall framing members to provide a surface for gluing shower panels.

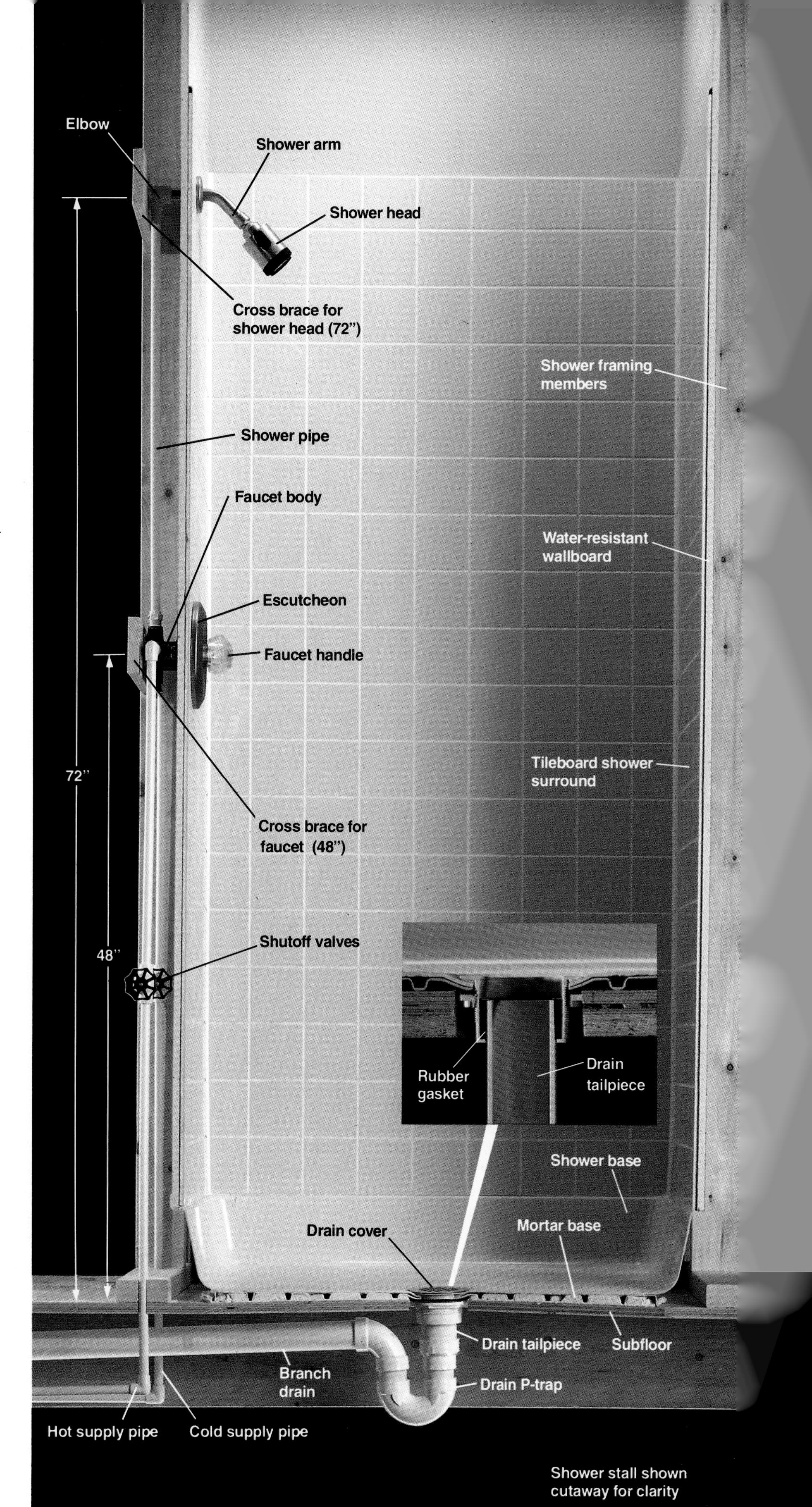

Shower stall shown cutaway for clarity

How to Frame a Shower Alcove

1 Measure shower base, and mark its dimensions on the floor. Measure from center of drain pipe to ensure that drain will be centered in the shower alcove. Install blocking between studs in existing walls to provide a surface for anchoring the alcove walls.

2 Build 2 × 4 alcove walls just outside the marked lines on the floor. Anchor the alcove walls to the existing wall and the subfloor. If necessary, drill holes or cut notches in the sole plate for plumbing pipes.

3 In the stud cavity that will hold the shower faucet and shower head, mark reference points 48" and 72" above the floor to indicate location of the faucet and shower head.

4 Attach 1 × 4 cross braces between studs to provide surfaces for attaching shower head and faucet. Center the cross braces on the marked reference points, and position them flush with the back edge of the studs to provide adequate space for the faucet body (inset) and shower head fittings.

5 Following manufacturer's directions, assemble plumbing pipes and attach faucet body and shower-head fitting to cross braces. (See pages 56 to 63 for information on working with plastic plumbing pipes.) Attach the faucet handle and shower head after the shower panels have been installed (pages 102 to 103).

How to Install a Shower Base

1 Trim the drain pipe in the floor to the height recommended by the manufacturer (usually near or slightly above floor level). Stuff a rag into the drain pipe, and leave it in until you are ready to make the drain connections.

2 Prepare the shower drain piece as directed by the manufacturer, and attach it to the drain opening in the shower base (see inset photo, page 99). Tighten locknut securely onto drain tailpiece to ensure a waterproof fit.

3 Mix a batch of dry-set mortar, then apply a 1" layer to the subfloor, covering the shower base area. Mortar stabilizes and levels the shower base.

4 Apply soap to the outside of the drain pipe in the floor, and to the inside of the rubber gasket in the drain tailpiece. Set the shower base onto the drain pipe, and press down slowly until the rubber gasket in the drain tailpiece fits snugly over the drain pipe.

5 Press the shower base down into the dry-set mortar, carefully adjusting it so it is level. If directed by manufacturer, anchor the shower base with screws driven through the edge flanges and into the wall studs. Let mortar dry for 6 to 8 hours.

How to Install a Shower Panel Kit

1 Attach the shower dome, if included with the shower panel kit, to the walls of the shower alcove. Refer to the manufacturer's directions for exact installation height for the dome.

2 Screw the mounting strips for the shower door frame to both sides of the shower opening.

3 Make a template for marking pipe hole locations in shower panels, using the panel kit shipping carton or heavy paper. Mark holes for the shower head and the faucet handle or handles.

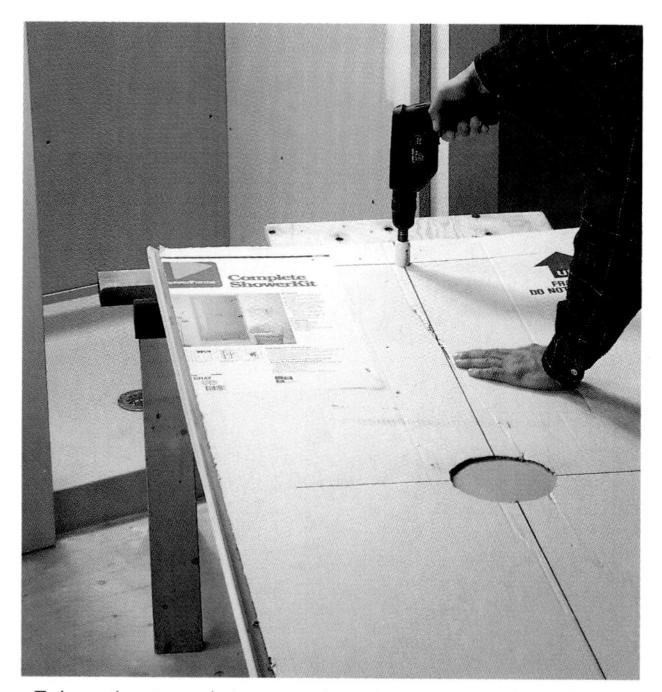

4 Lay the template over the shower panel, and drill out holes with a hole saw. Cut larger holes with a jig saw. For more accurate cuts, use plywood to support the panel near the cutout area.

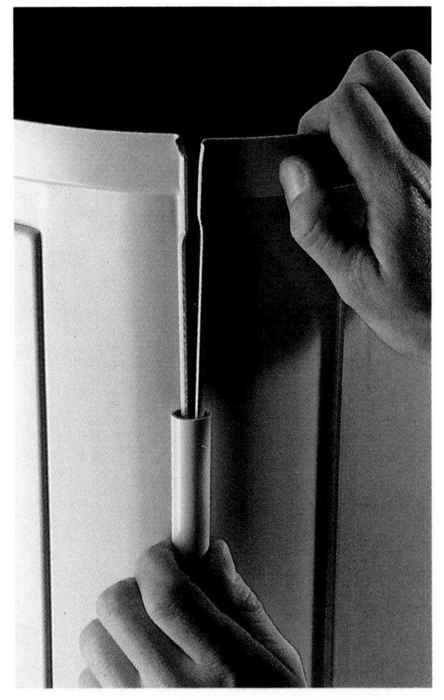

5 Assemble the shower panels according to manufacturer's directions. Most shower panels are joined together with clips or interlocking flanges before they are installed.

6 Apply heavy beads of panel adhesive to the walls of the shower alcove, and to the mounting strips for the door frame.

7 Slide shower panels into the alcove, and press them against the walls. Pull the panels away from the walls for about 1 minute to let adhesive begin to set, then press the panels back in place.

8 Wedge lengths of scrap lumber between the shower panels to hold the panels in place until the adhesive has set completely. Cover the lumber with scraps of carpeting or rags to avoid damaging the surfaces of the shower panels. Allow the panel adhesive to dry for at least 24 hours before removing the braces.

9 Attach faucet handles, shower arm, and shower head (pages 116 to 119), then caulk the seams between the panels and around the base, using tub & tile caulk.

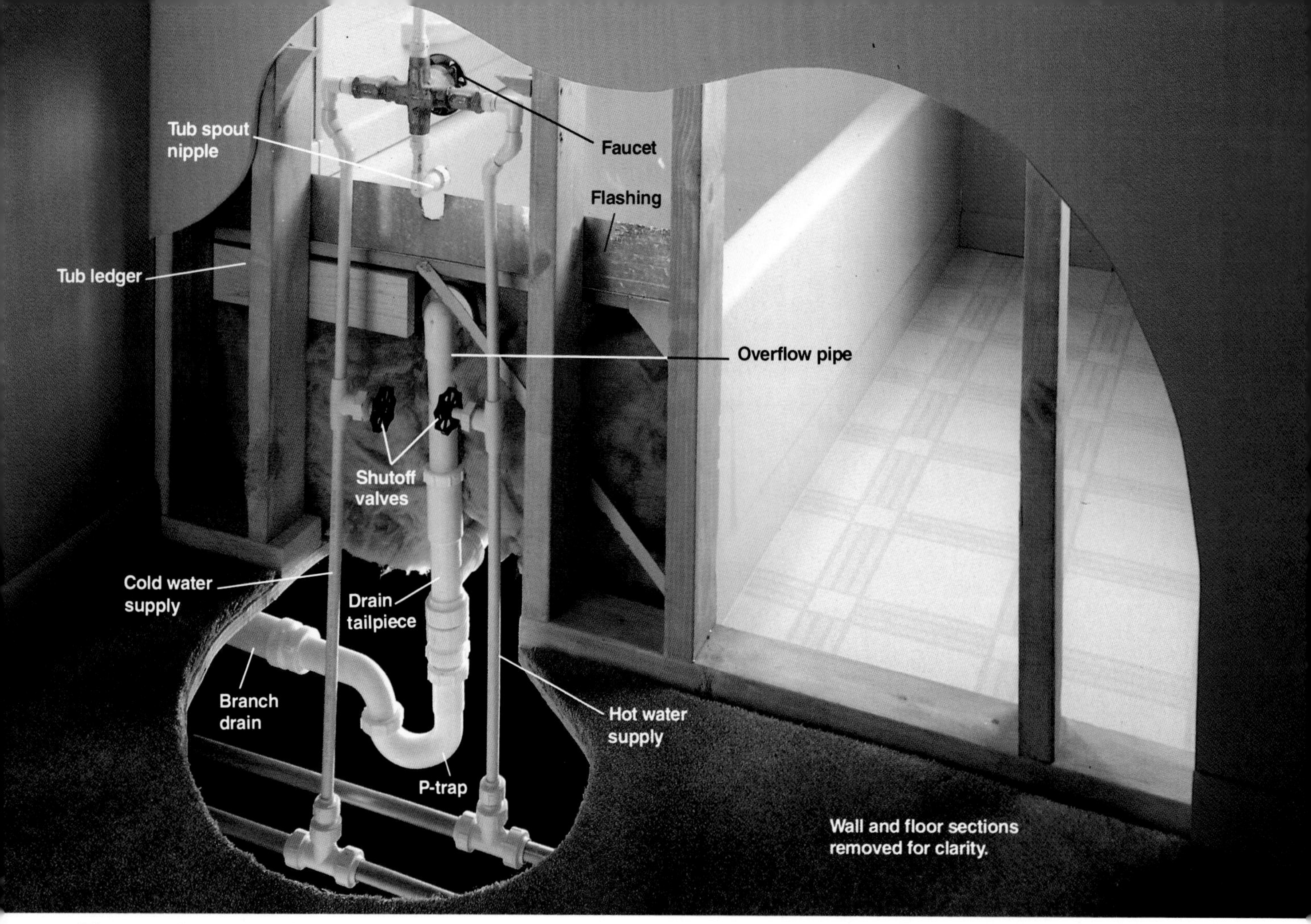

The supply system for a bathtub includes hot and cold supply pipes, shutoff valves, a faucet and handles, and a spout. Supply connections can be made before or after the tub is installed.

The drain-waste-overflow system for a bathtub includes the overflow pipe, drain T, P-trap, and branch drain. The overflow pipe assembly is attached to the tub before installation.

Installing Bathtubs

The development of modern plastic polymers and better construction techniques has created a new generation of tubs that are strong, light, and easy to clean. Even if your old fiberglass or cast-iron tub is in good condition, consider replacing it with a newer model that resists staining and rusting.

Take care when handling a new bathtub, since the greatest chance of damaging the tub occurs during installation. If the inside of your tub has a protective layer of removable plastic, leave it on during installation. Also set a layer of cardboard into the bottom of the tub for added protection while you work.

Combination tub-showers: Special faucets have a removable plug in the top. Replace the plug with a shower pipe connector.

Everything You Need:

Tools: channel-type pliers, hacksaw, carpenter's level, pencil, tape measure, saw, screwdriver, drill, adjustable wrench.

Materials: tub protector; shims; galvanized deck screws; drain-waste-overflow kit; 1 × 3, 1 × 4, and 2 × 4 lumber; galvanized roofing nails; galvanized roof flashing; tub & tile caulk.

Tips for Installing Bathtubs

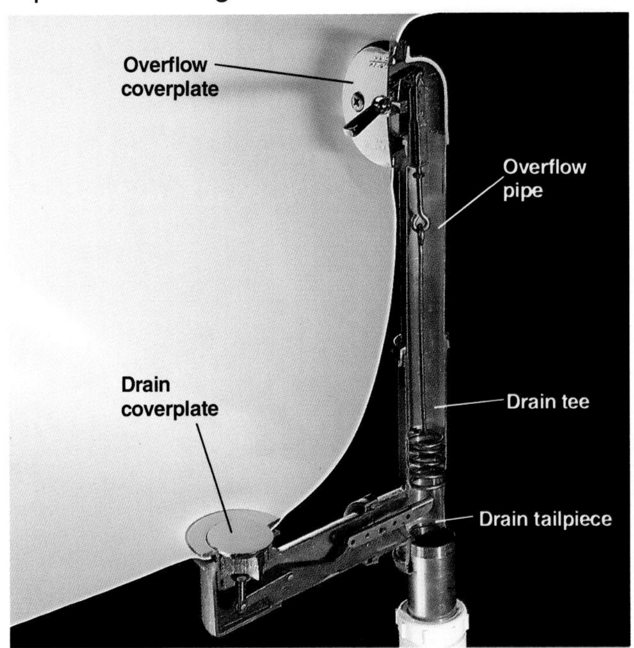

Drain-waste-overflow kit with stopper mechanism must be purchased separately and attached to the tub before it is installed (pages 106 to 107). Available in both brass and plastic types, most kits include an overflow coverplate, an overflow pipe that can be adjusted to different heights, a drain T-fitting, an adjustable drain tailpiece, and a drain coverplate that screws into the tailpiece.

Build a deck for a drop-in style tub or whirlpool (pages 112 to 113). Used frequently with whirlpools, most decks are finished with cementboard and ceramic tile after the tub or whirlpool is installed.

How to Install a Bathtub in an Alcove

1 Attach the faucet body and shower head to the water supply pipes, and attach the assembly to 1 × 4 cross braces before installing the tub. Trim the drain pipe to the height specified by the drain-waste-overflow kit manufacturer.

2 Place a tub-bottom protector, which can be cut from the shipping carton, into the tub. Test-fit the tub by sliding it into the alcove so it rests on the subfloor, flush against the wall studs.

(continued next page)

3 Check the tub rim with a carpenter's level, and shim below the tub to make it level. Mark the top of the nailing flange at each stud. Remove the bathtub from the alcove.

4 Measure the distance from the top of the nailing flange to the underside of the tub rim (inset), and subtract that amount (usually about 1") from the marks on the wall studs. Draw a line at that point on each wall stud.

5 Cut ledger board strips, and attach them to the wall studs just below the mark for the underside of the tub rim (step 4). You may have to install the boards in sections to make room for any structural braces at the ends of the tub.

6 Adjust the drain-waste-overflow assembly (usually sold as a separate kit) to fit the drain and overflow openings. Attach gaskets and washers as directed by the manufacturer, then position the assembly against the tub drain and overflow openings.

7 Apply a ring of plumber's putty to the bottom of the drain piece flange, then insert the drain piece through the drain hole in the bathtub. Screw the drain piece into the drain tailpiece, and tighten until snug. Insert pop-up drain plug.

8 Insert drain plug linkage into the overflow opening, and attach the overflow coverplate with long screws driven into the mounting flange on the overflow pipe. Adjust drain plug linkage as directed by manufacturer.

9 Apply a 1/2"-thick layer of dry-set mortar to the subfloor, covering the entire area where the tub will rest.

10 Lay soaped 1 × 4 runners across the alcove so they rest on the far sill plate. The runners will allow you to slide the tub into the alcove without disturbing the mortar base.

(continued next page)

How to Install a Bathtub in an Alcove (continued)

11 Slide the tub over the runners and into position, then remove the runners, allowing the tub to settle into the mortar. Press down evenly on the tub rims until they touch the ledger boards.

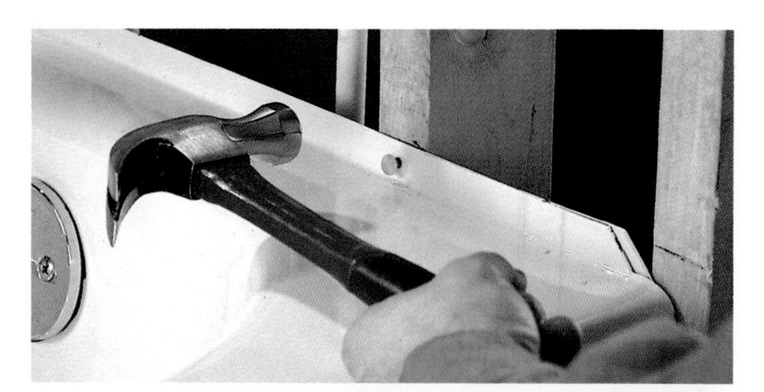

12 Before mortar sets, nail the tub rim flanges to the wall studs. Rim flanges are attached either by predrilling holes into the flanges and nailing with galvanized roofing nails (top), or by driving roofing nails into studs so the head of the nail covers the rim flange (bottom). After rim flanges are secured, allow the mortar to dry for 6 to 8 hours.

13 Attach 4"-wide strips of galvanized metal roof flashing over the tub flange to help keep water out of the wall. Leave a 1/4" expansion gap between the flashing and the tub rim. Nail the flashing to each wall stud, using 1" galvanized roofing nails.

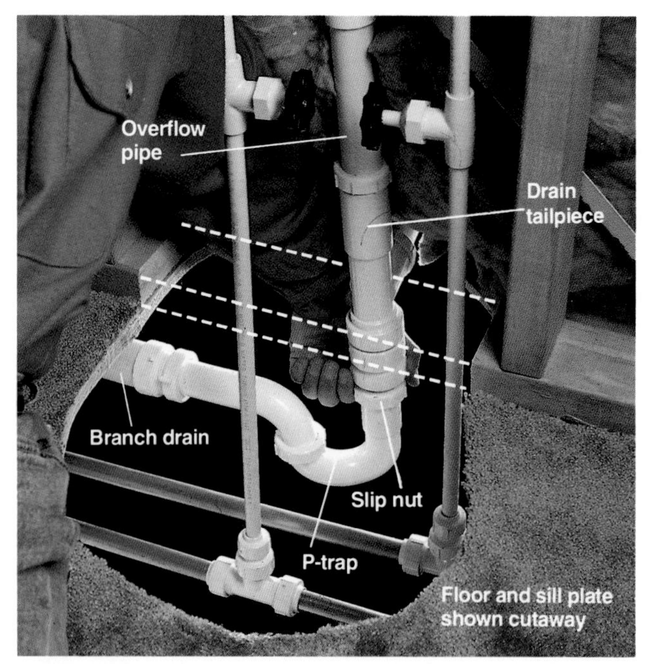

14 Adjust the drain tailpiece so it will fit into the P-trap (you may have to trim it with a hacksaw), then connect it, using a slip nut. Install wall surfaces, then install faucet handles and tub spout (pages 116 to 119). Finally, caulk all around the bathtub.

How to Install a Tub Surround

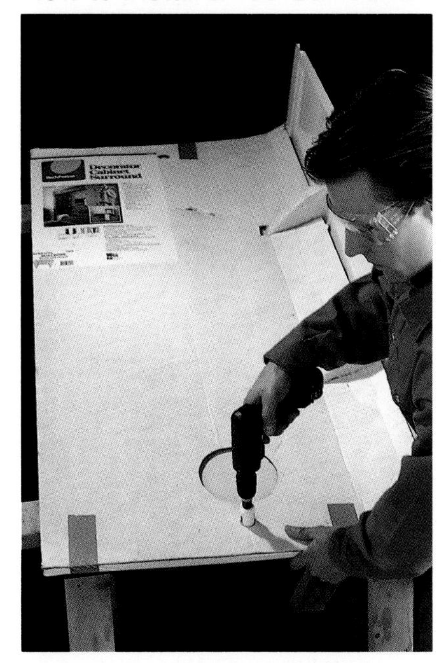

1 Mark a cardboard template for the plumbing cutouts, then tape it to the tub surround panel that will cover the plumbing wall. Make the cutouts in the panel with a hole saw or a jig saw.

2 Test-fit surround panels according to manufacturer's suggested installation sequence, and tape them in place. Draw lines along the tops of all the panels, at the outside edges of side panels, and on the tub rim, along the bottoms of panels.

3 Remove panels in reverse order, one at a time. As they are exposed, outline the inside edges of each panel on the surface of the wall.

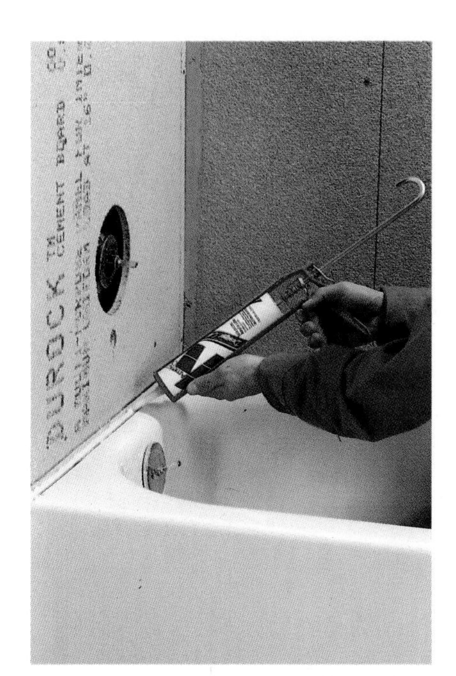

4 Apply a heavy bead of tub & tile caulk to the tub rim, following the marks made where the panels will rest.

5 Apply panel adhesive recommended by manufacturer to the wall in the outline area for the first panel. Carefully press the panel in place.

6 Install the rest of the panels in the proper sequence, following the manufacturer's directions for connecting panels and sealing the seams. Press all the panels in place, then brace for drying (page 103, step 8).

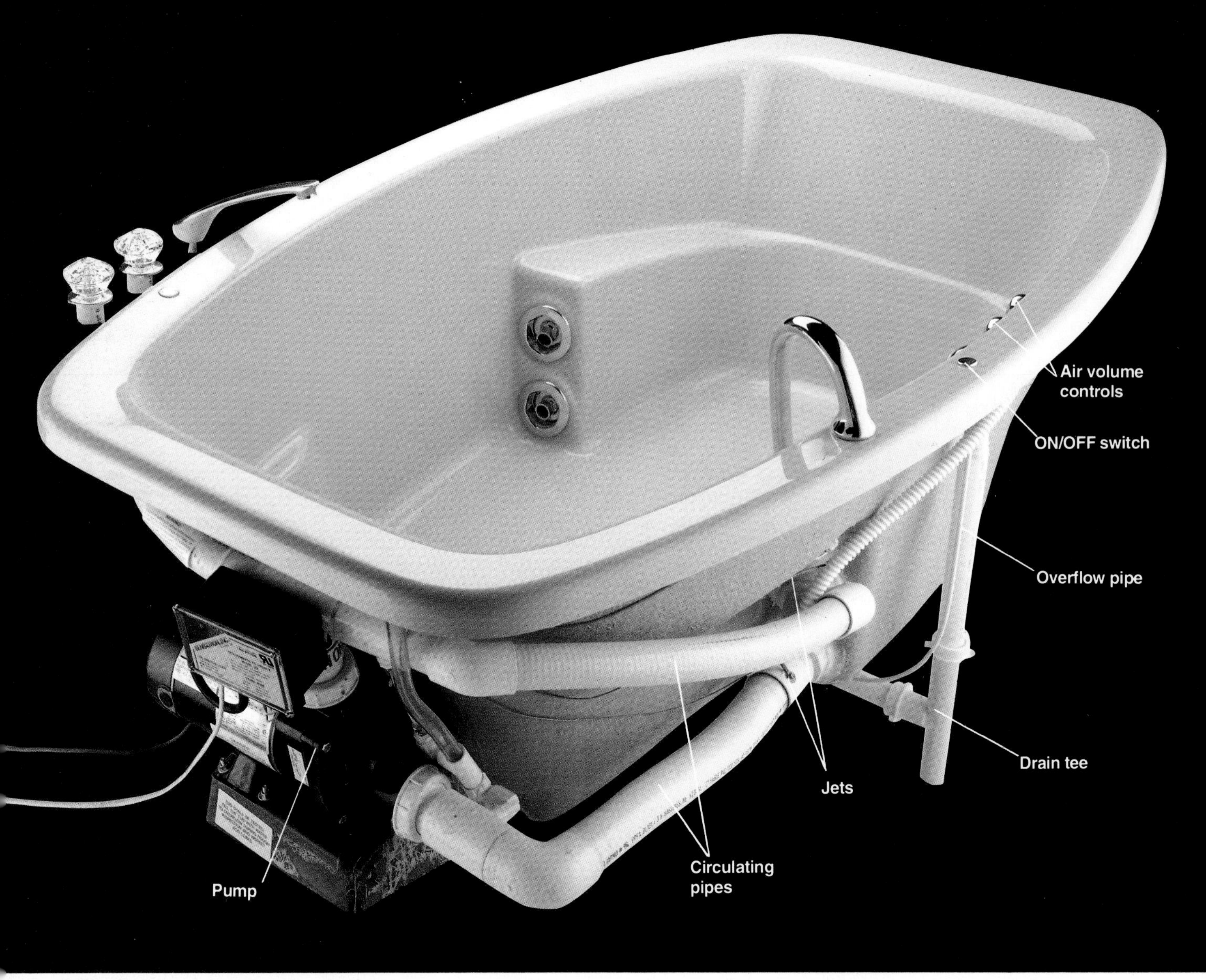

A whirlpool circulates aerated water through jets mounted in the body of tub. Whirlpool pumps move as much as 50 gallons of water per minute to create a relaxing "hydromassage" effect. The pump, pipes, jets, and most of the controls are installed at the factory, making the actual hookup in your home quite simple.

Installing Whirlpools

Installing a whirlpool is very similar to installing a bathtub, once the rough-in is completed. Completing a rough-in for a whirlpool requires that you install a separate electrical circuit for the pump motor. Some Building Codes specify that a licensed electrician be hired to wire whirlpools; check with your local building inspector.

Select your whirlpool before you do rough-in work, because exact requirements will differ from model to model. Select your faucet to match the trim kit that comes with your whirlpool. When selecting a faucet, make sure the spout is large enough to reach over the tub rim. Most whirlpools use "widespread" faucets because the handles and spout are separate, and can be positioned however you like, even on opposite sides of the tub. Most building centers carry flex tube in a variety of lengths for connecting faucet handles and spout.

Everything You Need:

Tools: marker, tape measure, circular saw, jig saw, drill & spade bits, hacksaw, screwdriver.

Materials: 2 × 4s, 10d nails, ¾" exterior grade plywood, deck screws, dry-set mortar, wood spacer blocks, 8-gauge insulated wire, grounding clamp.

Optional Whirlpool Accessories

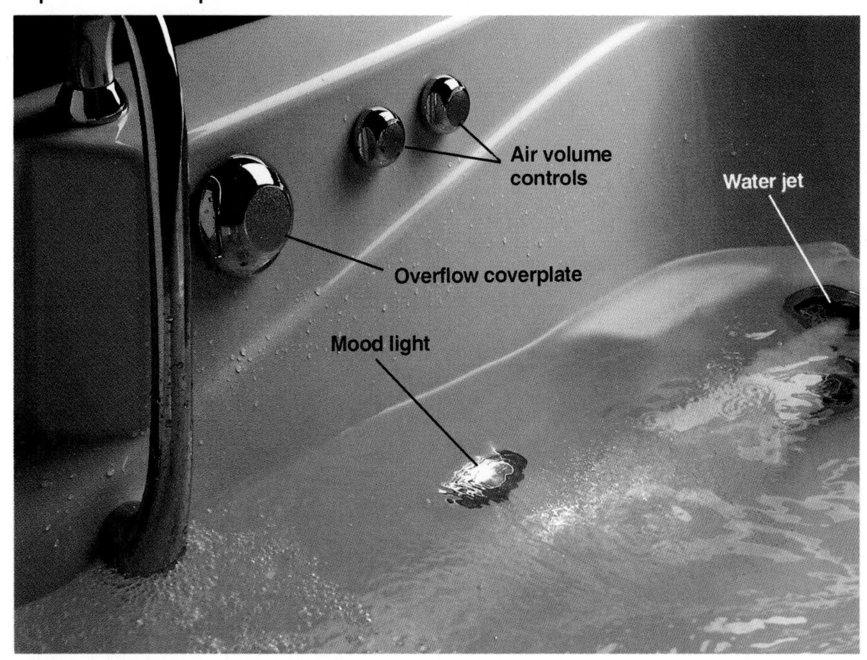

Mood lights are sold as factory-installed accessories by many manufacturers. Most are available with several filters to let you adjust the color to suit your mood. Mood lights are low-voltage fixtures wired through 12-volt transformers. Do not wire mood lights or other accessories into the electrical circuit that supplies the pump motor.

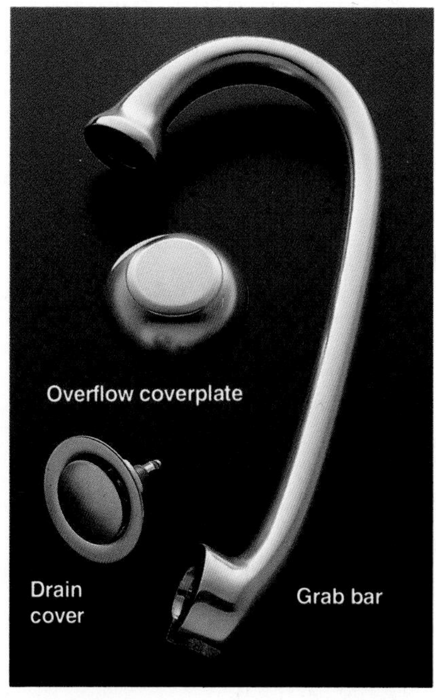

Trim kits for whirlpools are ordered at the time of purchase. Available in a variety of finishes, all of the trim pieces except the grab bar and overflow coverplate normally are installed at the factory.

Requirements for Making Electrical Hookups

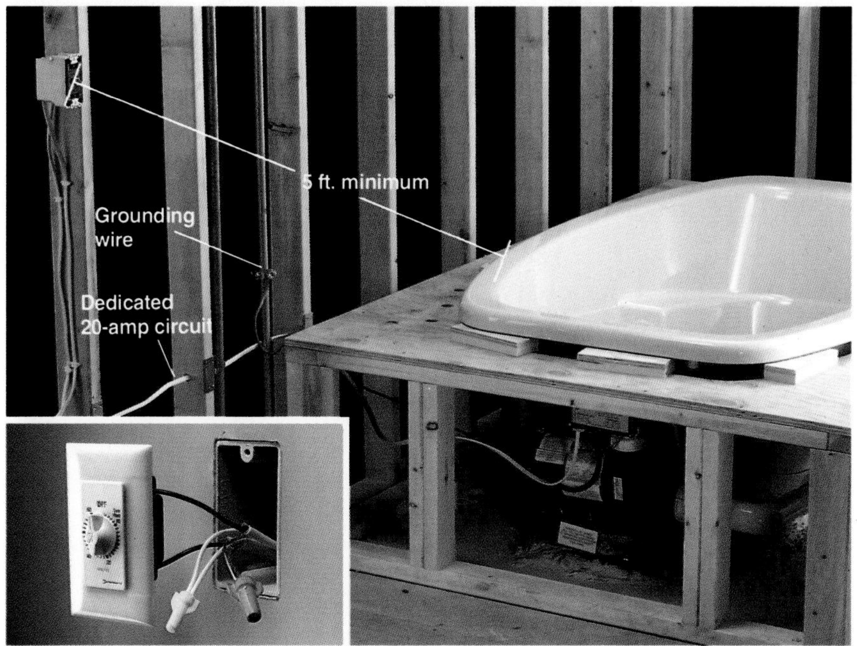

The electrical service for a whirlpool should be a dedicated 115 to 120-volt, 20-amp circuit. The pump motor should be grounded separately, normally to a metal cold water supply pipe. Most whirlpool motors are wired with 12-2 NM cable, but some local Codes require the use of conduit. **Remote timer switches** (inset), located at least 5 ft. from the tub, are required by some Codes, even for a tub with a built-in timer.

A GFCI circuit breaker at the main service panel is required with whirlpool installations. Always hire an electrician to connect new circuits at your service panel, even if you install the circuit cable yourself.

How to Install a Whirlpool

1 Outline the planned location of the deck frame on the subfloor. Use the plumbing stub-outs as starting points for measuring. Before you begin to build the deck, check the actual dimensions of your whirlpool tub to make sure they correspond to the dimensions listed in the manufacturer's directions. TIP: Plan your deck so it will be at least 4" wide at all points around the whirlpool.

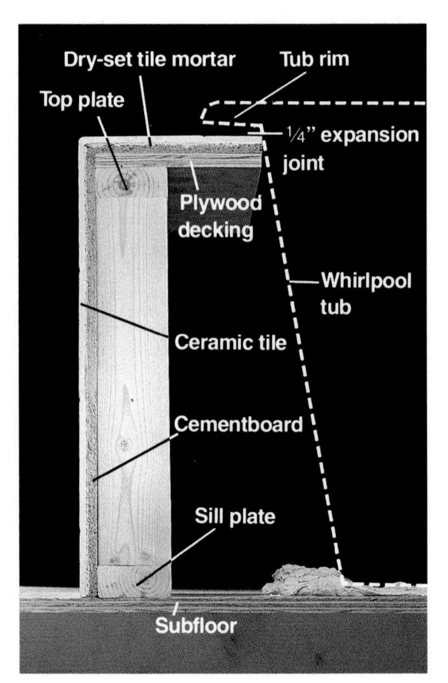

2 Cut top plates, sill plates, and studs for the deck frame. The height of frame should allow 3/4" for the plywood decking, 1/4" for an expansion gap between the deck and the tub rim, and 1" for cementboard, tile, and mortar.

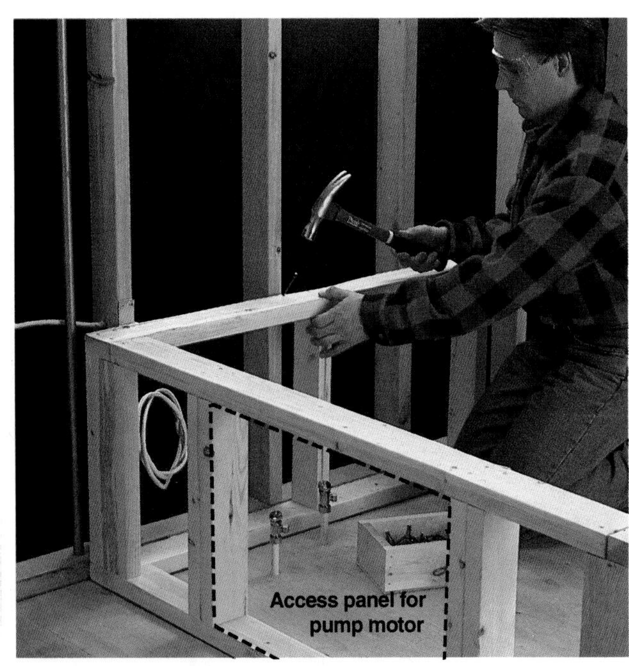

3 Assemble the deck frame. Make sure to leave a framed opening for access panels at the pump location and the drain location. Nail the frame to the floor joists and wall studs or blocking, using 10d nails.

4 Cover the deck frame with 3/4" exterior-grade plywood, and attach with screws spaced every 12". Using a template of the whirlpool cutout (usually included with the tub), mark the deck for cutting. If no template is included, make one from the shipping carton. Cutout will be slightly smaller than the outside dimensions of the whirlpool rim.

5 Make the cutout hole in the deck, using a jig saw. Drill a pilot hole along the cutout line to start the cut.

6 Measure and mark holes for faucet tailpieces and spout tailpiece according to the faucet manufacturer's suggestions. Drill holes with a spade bit or hole saw.

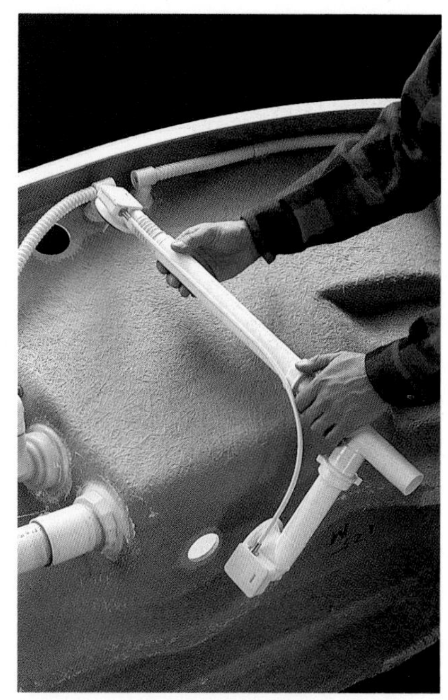

7 Attach drain-waste-overflow assembly (included with most whirlpools) at the drain and overflow outlets in the tub (pages 106 to 107). Trim the drain pipe in the floor to the proper height.

8 Apply a layer of dry-set mortar to subfloor where tub will rest. Make 12" spacer blocks, 1 1/4" thick (equal to expansion gap, tile, mortar, and cementboard; see step 2). Arrange blocks along the edges of the cutout.

9 With a helper, lift the tub by the rim and set it slowly into the cutout hole. Lower the tub, pressing it into the mortar base, until the rim rests on the spacers at the edges of the cutout area. Avoid moving or shifting the tub once it is in place, and allow the mortar to set for 6 to 8 hours before proceeding with the tub installation. Align the tailpiece of the drain-waste-overflow assembly with the P-trap as you set the tub in place.

(continued next page)

10 Adjust the length of the tailpiece for the drain-waste-overflow assembly, if necessary, then attach assembly to the P-trap in the drain opening, using a slip nut.

11 Inspect the seals on the built-in piping and hoses for loose connections. If you find a problem, contact your dealer for advice. Attempting to fix the problem yourself could void the whirlpool warranty.

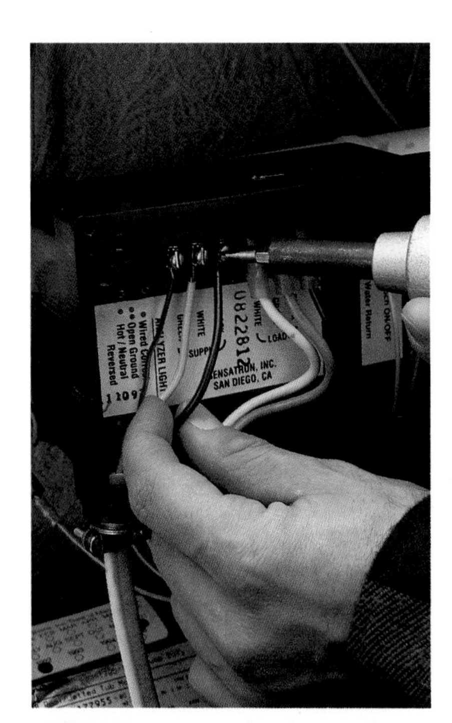

12 With power off, remove the wiring cover from the pump motor. Feed circuit wires from the power source or wall timer into the motor. Connect wires according to directions printed on the motor.

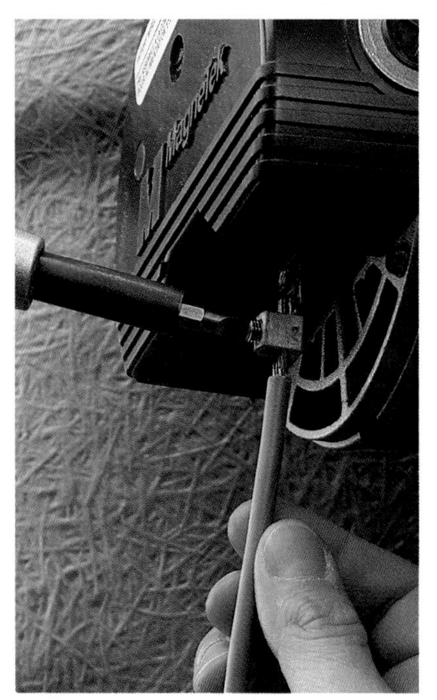

13 Attach an insulated 8-gauge wire to the grounding lug on the pump motor.

14 Attach the other end of the wire to a metal cold water supply pipe in the wall, using a grounding clamp. Test the GFCI circuit breaker.

15 Clean out the tub, then fill it so the water level is at least 3" above the highest water jet.

16 Turn on the pump, and allow it to operate for at least 20 minutes while you check for leaks. Contact your whirlpool dealer if leaks are detected.

17 Staple fiberglass insulation with attached facing to vertical frame supports. Facing should point inward, to keep fibers out of motor. Do not insulate within 6" of pumps, heaters, or lights.

18 Attach cementboard to the sides and top of the deck frame if you plan to install ceramic tile on the deck. Use 3⁄4" plywood for access panel coverings.

19 Attach finish surfaces to deck and deck frame, then install grab bar and the faucet and spout (pages 116 to 119). Fill the joints between floor and deck, and between the tub rim and deck surface, with tub & tile caulk.

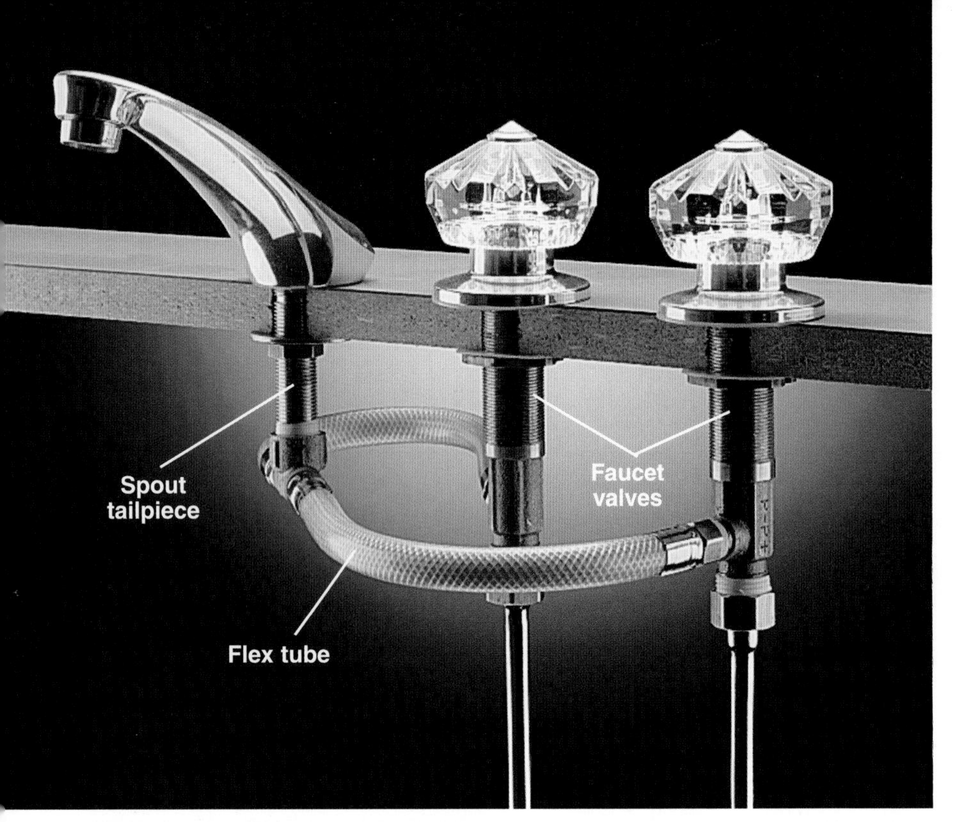

Widespread faucets have handles that are installed separately from the spout. Installed most often with whirlpools, widespread faucets also are becoming popular for sinks. The possibilities for arranging widespread faucets are limited only by the length of the flex tubes that connect the faucet handle valves to the spout.

Installing Faucets & Spouts

One-piece faucets, with either one or two handles, are the most popular fixtures for bathroom installations. "Widespread" faucets with separate spout and handles are being installed with increasing frequency, however. Because the handles are connected to the spout with flex tubes that can be 18" or longer, widespread faucets can be arranged in many ways.

Everything You Need:

Tools: drill with spade bit, basin wrench, adjustable wrench, screwdriver.

Materials: plumber's putty, Teflon tape, joint compound.

How to Install a Widespread Faucet

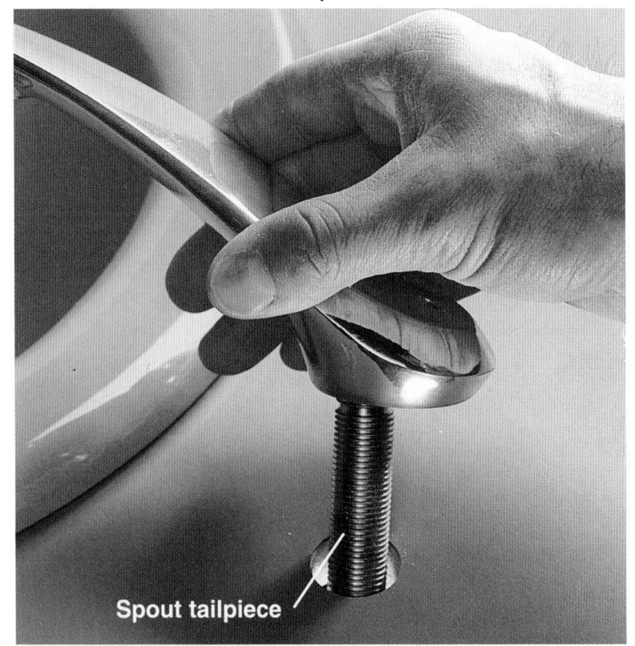

1 Drill holes for faucet handles and spout in deck or countertop, according to manufacturer's suggestions. Slide a protective washer onto the spout tailpiece, then insert the tailpiece into the spout hole.

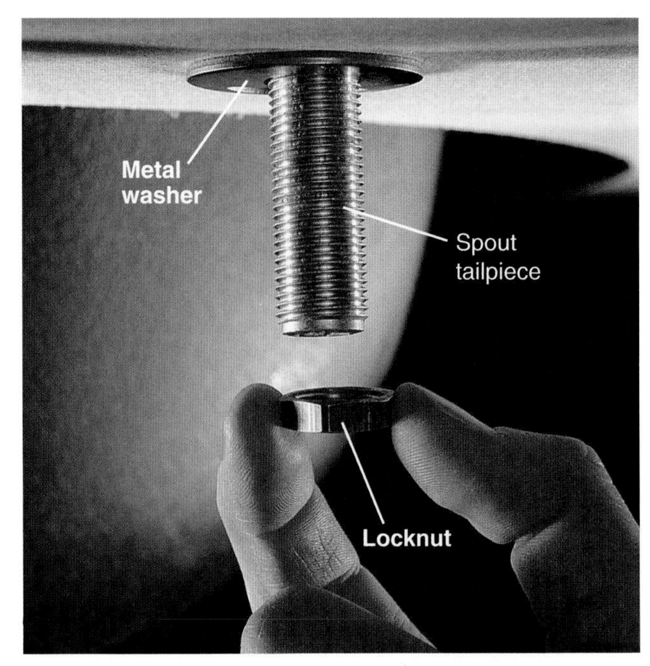

2 From beneath the deck, slide a metal washer onto spout tailpiece, then attach a locknut. Tighten the nut by hand, then check to make sure the spout is properly aligned. Tighten with a basin wrench until snug.

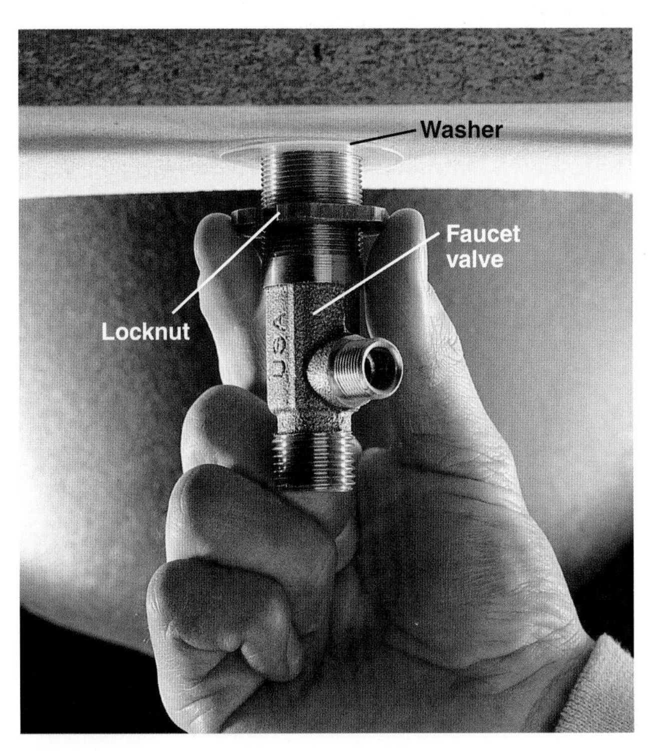

3 Attach faucet valves to deck, using washers and locknuts, as directed by the manufacturer. NOTE: Some widespread faucet valves, like the one above, are inserted up through the hole, and have locknuts above and below.

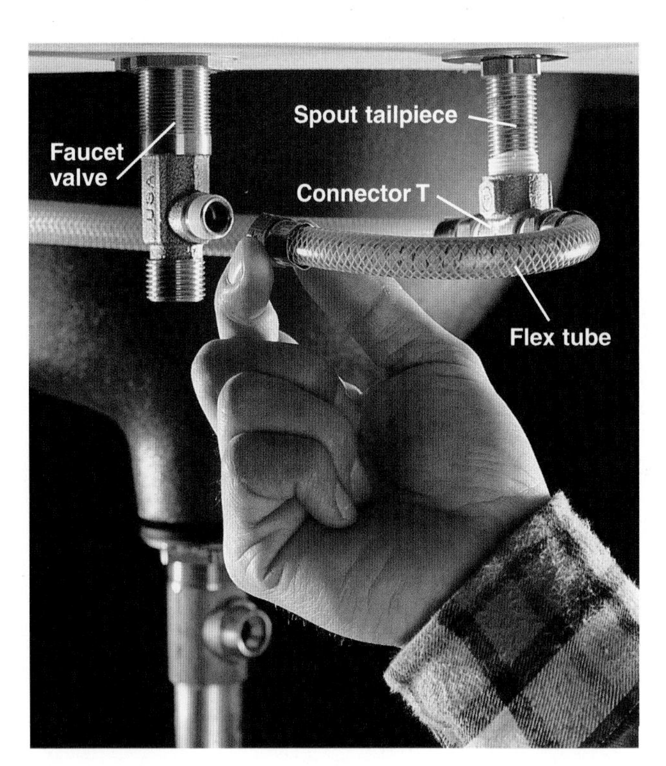

4 Wrap Teflon tape around spout tailpiece, then attach the connector T to the tailpiece. Attach one end of each flex tube to the T, and the other end to the proper faucet valve.

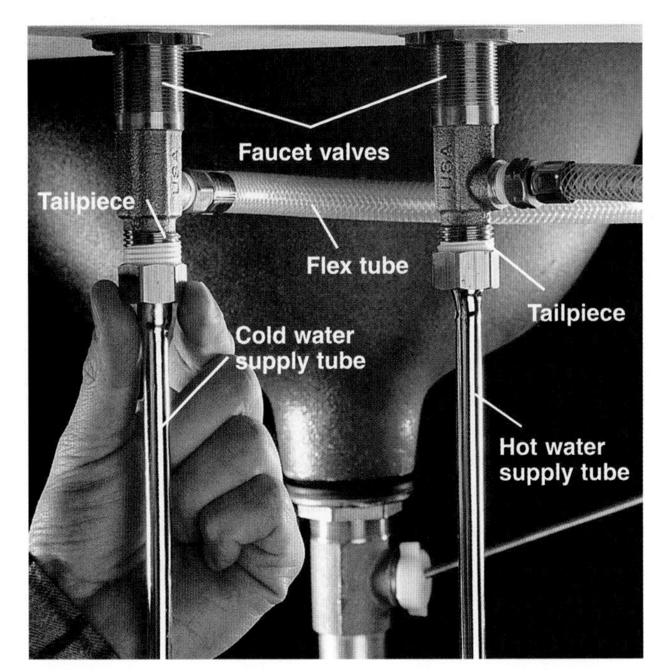

5 Wrap Teflon tape around faucet valve tailpieces, then attach hot and cold water supply tubes to tailpieces.

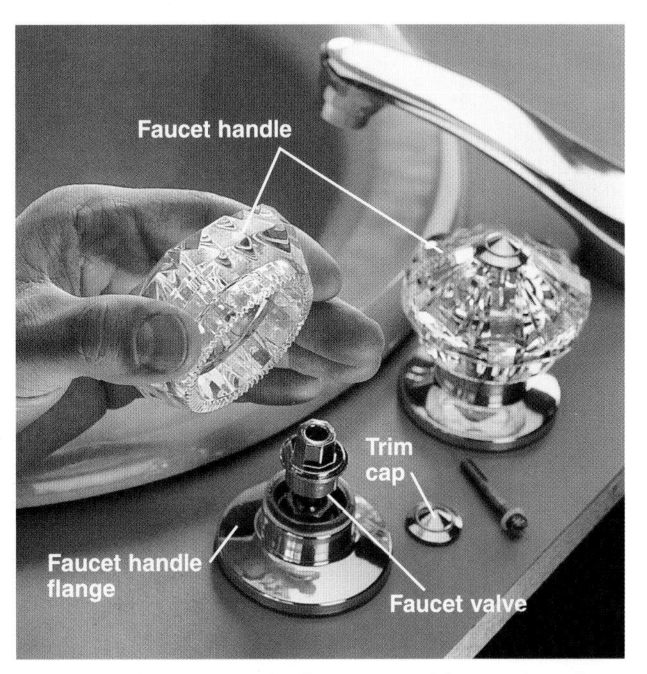

6 Attach faucet handle flanges and faucet handles, according to manufacturer's directions. Cover exposed screw heads with trim caps.

How to Install a One-piece Faucet

1 Apply a ring of plumber's putty around the base of the faucet body. (Some faucets use a gasket that does not require plumber's putty—read the manufacturer's directions carefully.)

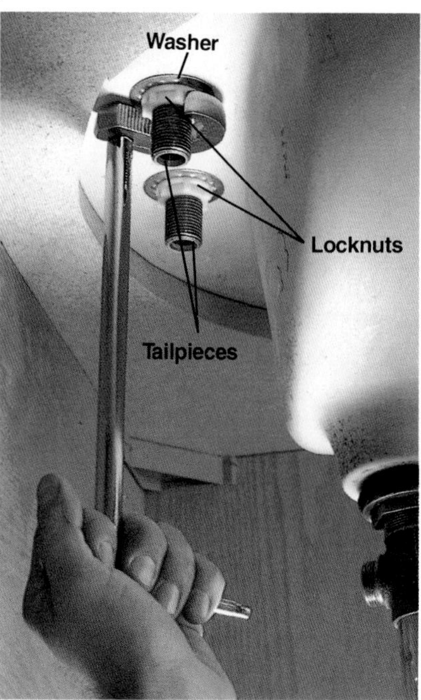

2 Insert the faucet tailpieces through holes in countertop or sink. From below, thread washers and locknuts over the tailpieces, then tighten the locknuts with a basin wrench until snug.

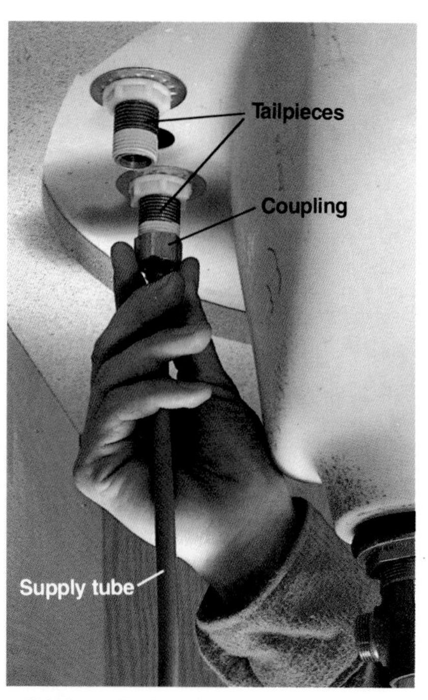

3 Wrap Teflon® tape around the tailpiece threads, then attach the supply tube couplings and tighten until snug. Connect drain linkage, then attach handles and trim caps.

How to Install Tub & Shower Faucets

Two-handled faucet: Screw handle flanges onto faucet valve stems, then attach handles to stems, using mounting screws. Attach spout (page 73) and trim caps. NOTE: Faucet body is attached before wall surface is installed (page 100).

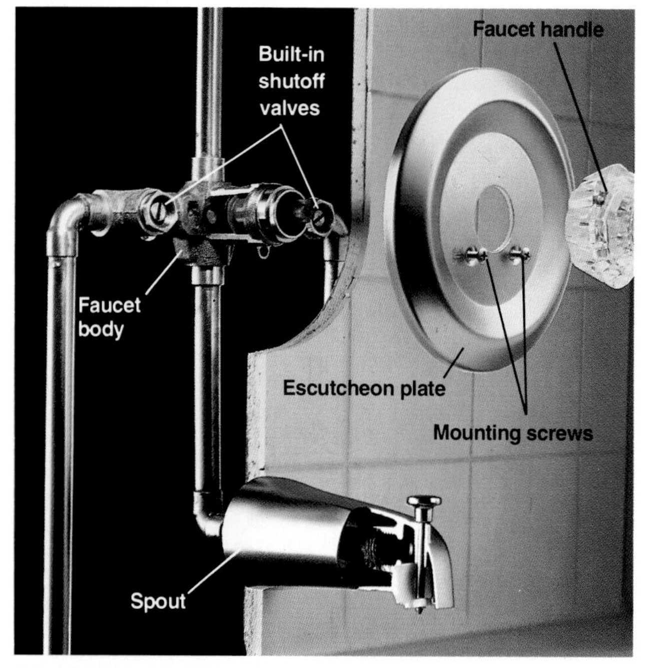

Single-handled faucets: Open built-in shutoff valves, using a screwdriver, then attach escutcheon plate to faucet body with mounting screws. Attach faucet handle with mounting screw, then attach spout and trim cap. NOTE: Faucet body is attached before wall surface is installed (page 100).

How to Connect Supply Tubes

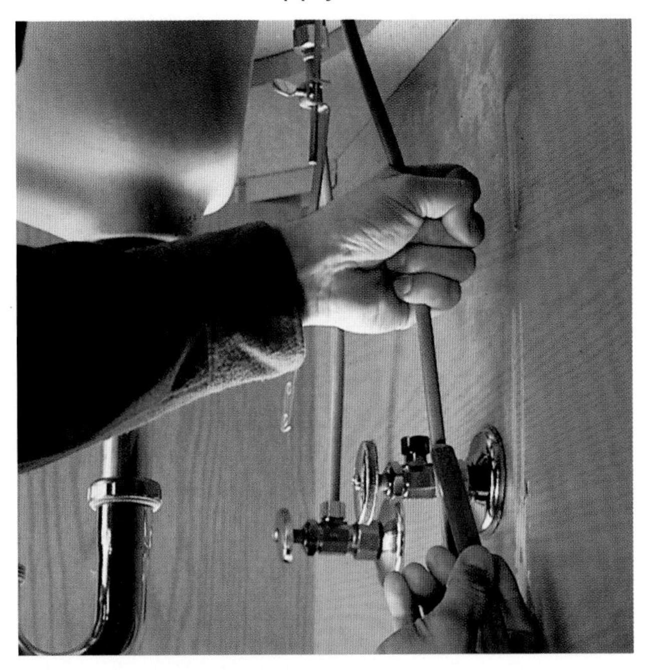

Connect supply tubes after the sink and faucet body are installed. Tubes should be slightly longer than the distance from the shutoff valves to the faucet tailpieces.

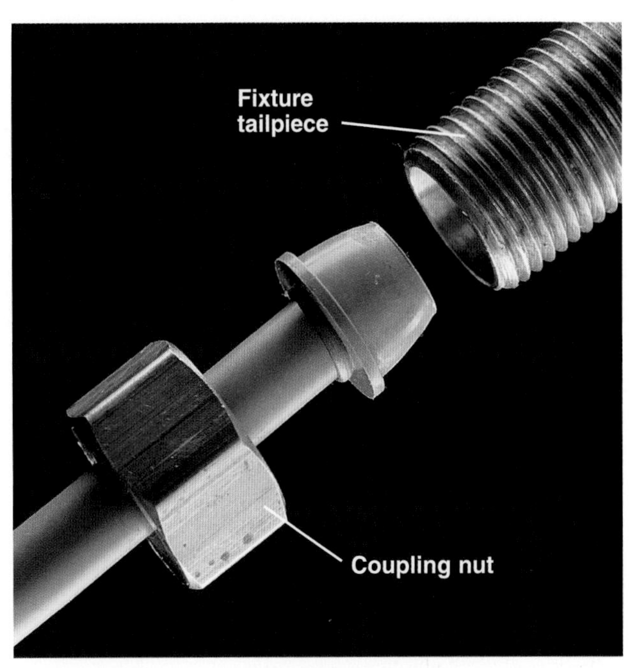

Most tubes have a flared end that fits into the faucet tailpiece. Wrap threads of tailpiece with Teflon tape before attaching tubes.

How to Install a Tub Spout

Connect tub spouts by applying joint compound of Teflon tape to the threaded end of the spout nipple that extends from the wall.

Screw the spout onto the nipple, using a long screwdriver as a lever. Some spouts have a set-screw on the underside that must be tightened.

Installing Toilets

Install a toilet by anchoring the bowl to the floor first, then mounting the tank onto the bowl. China fixtures crack easily, so use care when handling them.

Most toilets in the low-to-moderate price range are two-piece units, with a separate tank and bowl, made of vitreous china. One-piece toilets, with integral tank and bowl, also are available, but the cost is usually two or three times that of two-piece units.

Standard toilets have 3.5-gallon tanks, but water-saver toilets, with 1.6-gallon tanks, are becoming increasingly common. Most states now require water-saver toilets in new construction.

Everything You Need:

Tools: adjustable wrench, ratchet wrench or basin wrench, screwdriver.

Materials: wax ring & sleeve, plumber's putty, floor bolts, tank bolts with rubber washers, seat bolts and mounting nuts.

How to Install a Toilet

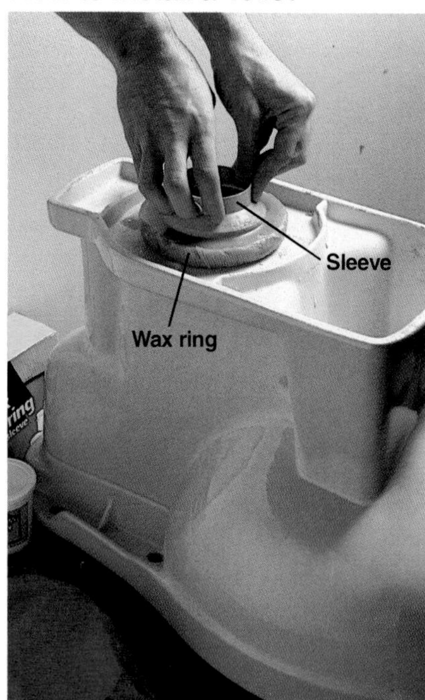

1 Turn the bowl upside down, and place a new wax ring and sleeve onto the toilet horn. Apply a ring of plumber's putty around the bottom edge of the toilet base.

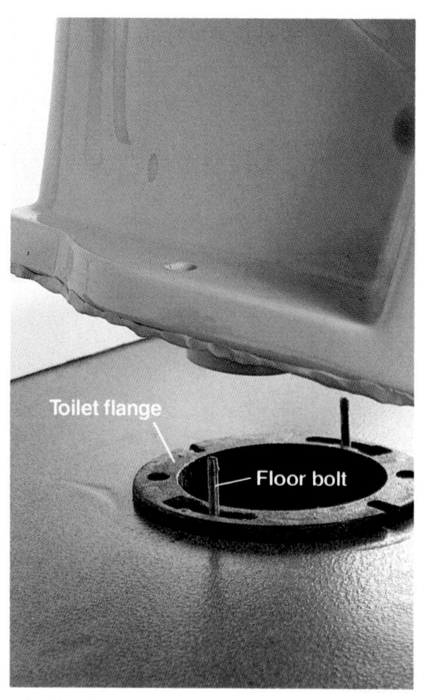

2 Position the toilet over the toilet flange so the floor bolts fit through the holes in the base of the toilet. The flange should be clean, and the floor bolts should point straight up.

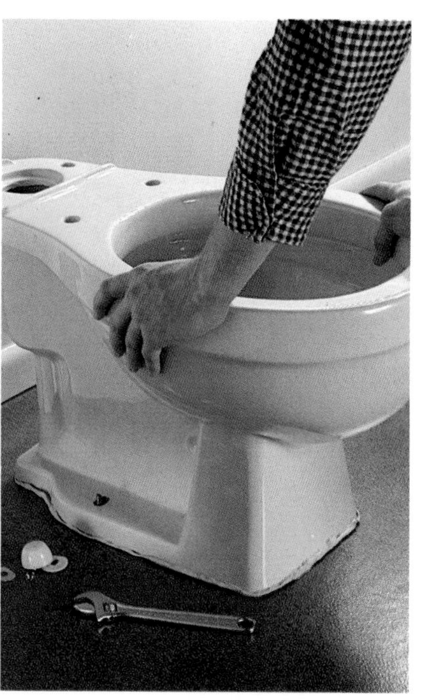

3 Press down on the toilet bowl to compress the wax ring and plumber's putty. Attach washers and nuts to the floor bolts, and tighten with an adjustable wrench until snug. Do not overtighten. Attach trim caps.

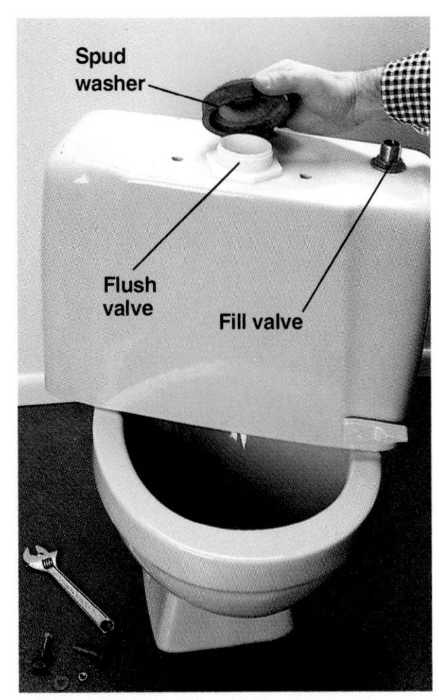

4 Turn the tank upside down, and set the spud washer over the tailpiece of the flush valve. Turn tank right-side up. NOTE: with some toilets, you will need to purchase a flush handle, fill valve, and flush valve separately.

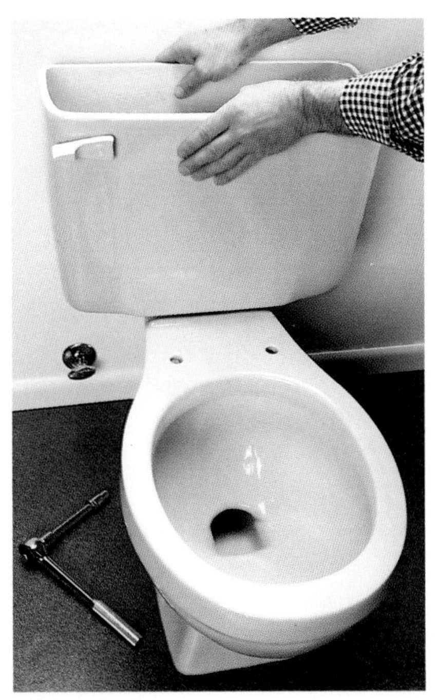

5 Set the tank onto the bowl, centering the spud washer over the water inlet opening near the back edge of the bowl.

6 Shift the tank gently until the tank bolt holes in the tank are aligned over the tank bolt holes in the bowl flange. Place rubber washers onto tank bolts, then insert the bolts down through the holes in the tank.

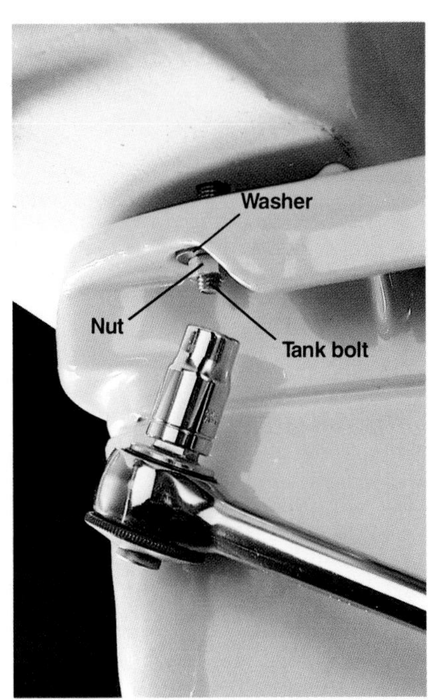

7 From beneath the bowl flange, attach washers and nuts to the tank bolts, and tighten with a ratchet wrench or basin wrench until snug. Do not overtighten.

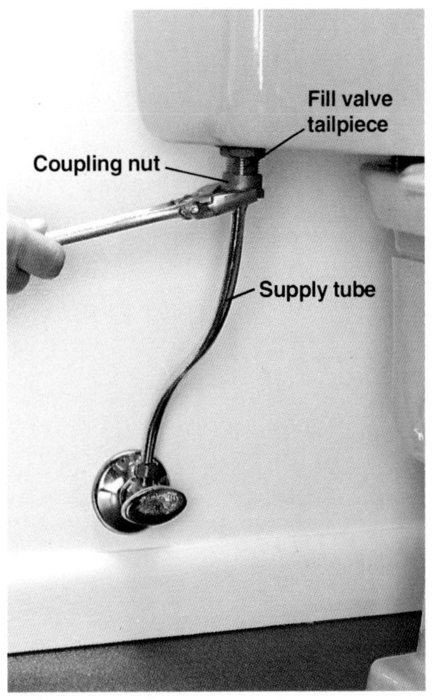

8 Cut a piece of supply tube to fit between the shutoff valve and the toilet tank. Attach the tube to the shutoff valve, then to the fill valve tailpiece. Use an adjustable wrench to tighten coupling nuts until they are snug.

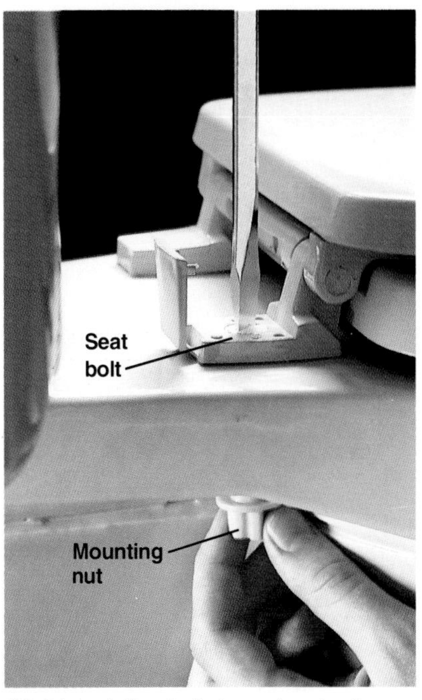

9 Mount the toilet seat onto the bowl by tightening the mounting nuts onto the seat bolts from below the seat flange.

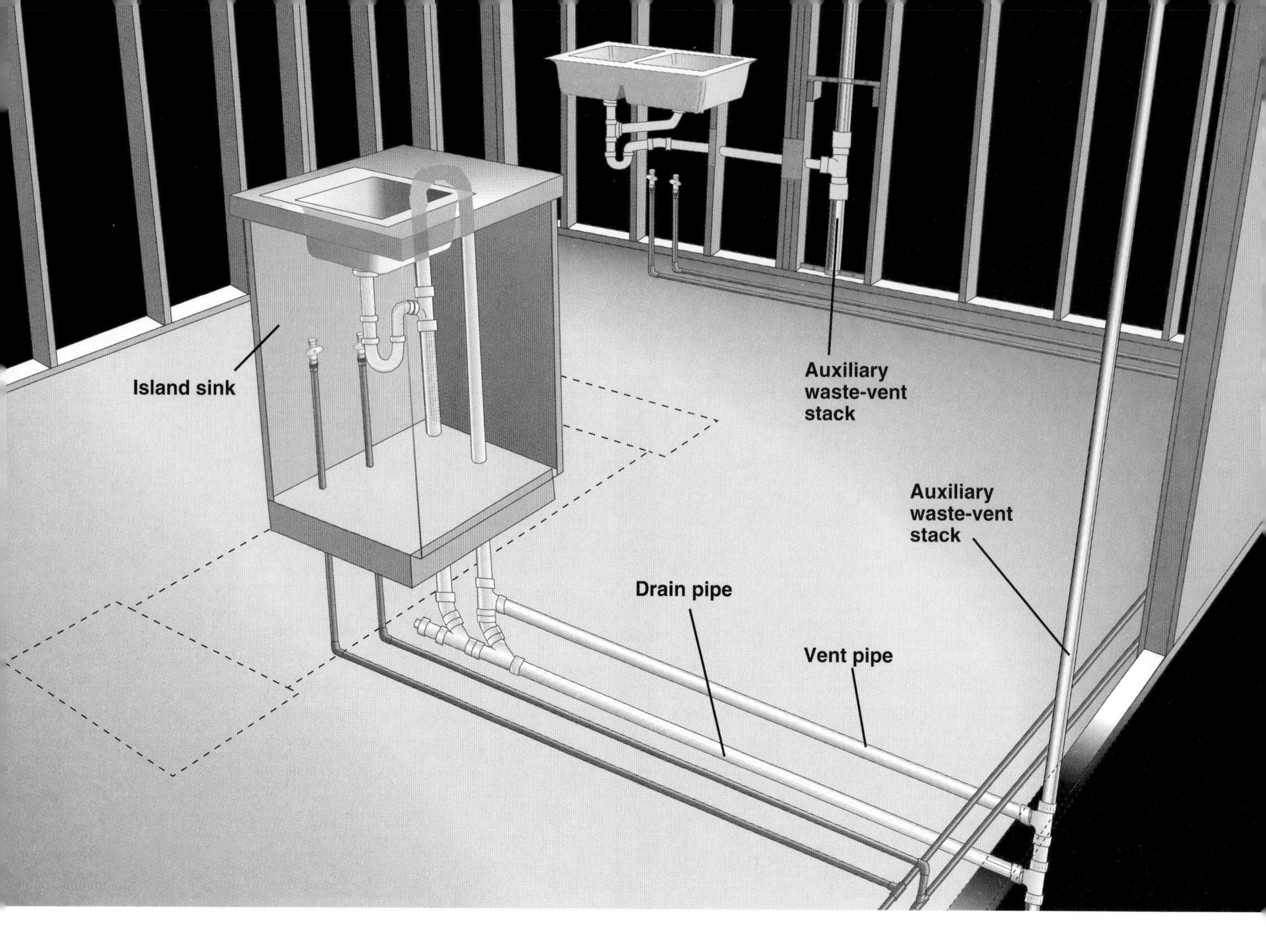

Plumbing a Kitchen

Plumbing a remodeled kitchen is a relatively easy job if your kitchen includes only a wall sink. If your project includes an island sink, however, the work becomes more complicated.

An island sink poses problems because there is no adjacent wall in which to run a vent pipe. For an island sink, you will need to use a special plumbing configuration known as a *loop vent*.

Each loop vent situation is different, and your configuration will depend on the location of existing waste-vent stacks, the direction of the floor joists, and the size and location of your sink base cabinet. Consult your local plumbing inspector for help in laying out the loop vent.

For our demonstration kitchen, we have divided the project into three phases:

- How to Install DWV Pipes for a Wall Sink (pages 124 to 126)
- How to Install DWV Pipes for an Island Sink (pages 127 to 131)
- How to Install New Supply Pipes (pages 132 to 133)

Our demonstration kitchen includes a double wall sink and an island sink. The 1½" drain for the wall sink connects to an existing 2" galvanized waste-vent stack; since the trap is within 3½ ft. of the stack, no vent pipe is required. The drain for the island sink uses a loop vent configuration connected to an auxiliary waste-vent stack in the basement.

Tips for Plumbing a Kitchen

Insulate exterior walls if you live in a region with freezing winter temperatures. Where possible, run water supply pipes through the floor or interior partition walls, rather than exterior walls.

Use existing waste-vent stacks to connect the new DWV pipes. In addition to a main waste-vent stack, most homes have one or more auxiliary waste-vent stacks in the kitchen that can be used to connect new DWV pipes.

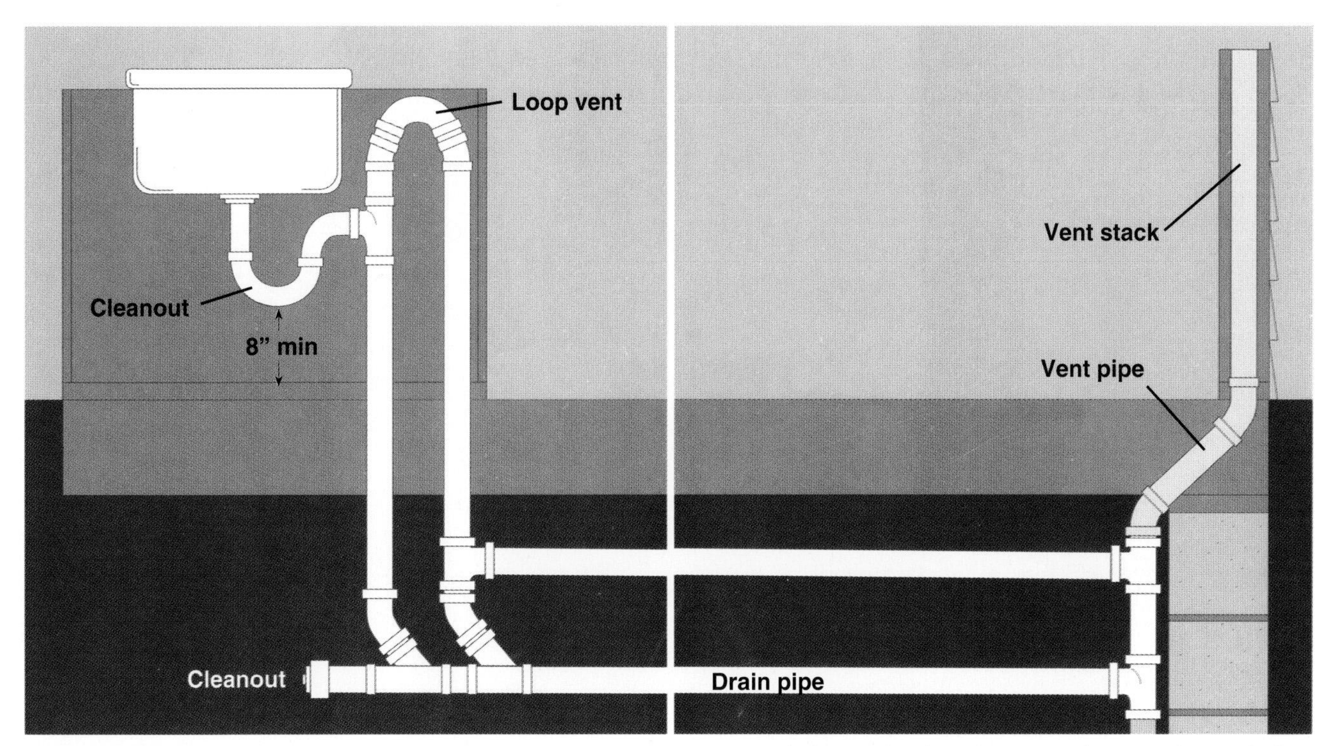

Loop vent makes it possible to vent a sink when there is no adjacent wall to house the vent pipe. The drain is vented with a loop of pipe that arches up against the countertop and away from the drain before dropping through the floor. The vent pipe then runs horizontally to an existing vent pipe. In our project, we have tied the island vent to a vent pipe extending up from a basement utility sink. NOTE: Loop vents are subject to local Code restrictions. Always consult your building inspector for guidelines on venting an island sink.

How to Install DWV Pipes for a Wall Sink

1 Determine the location of the sink drain by marking the position of the sink and base cabinet on the floor. Mark a point on the floor indicating the position of the sink drain opening. This point will serve as a reference for aligning the sink drain stub-out.

2 Mark a route for the new drain pipe through the studs behind the wall sink cabinet. The drain pipe should angle ¼" per foot down toward the waste-vent stack.

3 Use a right-angle drill and hole saw to bore holes for the drain pipe (page 76). On non-loadbearing studs, such as the cripple studs beneath a window, you can notch the studs with a reciprocating saw to simplify the installation of the drain pipe. If the studs are loadbearing, however, you must thread the run though the bored holes, using couplings to join short lengths of pipe as you create the run.

4 Measure, cut, and dry-fit a horizontal drain pipe to run from the waste-vent stack to the sink drain stub-out. Create the stub-out with a 45° elbow and 6" length of 1½" pipe. NOTE: If the sink trap in your installation will be more than 3½ ft. from the waste-vent pipe, you will need to install a waste T and run a vent pipe up the wall, connecting it to the vent stack at a point at least 6" above the lip of the sink.

5 Remove the neoprene sleeve from a banded coupling, then roll the lip back and measure the thickness of the separator ring.

6 Attach two lengths of 2" pipe, at least 4" long, to the top and bottom openings on a 2" × 2" × 1½" waste T. Hold the fitting alongside the waste-vent stack, then mark the stack for cutting, allowing space for the separator rings on the banded couplings.

(continued next page)

7 Use riser clamps and 2 × 4 blocking to support the waste-vent stack above and below the new drain pipe, then cut out the waste-vent stack along the marked lines, using a reciprocating saw and metal-cutting blade.

8 Slide banded couplings onto the cut ends of the waste-vent stack, and roll back the lips of the neoprene sleeves. Position the waste T assembly, then roll the sleeves into place over the plastic pipes.

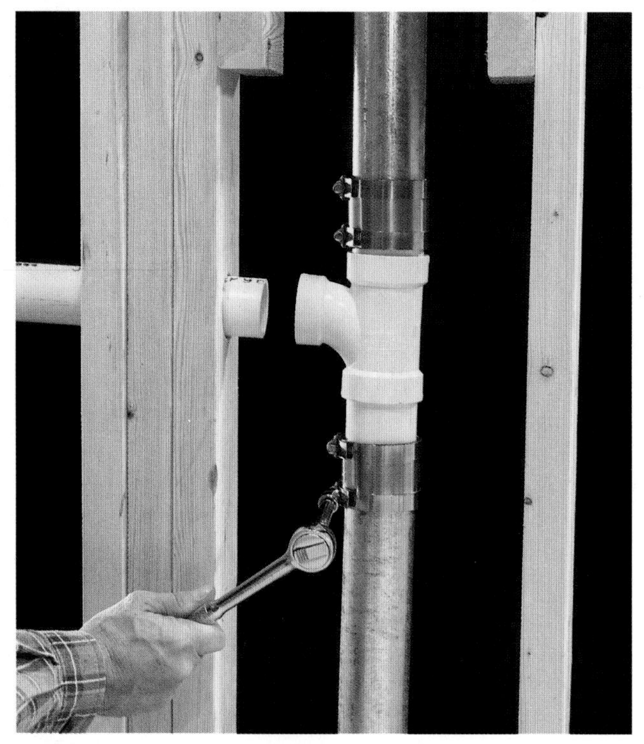

9 Slide the metal bands into place over the neoprene sleeves, and tighten the clamps with a ratchet wrench or screwdriver.

10 Solvent-glue the drain pipe, beginning at the waste-vent stack. Use a 90° elbow and a short length of pipe to create a drain stub-out extending about 4" out from the wall.

How to Install DWV Pipes for an Island Sink

1 Position the base cabinet for the island sink, according to your kitchen plans. Mark the cabinet position on the floor with tape, then move the cabinet out of the way.

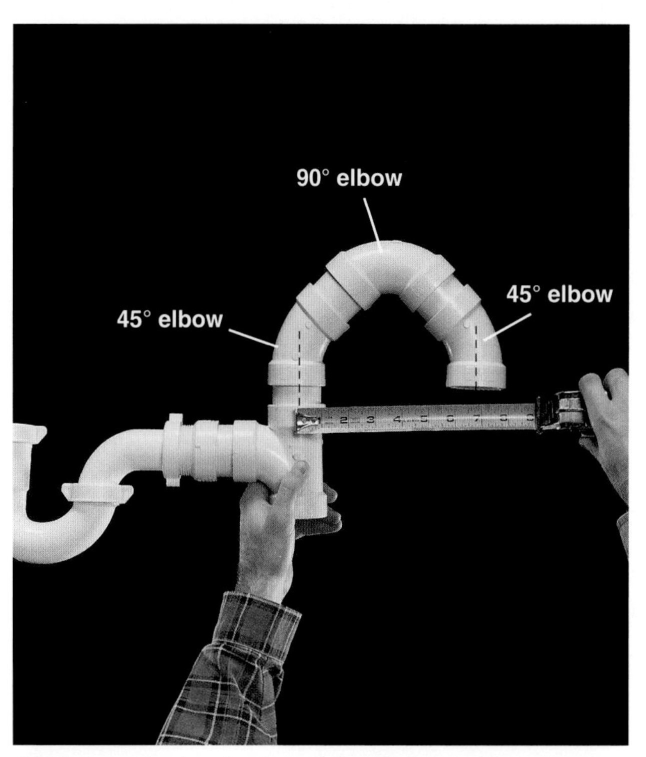

2 Create the beginning of the drain and loop vent by test-fitting a drain trap, waste T, two 45° elbows, and a 90° elbow, linking them with 2" lengths of pipe. Measure the width of the loop between the centerpoints of the fittings.

3 Draw a reference line perpendicular to the wall to use as a guide when positioning the drain pipes. A cardboard template of the sink can help you position the loop vent inside the outline of the cabinet.

4 Position the loop assembly on the floor, and use it as a guide for marking hole locations. Make sure to position the vent loop so the holes are not over joists.

(continued next page)

5 Use a hole saw with a diameter slightly larger than the vent pipes to bore holes in the subfloor at the marked locations. Note the positions of the holes by carefully measuring from the edges of the taped cabinet outline; these measurements will make it easier to position matching holes in the floor of the base cabinet.

6 Reposition the base cabinet, and mark the floor of the cabinet where the drain and vent pipes will run. (Make sure to allow for the thickness of the cabinet sides when measuring.) Use the hole saw to bore holes in the floor of the cabinet, directly above the holes in the subfloor.

7 Measure, cut, and assemble the drain and loop vent assembly. Tape the top of the loop in place against a brace laid across the top of the cabinet, then extend the drain and vent pipes through the holes in the floor of the cabinet. The waste T should be about 18" above the floor, and the drain and vent pipes should extend about 2 ft. through the floor.

8 In the basement, establish a route from the island vent pipe to an existing vent pipe. (In our project, we are using the auxiliary waste-vent stack near a utility sink.) Hold a long length of pipe between the pipes, and mark for T-fittings. Cut off the plastic vent pipe at the mark, then dry-fit a waste T-fitting to the end of the pipe.

9 Hold a waste T against the vent stack, and mark the horizontal vent pipe at the correct length. Fit the horizontal pipe into the waste T, then tape the assembly in place against the vent stack. The vent pipe should angle ¼" per foot down toward the drain.

10 Fit a 3" length of pipe in the bottom opening on the T-fitting attached to the vent pipe, then mark both the vent pipe and the drain pipe for 45° elbows. Cut off the drain and vent pipes at the marks, then dry-fit the elbows onto the pipes.

11 Extend both the vent pipe and drain pipe by dry-fitting 3" lengths of pipe and Y-fittings to the elbows. Using a carpenter's level, make sure the horizontal drain pipe will slope toward the waste-vent at a pitch of ¼" per ft. Measure and cut a short length of pipe to fit between the Y-fittings.

(continued next page)

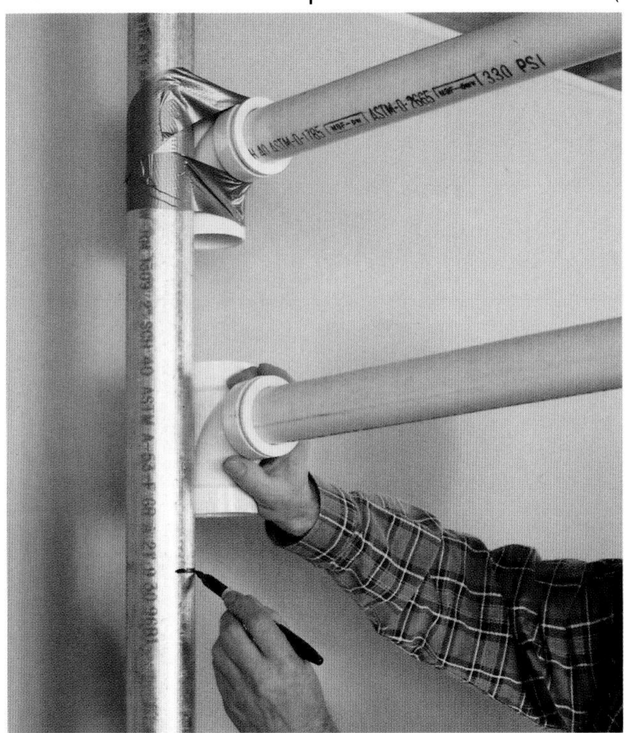

12 Cut a horizontal drain pipe to reach from the vent Y-fitting to the auxiliary waste-vent stack. Attach a waste T to the end of the drain pipe, then position it against the drain stack, maintaining a downward slope of ¼" per ft. Mark the auxiliary stack for cutting above and below the fittings.

13 Cut out the auxiliary stack at the marks. Use the T-fittings and short lengths of pipe to assemble an insert piece to fit between the cutoff ends of the auxiliary stack. The insert assembly should be about ½" shorter than the removed section of stack.

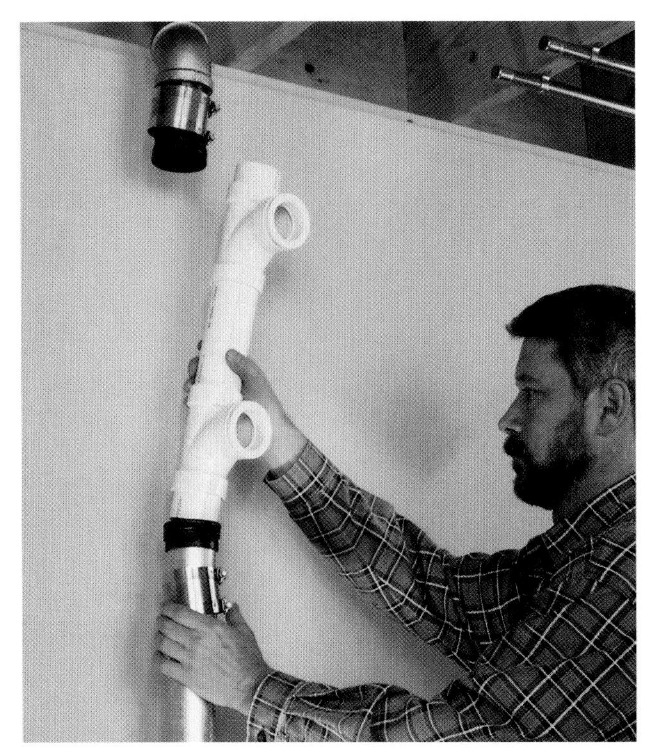

14 Slide banded couplings onto the cut ends of the auxiliary stack, then insert the plastic pipe assembly and loosely tighten the clamps.

15 At the open inlet on the drain pipe Y-fitting, insert a cleanout fitting.

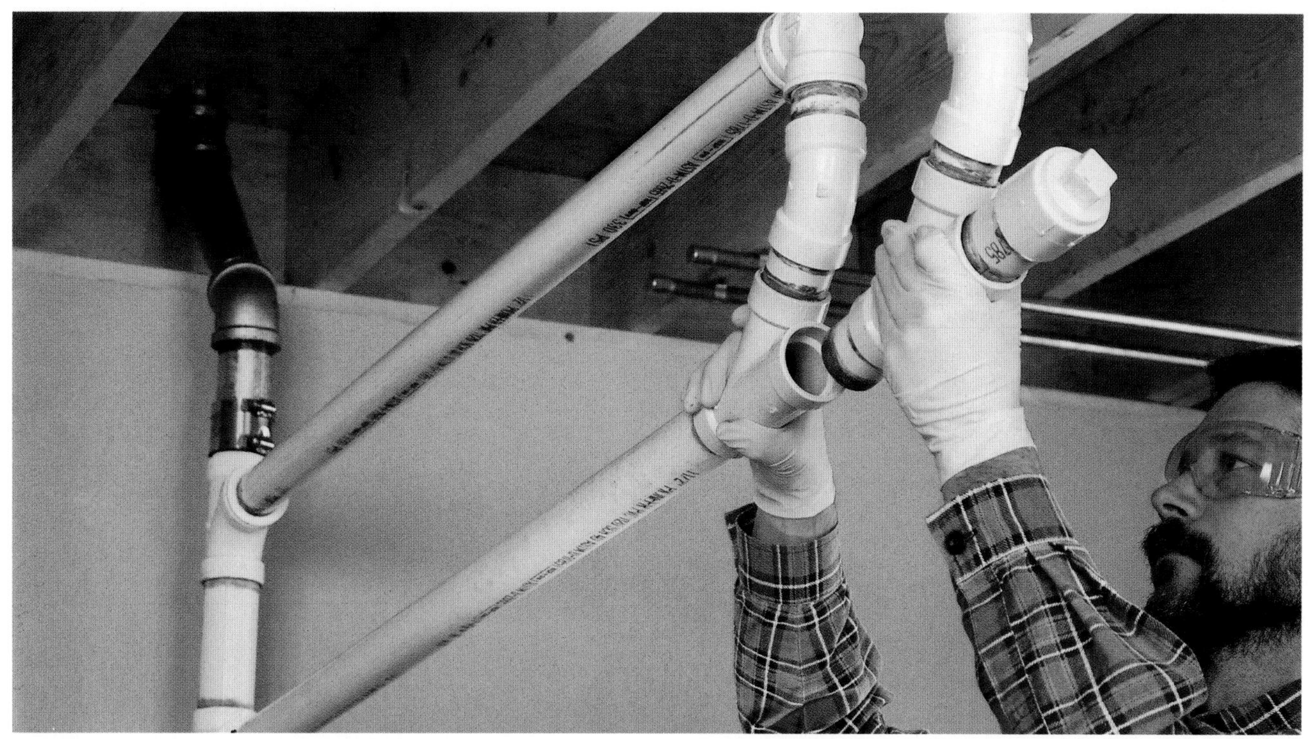

16 Solvent-glue all pipes and fittings found in the basement, beginning with the assembly inserted into the existing waste-vent stack, but do not glue the vertical drain and vent pipes running up into the cabinet. Tighten the banded couplings at the auxiliary stack. Support the horizontal pipes every 4 ft. with strapping nailed to the joists, then detach the vertical pipes extending up into the island cabinet. The final connection for the drain and vent loop will be completed as other phases of the kitchen remodeling project are finished.

17 After installing flooring and attaching cleats for the island base cabinet, cut away the flooring covering the holes for the drain and vent pipes.

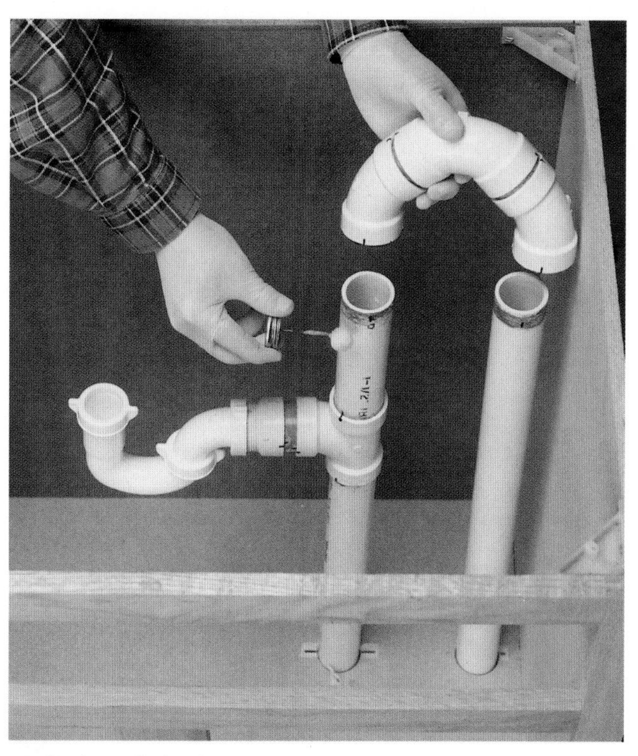

18 Install the base cabinet, then insert the drain and vent pipes through the holes in the cabinet floor and solvent-glue the pieces together.

How to Install New Supply Pipes

1 Drill two 1"-diameter holes, spaced about 6" apart, through the floor of the island base cabinet and the underlying subfloor. Position the holes so they are not over floor joists. Drill similar holes in the floor of the base cabinet for the wall sink.

2 Turn off the water at the main shutoff, and drain the pipes. Cut out any old water supply pipes that obstruct new pipe runs, using a tubing cutter or hacksaw. In our project, we are removing the old pipe back to a point where it is convenient to begin the new branch lines.

3 Dry-fit T-fittings on each supply pipe (we used ¾" × ½" × ½" reducing T-fittings). Use elbows and lengths of copper pipe to begin the new branch lines running to the island sink and the wall sink. The parallel pipes should be routed so they are between 3" and 6" apart.

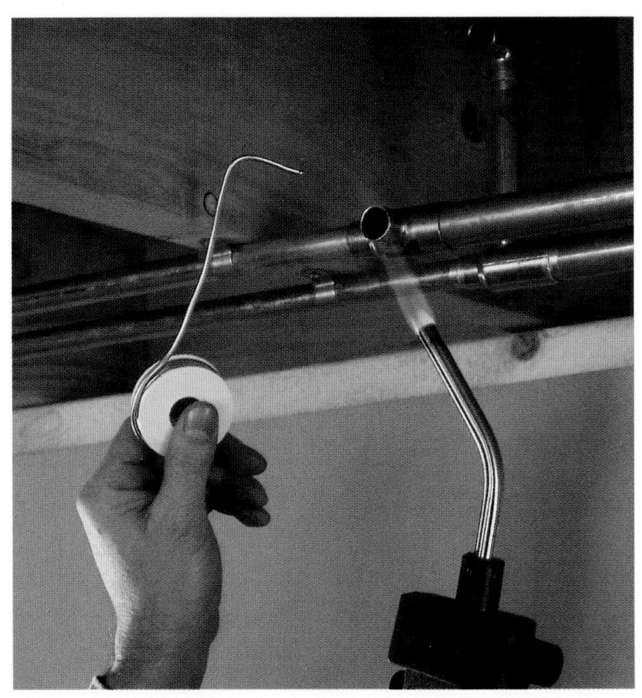

4 Solder the pipes and fittings together, beginning at the T-fittings. Support the horizontal pipe runs every 6 ft. with strapping attached to joists.

5 Extend the branch lines to points directly below the holes leading up into the base cabinets. Use elbows and lengths of pipe to form vertical risers extending at least 12" into the base cabinets. Use a small level to position the risers so they are plumb, then mark the pipe for cutting.

6 Fit the horizontal pipes and risers together, and solder them in place. Install blocking between joists, and anchor the risers to the blocking with pipe straps.

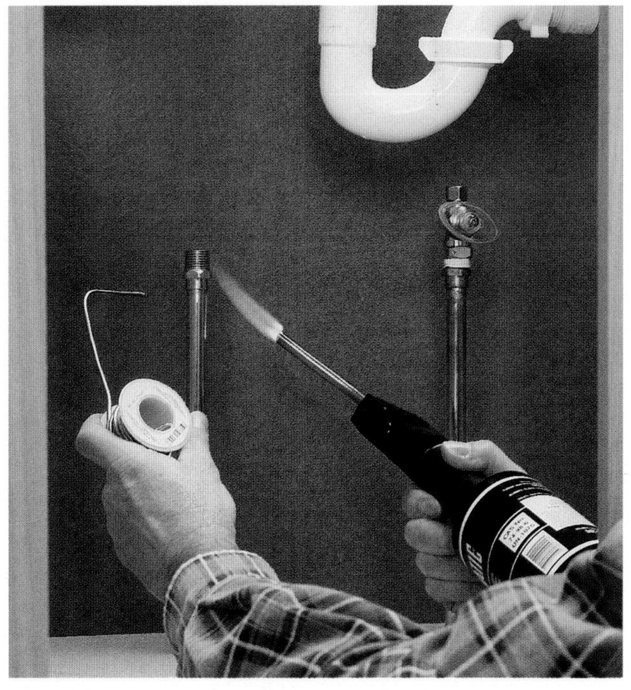

7 Solder male threaded adapters to the tops of the risers, then screw threaded shutoff valves onto the fittings.

Installing a Faucet & Drain

Most new kitchen faucets feature single-handle control levers and washerless designs. They rarely require maintenance. More expensive designer styles offer added features, like colorful enameled finishes, detachable spray nozzles, or even digital temperature readouts.

Connect the faucet to hot and cold water lines with easy-to-install flexible supply tubes made from vinyl or braided steel.

Where local Codes allow, use plastic piping for drain hookups. Plastic is inexpensive and easy to install.

A wide selection of extensions and angle fittings let you easily plumb any sink configuration. Manufacturers offer kits that contain all the fittings needed for attaching a food disposer or dishwasher to the sink drain system.

Everything You Need:

Basic Materials: faucet, flexible vinyl or braided steel supply tubes, drain components.

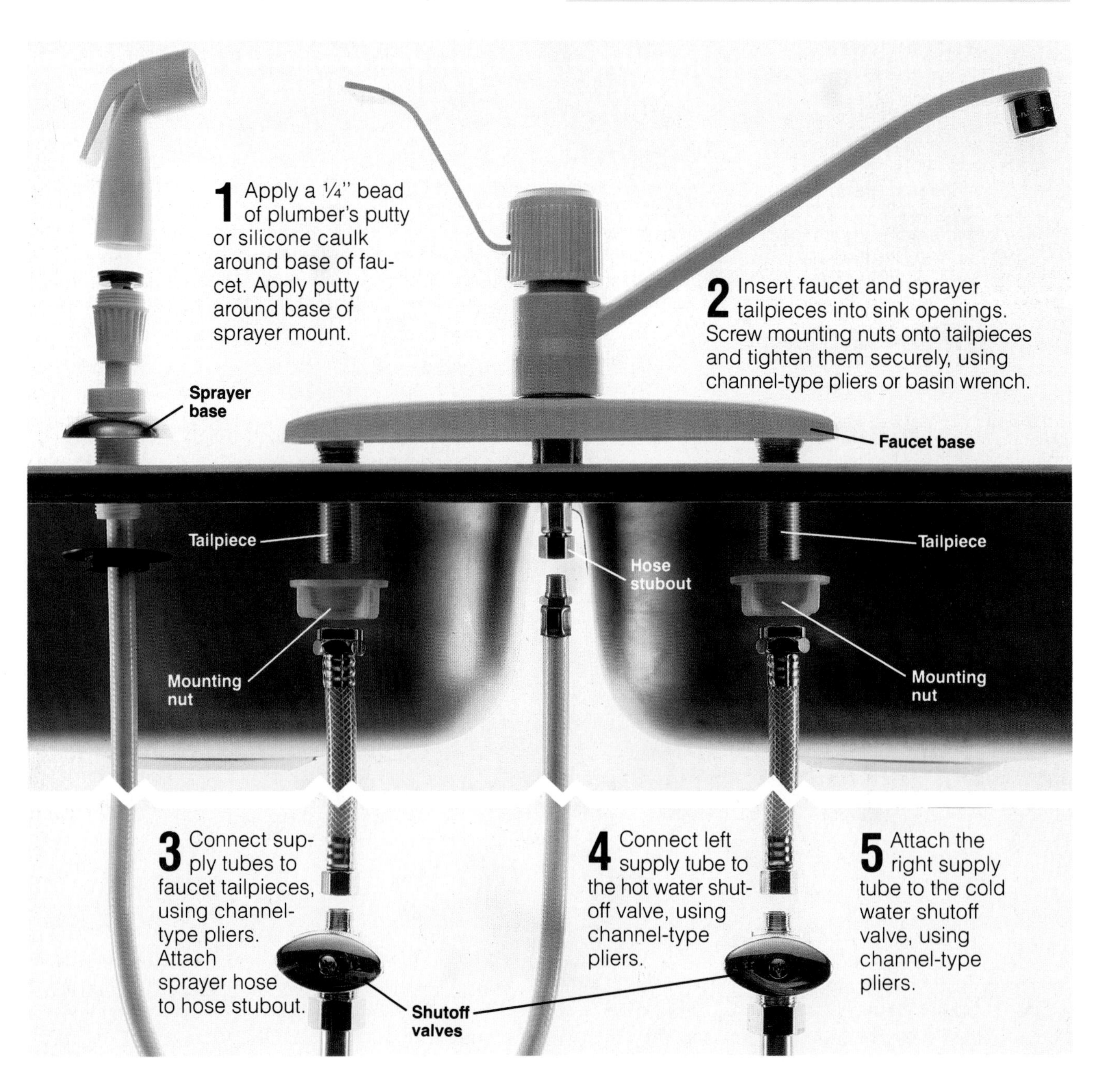

1 Apply a ¼" bead of plumber's putty or silicone caulk around base of faucet. Apply putty around base of sprayer mount.

2 Insert faucet and sprayer tailpieces into sink openings. Screw mounting nuts onto tailpieces and tighten them securely, using channel-type pliers or basin wrench.

3 Connect supply tubes to faucet tailpieces, using channel-type pliers. Attach sprayer hose to hose stubout.

4 Connect left supply tube to the hot water shutoff valve, using channel-type pliers.

5 Attach the right supply tube to the cold water shutoff valve, using channel-type pliers.

How to Attach Drain Lines

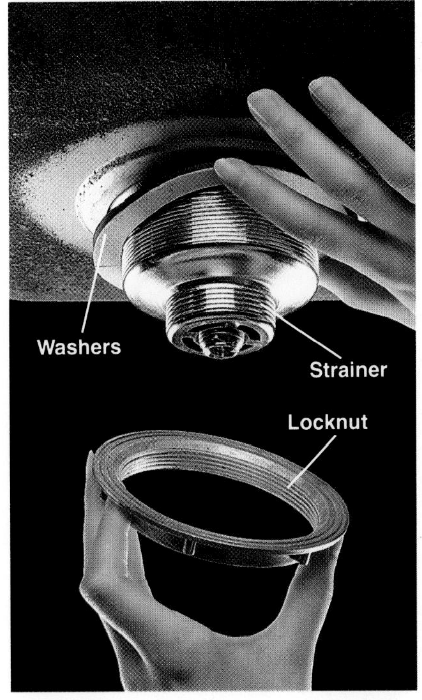

1 Install sink strainer in each sink drain opening. Apply ¼" bead of plumber's putty around bottom of flange. Insert strainer into drain opening. Place rubber and fiber washers over neck of strainer. Screw locknut onto neck and tighten with channel-type pliers.

2 Attach tailpiece to strainer. Place insert washer in flared end of tailpiece, then attach tailpiece by screwing a slip nut onto sink strainer. If necessary, tailpiece can be cut to fit with a hacksaw.

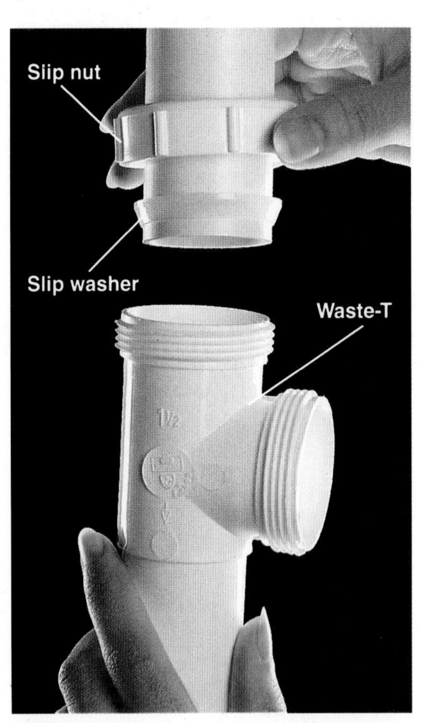

3 On sinks with two basins, use a continuous waste T-fitting to join the tailpieces (pages 114 to 115). Attach the fitting with slip washers and nuts. Beveled side of washers face threaded portion of pipes.

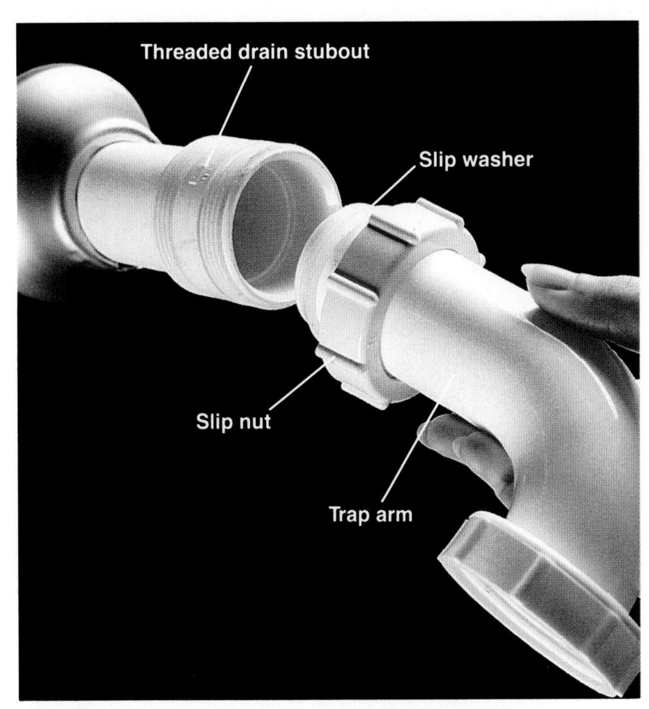

4 Attach the trap arm to the drain stubout, using a slip nut and washer. Beveled side of washer should face threaded drain stubout. If necessary, trap arm can be cut to fit with a hacksaw.

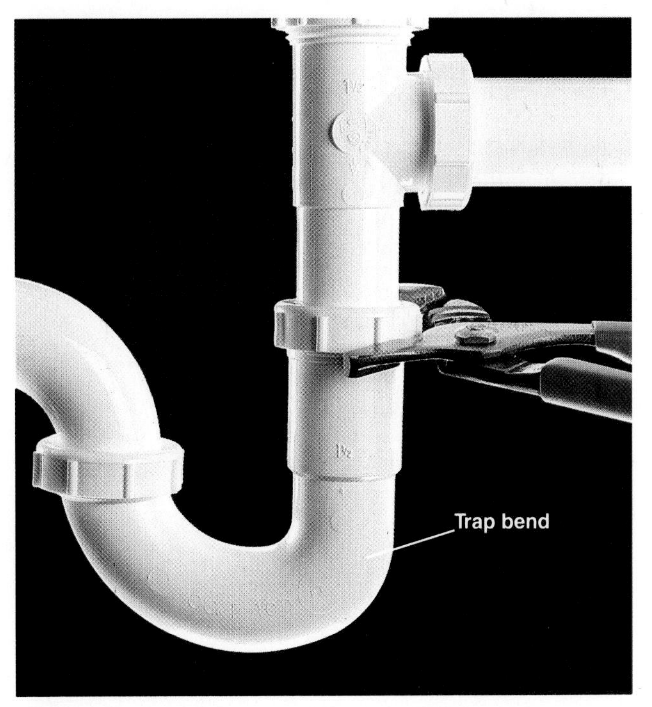

5 Attach trap bend to trap arm, using slip nuts and washers. Beveled side of washers should face trap bend. Tighten all nuts with channel-type pliers.

Kitchen sink with two basins is connected to the drain system by means of a P-trap and continuous waste T-fitting. One end of the continuous waste T connects to the food disposer, the other end to a drain tailpiece running to the other sink basin. Supply tubes controlled by shutoff valves bring water from the hot and cold supply risers to the faucet. The hot water riser may have a supply tube that runs to a dishwasher adjacent to the sink, and the cold water riser may have a saddle valve and tubing running to a refrigerator icemaker.

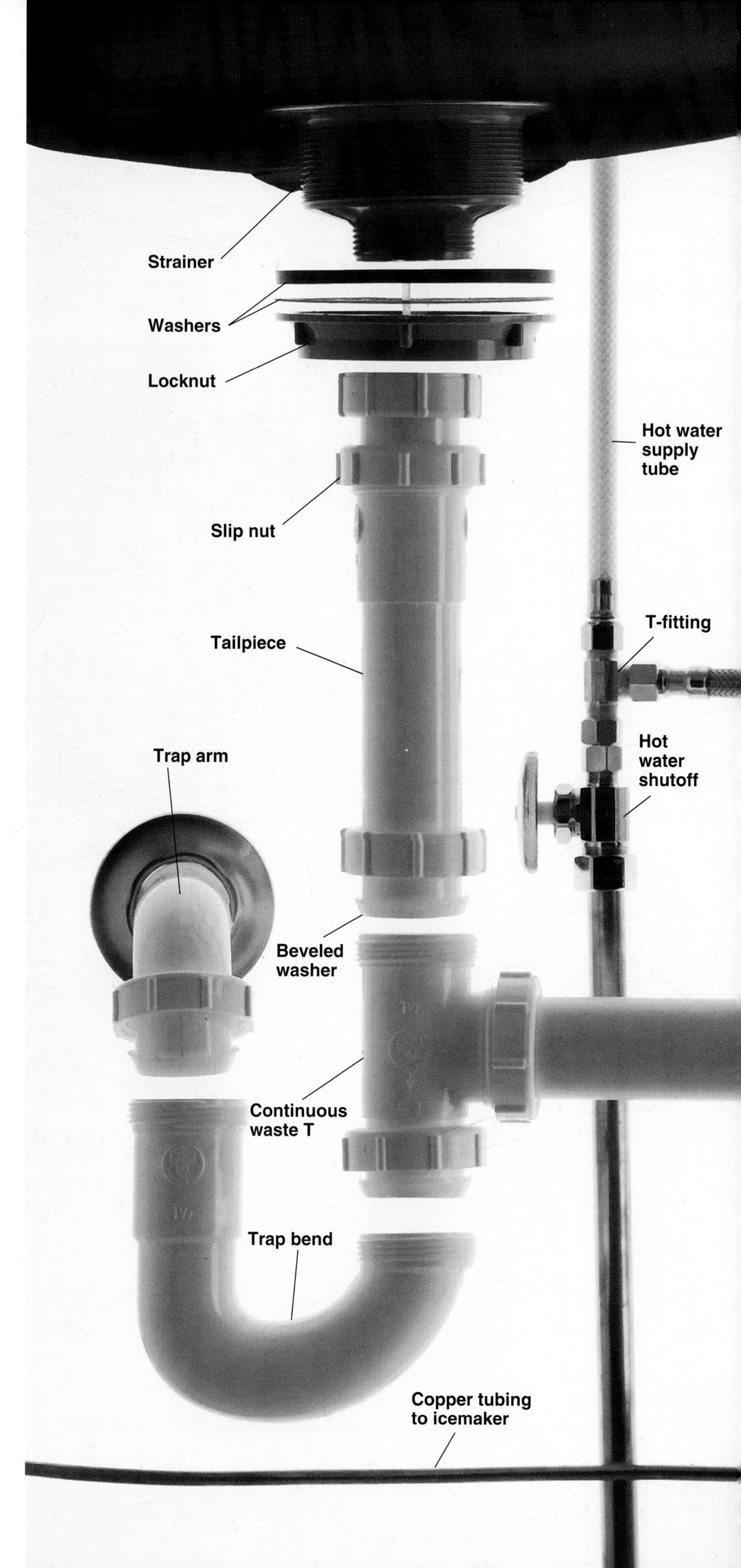

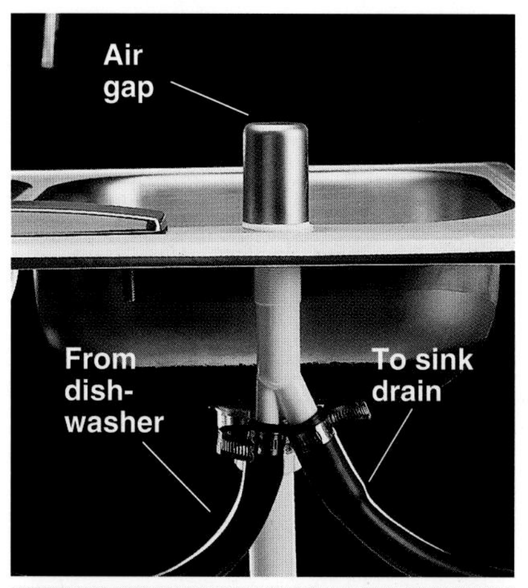

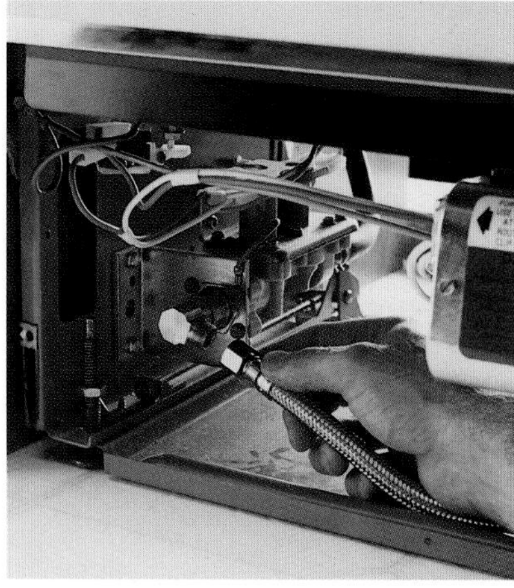

Dishwasher drain is looped up through an air gap device on the countertop (top), which prevents a plugged sink drain from backing up into the dishwasher. The supply tube running from the hot water supply pipe is connected to the water valve inside the access panel at the bottom of the dishwasher (bottom).

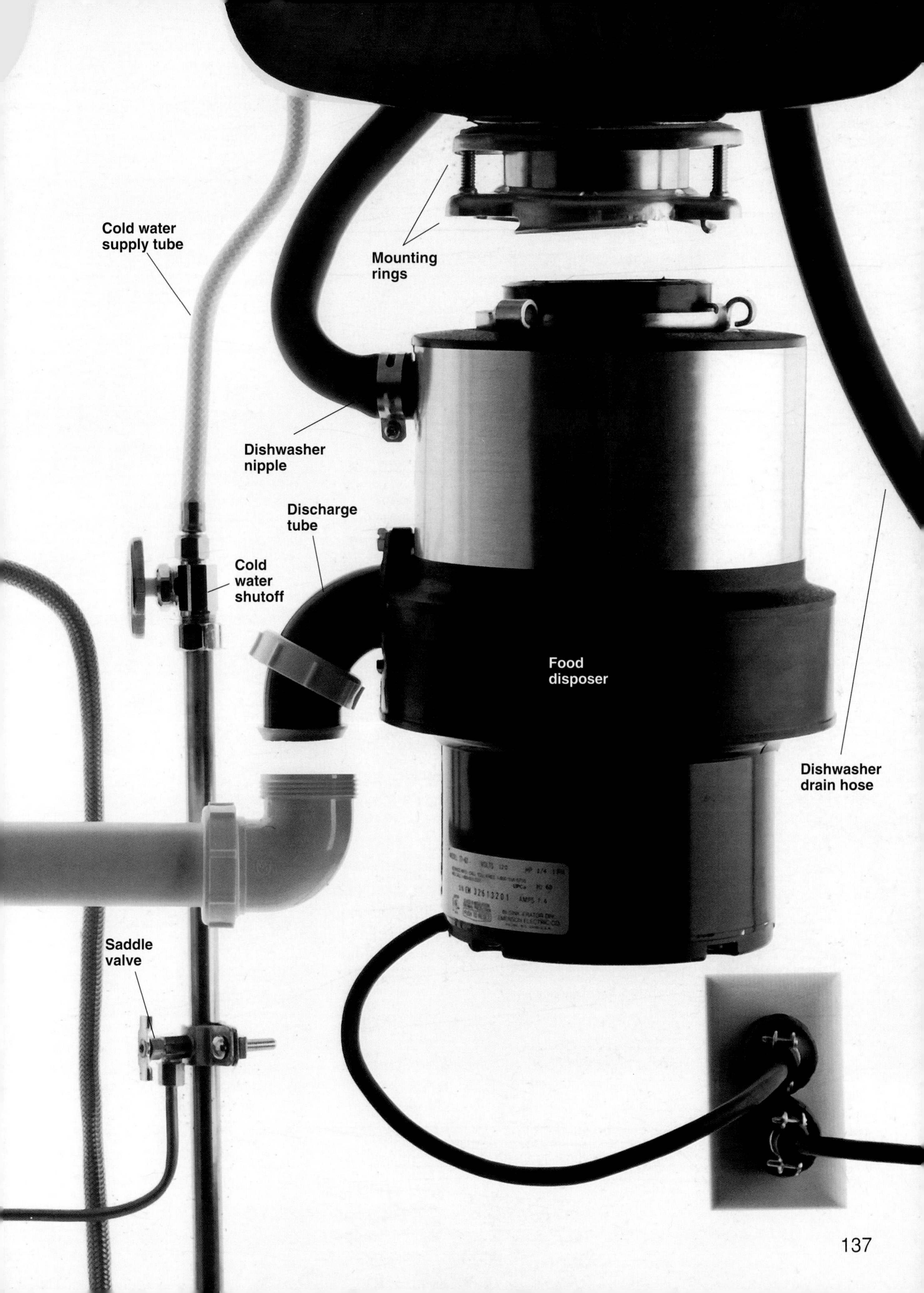
Cold water supply tube
Mounting rings
Dishwasher nipple
Discharge tube
Cold water shutoff
Food disposer
Dishwasher drain hose
Saddle valve

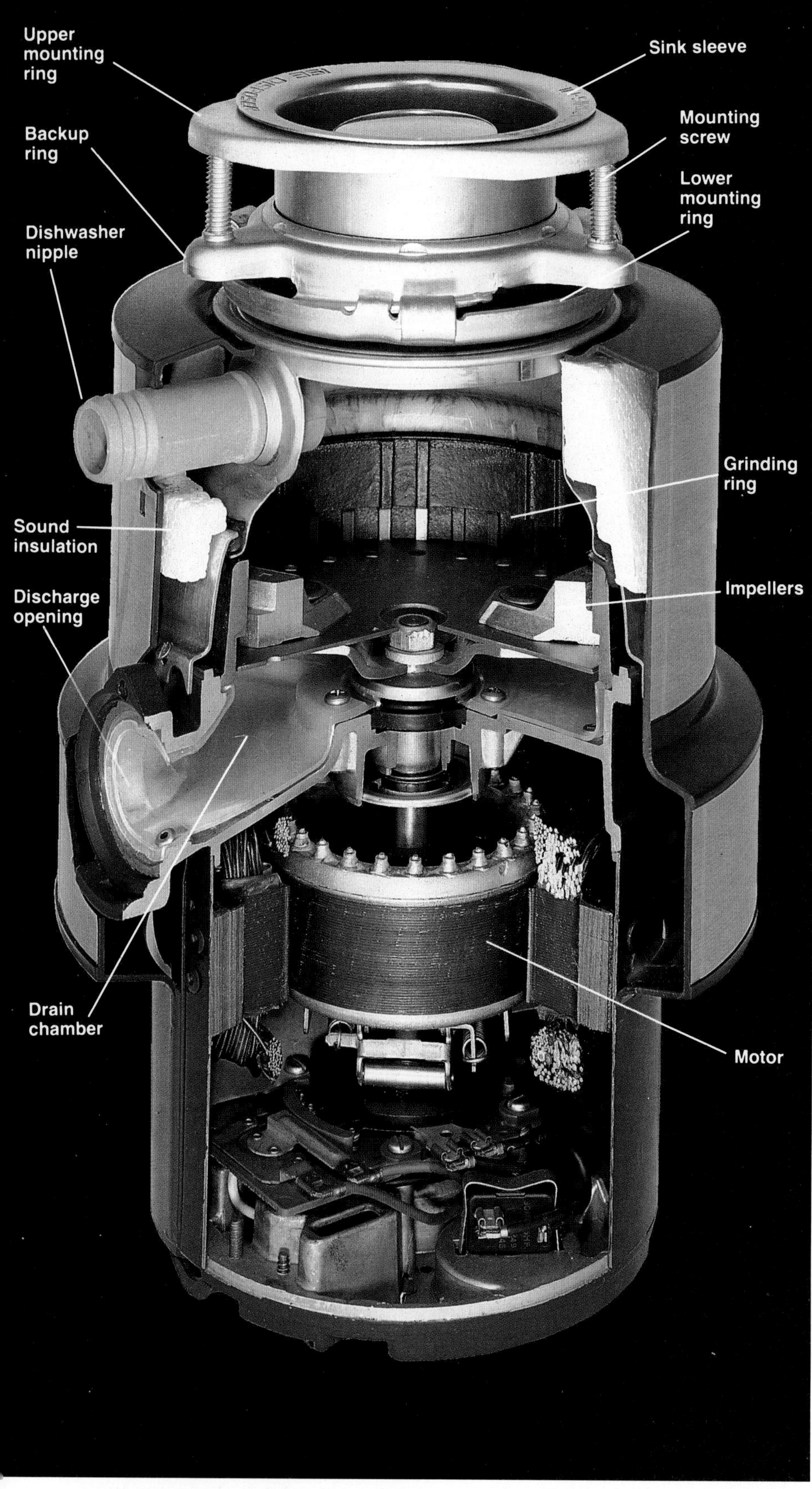

Food disposer grinds food waste so it can be flushed away through the sink drain system. A quality disposer has a ½-horsepower, self-reversing motor that will not jam. Other features to look for include foam sound insulation, a cast-iron grinding ring, and overload protection that allows the motor to be reset if it overheats. Better food disposers have a 5-year manufacturer's warranty.

Installing a Food Disposer

Choose a food disposer with a motor rated at ½ horsepower or more. Look for a self-reversing feature that prevents the disposer from jamming. Better models carry a manufacturer's warranty of up to five years.

Local Plumbing Codes may require that a disposer be plugged into a grounded outlet controlled by a switch above the sink.

Everything You Need:

Basic Tools (pages 34 to 37): screwdriver.

Basic Materials: 12-gauge appliance cord with grounded plug, wire nuts.

How to Install a Food Disposer

1 Remove plate on bottom of disposer. Use combination tool to strip about ½" of insulation from each wire in appliance cord. Connect white wires, using a wire nut. Connect black wires. Attach green insulated wire to green ground screw. Gently push wires into opening. Replace bottom plate.

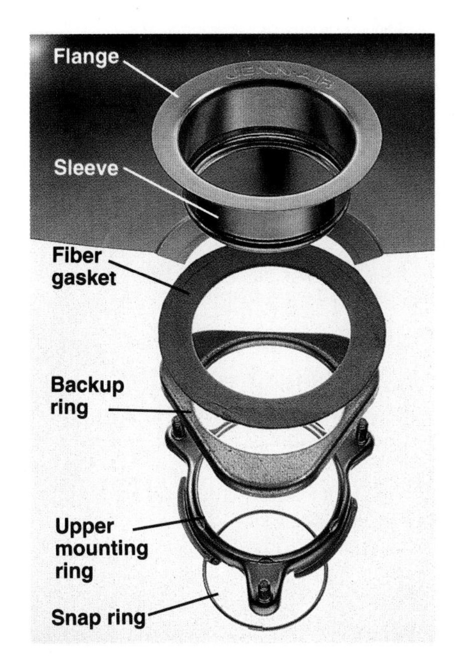

2 Apply ¼-inch bead of plumber's putty under the flange of the disposer sink sleeve. Insert sleeve in drain opening, and slip the fiber gasket and the backup ring onto the sleeve. Place upper mounting ring on sleeve and slide snap ring into groove.

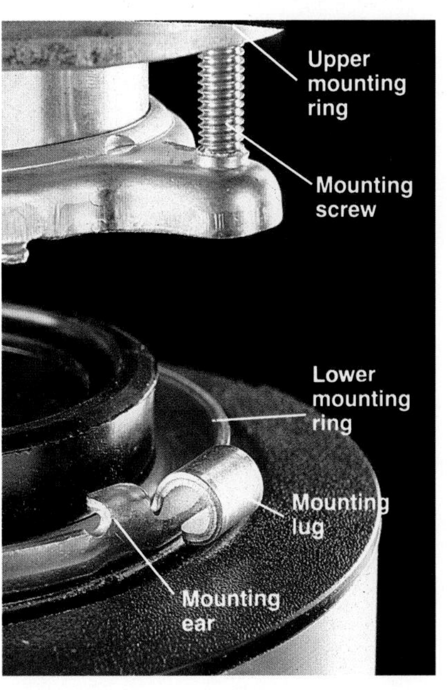

3 Tighten the three mounting screws. Hold disposer against upper mounting ring so that the mounting lugs on the lower mounting ring are directly under the mounting screws. Turn the lower mounting ring clockwise until the disposer is supported by the mounting assembly.

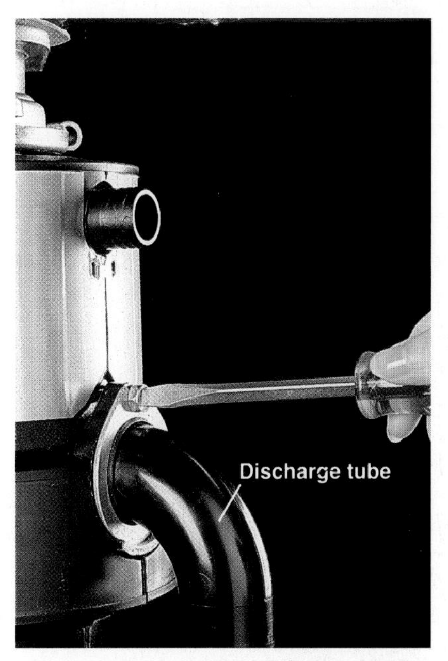

4 Attach the discharge tube to the discharge opening on the side of the disposer, using the rubber washer and metal flange.

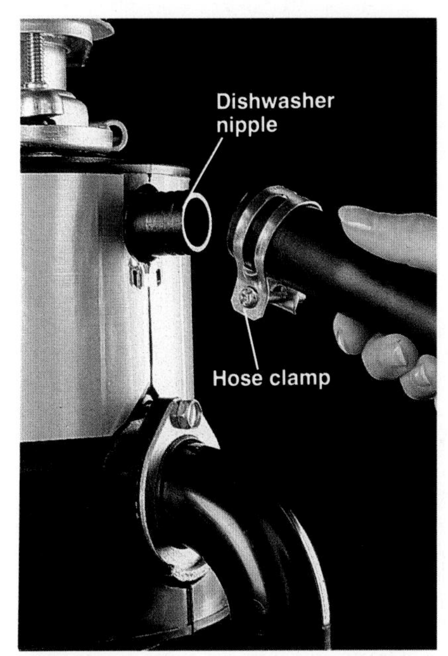

5 If dishwasher will be attached, knock out the plug in the dishwasher nipple, using a screwdriver. Attach the dishwasher drain hose to nipple with hose clamp.

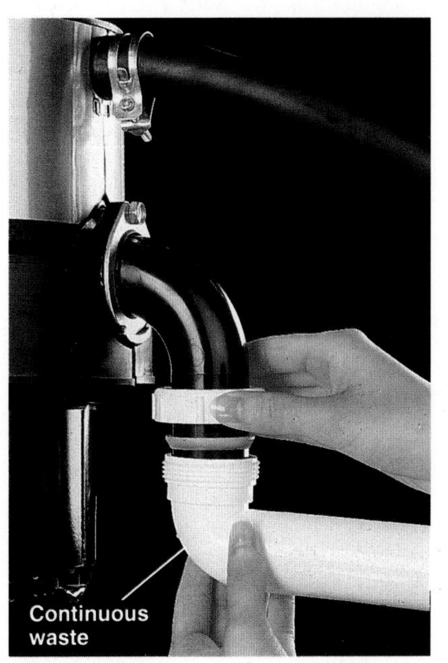

6 Attach the discharge tube to continuous waste pipe with slip washer and nut. If discharge tube is too long, cut it with a hacksaw or tubing cutter.

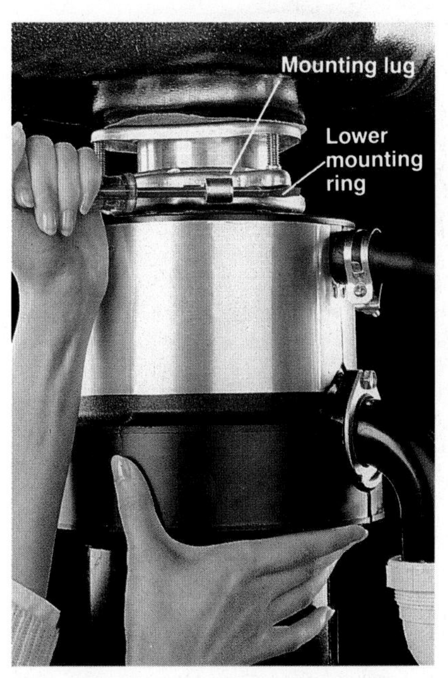

7 Lock disposer into place. Insert a screwdriver or disposer wrench into a mounting lug on the lower mounting ring, and turn clockwise until the mounting ears are locked. Tighten all drain slip nuts with channel-type pliers.

Installing a Dishwasher

A dishwasher requires a hot water supply connection, a drain connection, and an electrical hookup. These connections are easiest to make when the dishwasher is located next to the sink.

Hot water reaches the dishwasher through a supply tube. With a multiple-outlet shutoff valve or brass T-fitting on the hot water pipe, you can control water to the sink and dishwasher with the same valve.

For safety, loop the dishwasher drain hose up through an air gap mounted on the sink or countertop. An air gap prevents a clogged drain from backing up into the dishwasher.

A dishwasher requires its own 20-amp electrical circuit. For convenience, have this circuit wired into one-half of a split duplex receptacle. The other half of the receptacle powers the food disposer.

Everything You Need:

Basic Tools (pages 34 to 37): screwdriver, utility knife, drill with 2" hole saw.

Basic Materials: air gap, drain hose, waste-T tailpiece, braided steel supply tube, rubber connector for food disposer, brass L-fitting, 12-gauge appliance power cord.

How to Install a Dishwasher

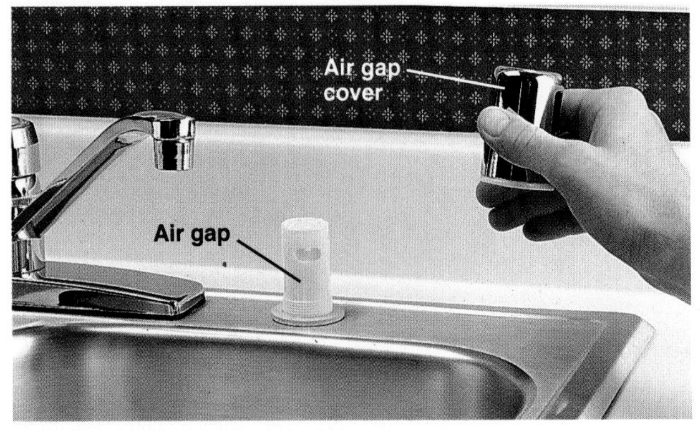

1 Mount air gap, using one of the predrilled sink openings. Or, bore a hole in the countertop with a drill and hole saw. Attach the air gap by tightening mounting nut over the tailpiece with channel-type pliers.

2 Cut openings in side of sink base cabinet for electrical and plumbing lines, using a drill and hole saw. Dishwasher instructions specify size and location of openings. Slide dishwasher into place, feeding rubber drain hose through hole in cabinet. Level the dishwasher.

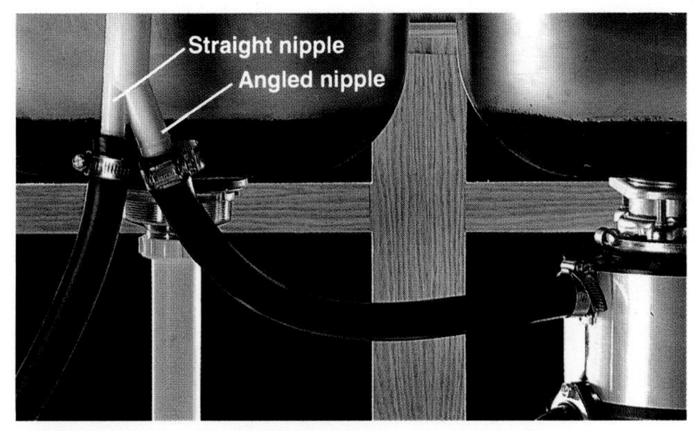

3 Attach the dishwasher drain hose to the smaller, straight nipple on the air gap, using a hose clamp. If hose is too long, cut to correct length with a utility knife. Cut another length of rubber hose to reach from the larger, angled nipple to the food disposer. Attach hose to the air gap and to the nipple on disposer with hose clamps.

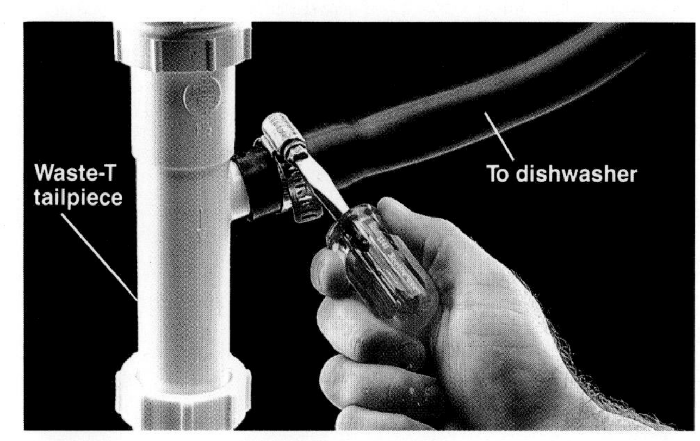

On sinks without food disposer, attach a special waste-T sink tailpiece to sink strainer. Attach the drain hose to the waste-T nipple with a hose clamp.

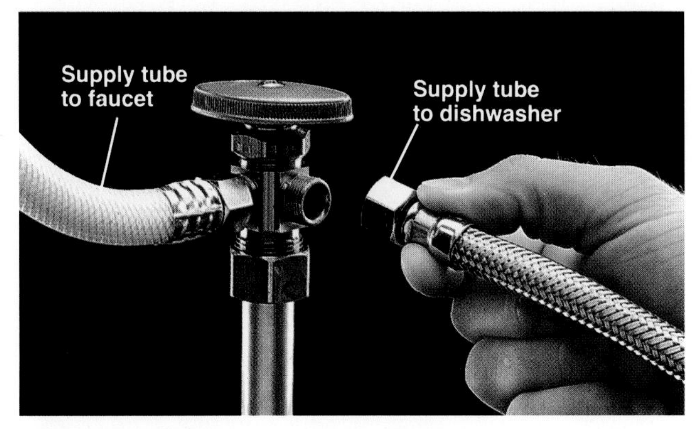

4 Connect dishwasher supply tube to hot water shutoff, using channel-type pliers. This connection is easiest with a multiple-outlet shutoff valve or a brass T-fitting (page 136).

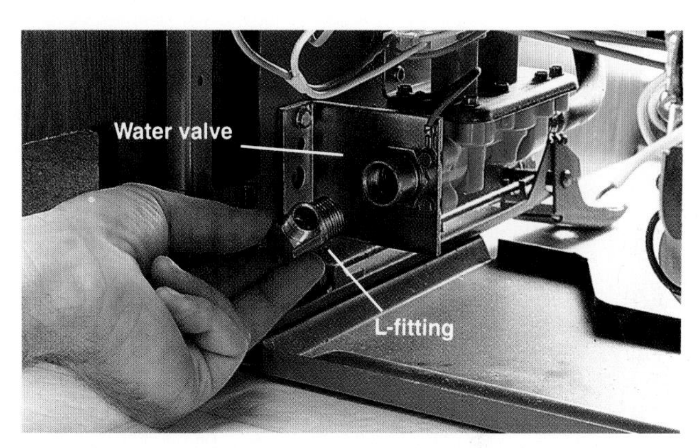

5 Remove access panel on front of dishwasher. Connect a brass L-fitting to the threaded opening on the dishwasher water valve, and tighten with channel-type pliers.

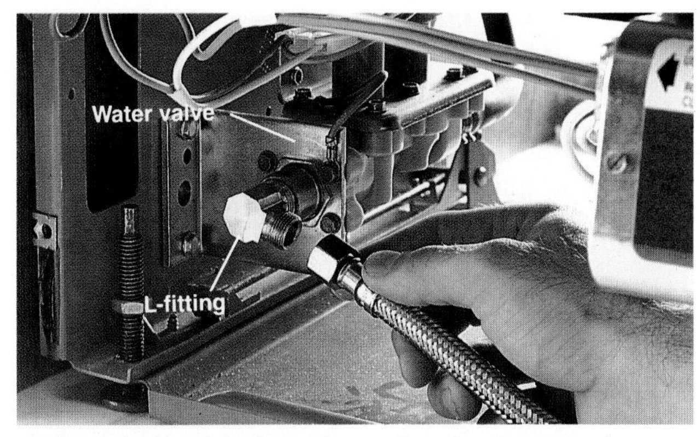

6 Run the braided steel supply tube from the hot water pipe to the dishwasher water valve. Attach supply tube to L-fitting, using channel-type pliers.

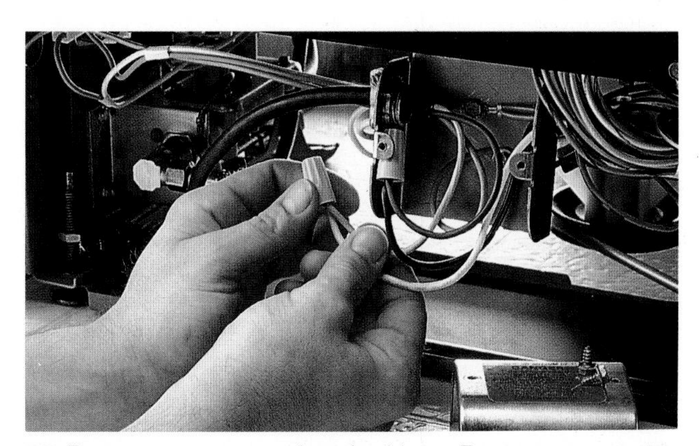

7 Remove cover on electrical box. Run power cord from outlet through to electrical box. Strip about ½" of insulation from each cord wire, using combination tool. Connect black wires, using a wire nut. Connect white wires. Connect green insulated wire to ground screw. Replace box cover and dishwasher access panel.

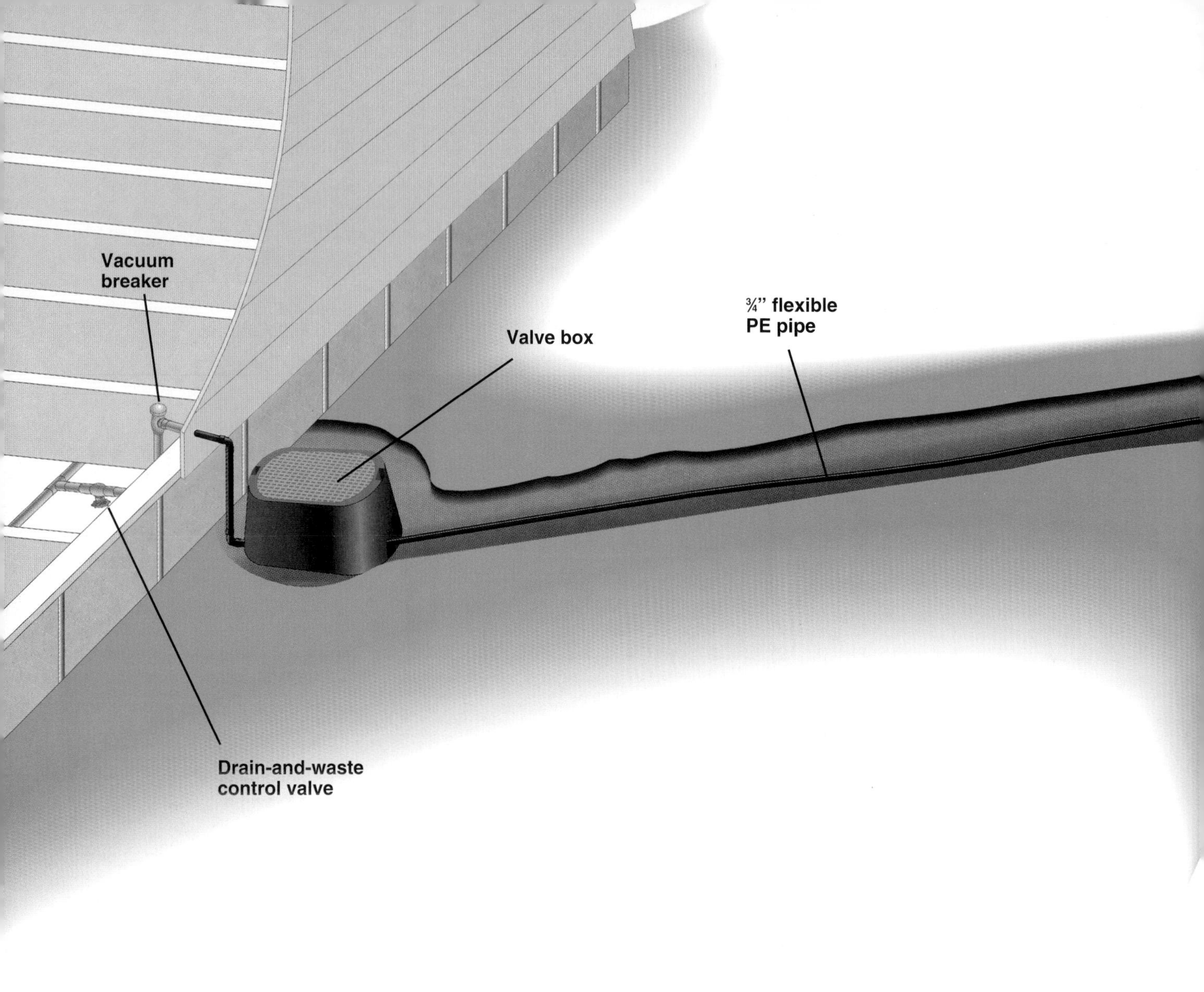

Installing Outdoor Plumbing

Flexible polyethylene (PE) pipe (pages 62 to 63) is used to extend cold-water plumbing lines to outdoor fixtures, such as a sink located in a shed or detached garage, a lawn sprinkler system, or garden spigot. In mild climates, outdoor plumbing can remain in service year-round, but in regions with a frost season, the outdoor supply pipes must be drained or blown empty with pressurized air to prevent the pipes from rupturing when the ground freezes in winter.

On the following pages you will see how to run supply pipes from the house to a utility sink in a detached garage. The utility sink drains into a rock-filled *dry well* installed in the yard. A dry well is designed to handle only "gray water" waste, such as the soapy rinse water created by washing tools or work clothes. Never use a dry well drain for septic materials, such as animal waste or food scraps. Never pour paints, solvent-based liquids, or solid materials into a sink that drains into a dry well. Such materials will quickly clog up your system and will eventually filter down into the groundwater supply.

Like an indoor sink, the garage utility sink has a vent pipe running up from the drain trap. This vent can extend through the roof (page 165), or it can be extended through the side wall of the garage and covered with a screen to keep birds and insects out.

Before digging a trench for an outdoor plumbing line, contact your local utility companies and ask them to mark the locations of underground gas, power, telephone, and water lines.

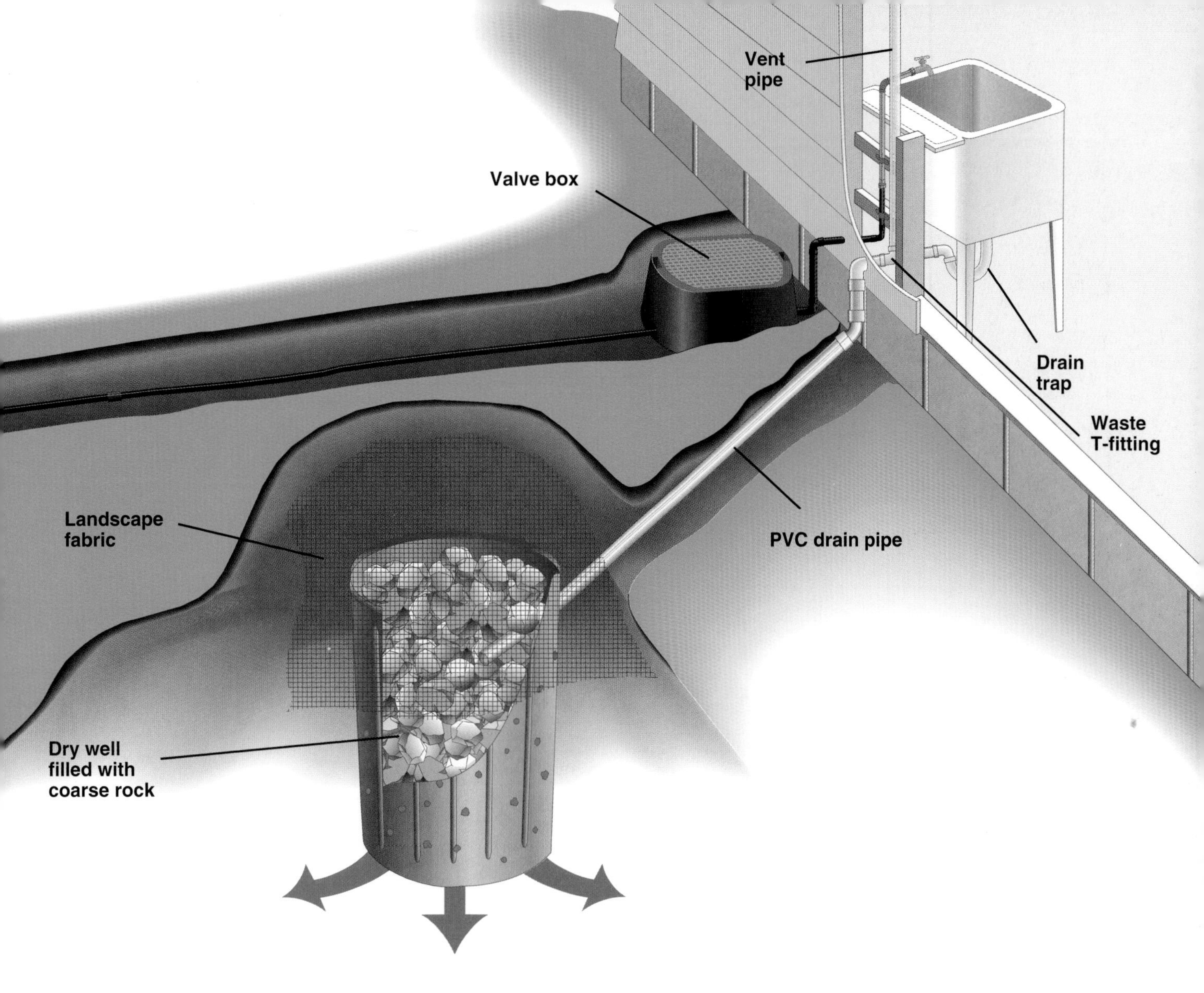

Underground lawn sprinkler systems can be installed using the same basic techniques used for plumbing an outdoor utility sink. Sprinkler systems vary from manufacturer to manufacturer, so make sure to follow the product recommendations when planning your system. See page 147.

How to Install Outdoor Plumbing for a Garage Sink

1 Plan a convenient route from a basement ¾" cold-water supply pipe to the outdoor sink location, then drill a 1"-diameter hole through the sill plate. Drill a similar hole where the pipe will enter the garage. On the ground outside, lay out the pipe run with spray paint or stakes.

2 Use a flat spade to remove sod for an 8"- to 12"-wide trench along the marked route from the house to the garage. Set the sod aside and keep it moist so it can be reused after the project is complete. Dig a trench that slopes slightly (⅛" per foot) toward the house and is at least 10" deep at its shallowest point. Use a long straight 2 × 4 and level to ensure that the trench has the correct slope.

3 Below the access hole in the rim joist, dig a small pit and install a plastic valve box so the top is flush with the ground. Lay a thick layer of gravel in the bottom of the box. Dig a similar pit and install a second valve box at the opposite end of the trench, where the water line will enter the garage.

4 Run ¾" PE pipe along the bottom of the trench from the house to the utility sink location. Use insert couplings and stainless steel clamps when it is necessary to join two lengths of pipe.

TIP: To run pipe under a sidewalk, attach a length of rigid PVC pipe to a garden hose with a pipe-to-hose adapter. Cap the end of the pipe, and drill a ⅛" hole in the center of the cap. Turn on the water, and use the high-pressure stream to bore a tunnel.

5 At each end of the trench, extend the pipe through the valve box and up the foundation wall, using a barbed elbow fitting to make the 90° bend. Install a barbed T-fitting with a threaded outlet in the valve box, so the threaded portion of the fitting faces down. Insert a male threaded plug in the bottom outlet of the T-fitting.

6 Use barbed elbow fittings to extend the pipe into the basement and garage, then use pipe straps and masonry screws to anchor the PE pipe to the foundation.

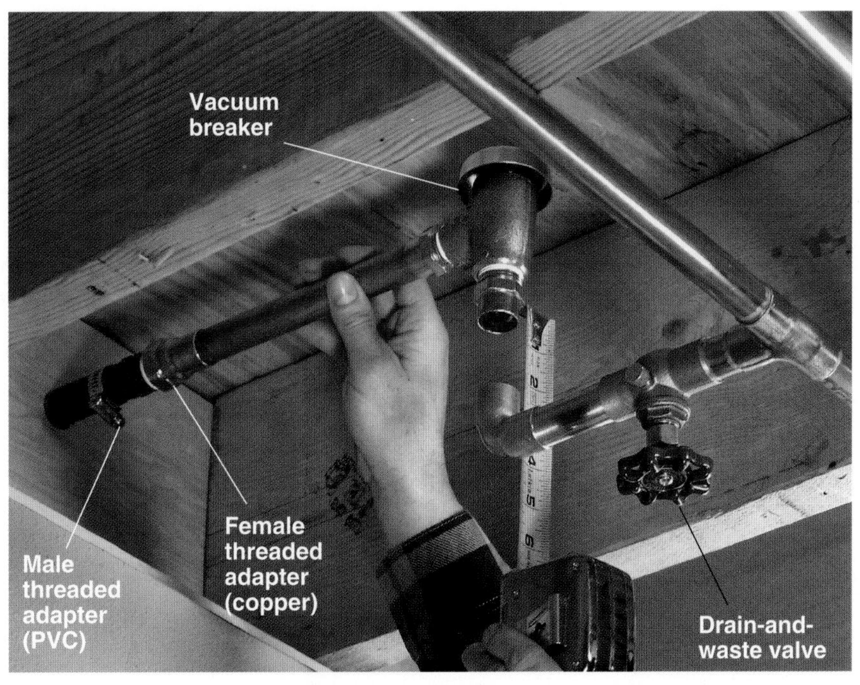

7 Inside the house, make the transition between the PE pipe and the copper cold-water supply pipe, using a threaded male PVC adapter, a female threaded copper adapter, a vacuum breaker, a drain-and-waste valve, and a copper T-fitting, as shown. The drain-and-waste valve includes a threaded cap, which can be removed to blow water from the lines when you are winterizing the system.

(continued next page)

How to Install Outdoor Plumbing for a Garage Sink (continued)

8 In the garage, attach a male threaded PVC adapter to the end of the PE pipe, then use a copper female threaded adapter, elbow, and male threaded adapter to extend a copper riser up to a brass hose bib. After completing the supply pipe installation, fill in the trench, tamping the soil firmly. Install the utility sink, complete with 1½" drain trap and waste T-fitting (page 143). Bore a 2" hole in the wall where the sink drain will exit the garage.

9 At least 6 ft. from the garage, dig a pit about 2 ft. in diameter and 3 ft. deep. Punch holes in the sides and bottom of an old trash can, and cut a 2" hole in the side of the can, about 4" from the top edge. Insert the can into the pit; the top edge should be about 6" below ground level. Run 1½" PVC drain pipe from the utility sink to the dry well (page 143). Fill the dry well with coarse rock, drape landscape fabric over it, then cover the trench and well with soil and reinstall the sod. Extend a vent pipe up from the waste T through the roof or side wall of the garage (page 165).

How to Winterize Outdoor Plumbing Pipes in Cold Climates

Close the drain-and-waste valve for the outdoor supply pipe, then remove the cap on the drain nipple. With the hose bib on the outdoor sink open, attach an air compressor to the valve nipple, then blow water from the system using no more than 50 pounds per square inch (psi) of air pressure. Remove the plugs from the T-fittings in each valve box, and store them for the winter.

Components of an Underground Sprinkler System

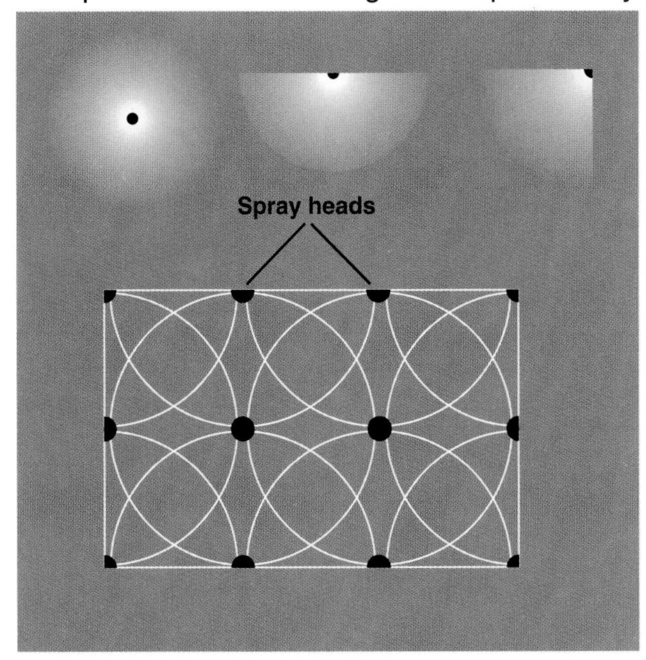

System layout is crucial to ensure proper irrigation of all parts of your yard. Spray heads are available in full-circle, semicircle, or quarter-circle patterns to cover the entire space. In most cases, you will divide your landscape into several individual zones, each controlled by its own valve. With a timer (photo below), you can program precise start and stop times for each irrigation zone.

Photo courtesy of Hunter Industries

Valve manifold is a group of valves used to control the various sprinkler zones. Some models are installed below ground in a valve box, while others extend above the ground. When a timer is used, each control valve in the manifold is wired separately into the timer.

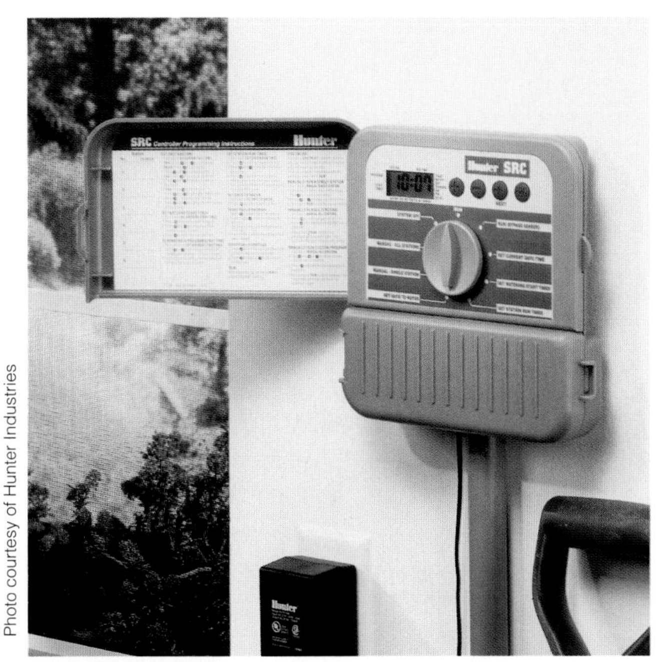

Photo courtesy of Hunter Industries

Sprinkler timers can be programmed to provide automatic control of all zones in an underground sprinkler system. Deluxe models control up to 16 different zones and have rain sensors that shut off the system when irrigation is not needed.

Photo courtesy of Hunter Industries

Sprinkler heads come in many styles to provide a variety of spray patterns. Flexible underground tubing, sometimes called *funny pipe*, links the spray head to saddle fittings on the underground water supply pipes.

Replacing & Repairing Old Plumbing

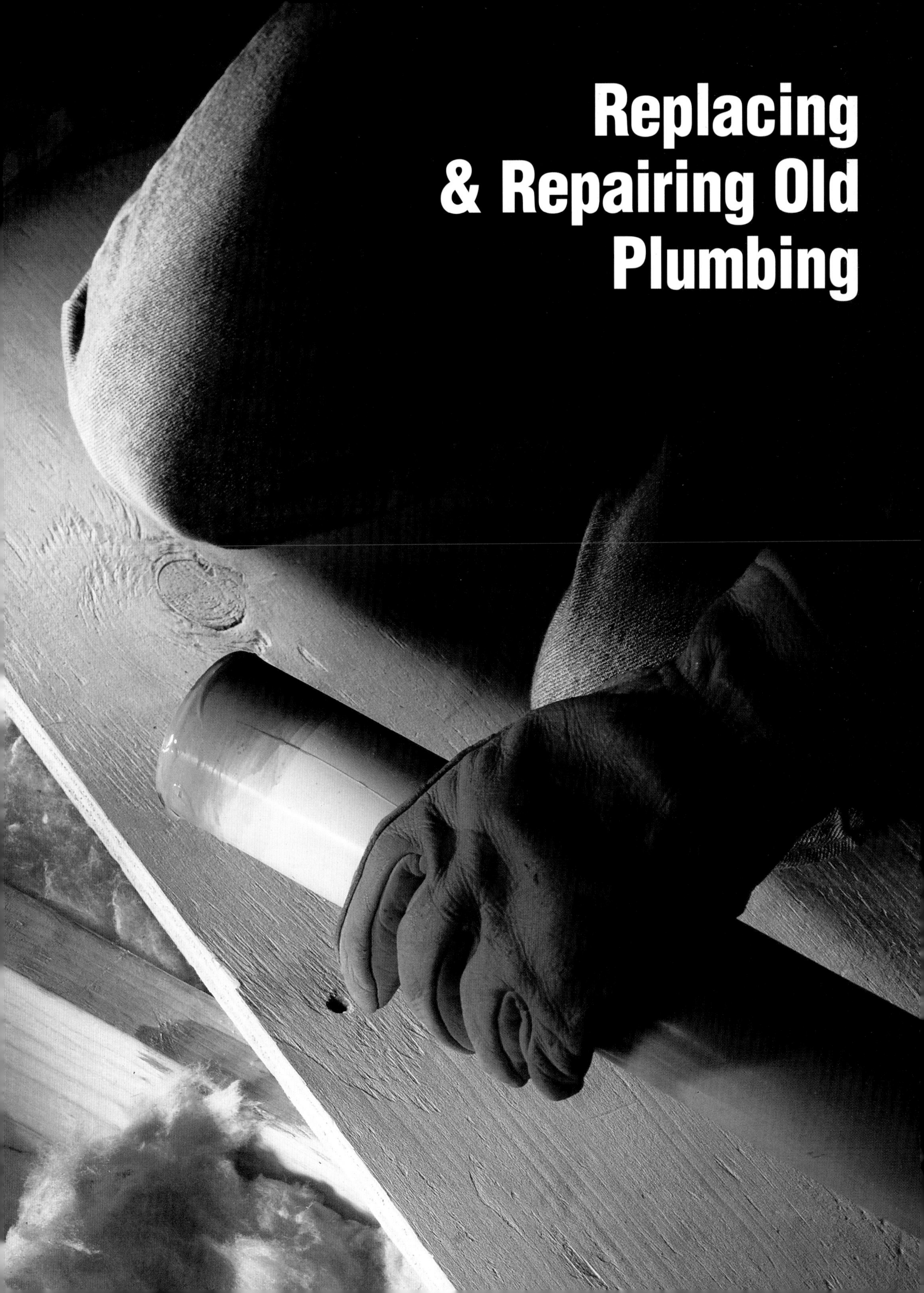

Leave old plumbing pipes in place, if possible. To save time, professional plumbing contractors remove old plumbing pipes only when they interfere with the routing of the new plumbing lines.

Replacing Old Plumbing

Plumbing pipes, like all building materials, eventually wear out and have to be replaced. If you find yourself repairing leaky, corroded pipes every few months, it may be time to consider replacing the old system entirely—and soon. A corroded water pipe that bursts while you are away can cost you many thousands of dollars in damage to wall surfaces, framing members, and furnishings.

Identifying the materials used in your plumbing system can also tell you if replacement is advised. If you have galvanized steel pipes, for example, it is a good bet that they will need to be replaced in the near future. Most galvanized steel pipes were installed before 1960, and since steel pipes have a maximum life expectancy of 30 to 35 years, such a system is probably living on borrowed time. On the other hand, if your system includes copper supply pipes and plastic drain pipes, you can relax; these materials were likely installed within the last 40 years, and they are considerably more durable than steel, provided they were installed correctly.

Unless you live in a rambler with an exposed basement ceiling, replacing old plumbing nearly always involves some demolition and carpentry work. Even in the best scenario, you probably will find it necessary to open walls and floors in order to run new pipes. For this reason, replacing old plumbing is often done at the same time as a kitchen or bathroom remodeling project, when wall and floor surfaces have to be removed and replaced.

A plumbing renovation project is subject to the same Code regulations as a new installation. Always work in conjunction with your local inspector (pages 24 to 29) when replacing old plumbing.

This section shows:

Replacement Options

Partial replacement involves replacing only those sections of your plumbing system that are currently causing problems. This is a quick, less expensive option than a complete renovation, but it is only a temporary solution. Old plumbing will continue to fail until you replace the entire system.

Complete replacement of all plumbing lines is an ambitious job, but doing this work yourself can save you thousands of dollars. To minimize the inconvenience, you can do this work in phases, replacing one branch of the plumbing system at a time.

Evaluating Your Plumbing

By the time you spot the telltale evidence of a leaky drain pipe or water supply pipe, the damage to the walls and ceilings of your home can be considerable. The tips on the following pages show early warning signals that indicate your plumbing system is beginning to fail.

Proper evaluation of your plumbing helps you identify old, suspect materials and anticipate problems. It also can save you money and frustration. Replacing an old plumbing system at your convenience before it reaches the disaster stage is preferable to hiring a plumbing contractor to bail you out of an emergency situation.

Remember that the network of pipes running through the walls of your home is only one part of the larger system. You should also evaluate the main water supply and sewer pipes that connect your home to the city utility system and make sure they are adequate before you replace your plumbing.

Fixture Units	Minimum Gallons per Minute (GPM)
10	8
15	11
20	14
25	17
30	20

Minumum recommended water capacity is based on total demand on the system, as measured by fixture units, a standard of measurement assigned by the Plumbing Code. First, add up the total units of all the fixtures in your plumbing system (page 26). Then, perform the water supply capacity test described below. Finally, compare your water capacity with the recommended minimums listed above. If the capacity falls below that recommended in the table above, then the main water supply pipe running from the city water main to your home is inadequate and should be replaced with a larger pipe by a licensed contractor.

How to Determine Your Water Supply Capacity

1 Shut off the water at the valve on your water meter, then disconnect the pipe on the house side of the meter. Construct a test spout using a 2" PVC elbow and two 6" lengths of 2" PVC pipe, then place the spout on the exposed outlet on the water meter. Place a large watertight tub under the spout to collect water.

2 Open the main supply valve and let the water run into the container for 30 seconds. Shut off the water, then measure the amount of water in the container and multiply this figure by 2. This number represents the gallons-per-minute (gpm) rate of your main water supply. Compare this measurement with the recommended capacity in the table above.

Rust stains on the surfaces of toilet bowls and sinks may indicate severe corrosion inside iron supply pipes. This symptom generally means your water supply system is likely to fail in the near future. NOTE: Rust stains can also be caused by a water heater problem or by a water supply with a high mineral content. Check for these problems before assuming your pipes are bad.

Low water pressure at fixtures suggests that the supply pipes either are badly clogged with rust and mineral deposits, or are undersized. To measure water pressure, plug the fixture drain and open the faucets for 30 seconds. Measure the amount of water and multiply by 2; this figure is the rating in gallons per minute (gpm). Vanity faucets should supply $1\frac{3}{4}$ gpm; bathtub faucets, 6 gpm; kitchen sink faucets, $4\frac{1}{2}$ gpm.

Slow drains throughout the house may indicate that DWV pipes are badly clogged with rust and mineral deposits. When a fixture faucet is opened fully with the drains unstopped, water should not collect in tubs and basins. NOTE: Slow drains may also be the result of inadequate venting. Check for this problem before assuming the drain pipes are bad.

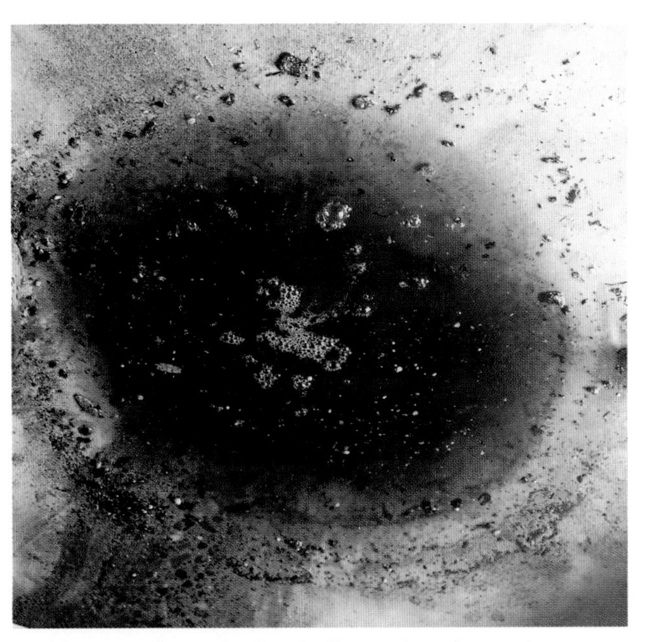

Backed-up floor drains indicate that the main sewer service to the street is clogged. If you have this problem regularly, have the main sewer lines evaluated by a plumbing contractor before you replace your house plumbing. The contractor will be able to determine if your sewer problem is a temporary clog or a more serious problem that requires major work.

Replacing Old Plumbing: A Step-by-step Overview

The overview sequence shown here represents the basic steps you will need to follow when replacing DWV and water supply pipes. On the following pages, you will see these steps demonstrated in complete detail, as we replace all the water supply pipes and drain pipes for a bathroom, including the main waste-vent stack running from basement to roofline.

Remember that no two plumbing jobs are ever alike, and your own project will probably differ from the demonstration projects shown in this section. Always work in conjunction with your local plumbing inspector, and organize your work around a detailed plumbing plan that shows the particulars of your project.

1 Plan the routes for the new plumbing pipes. Creating efficient pathways for new pipes is crucial to a smooth installation. In some cases this requires removing wall or floor surfaces. Or, you can frame a false wall, called a *chase* (page 156), to create space for running the new pipes.

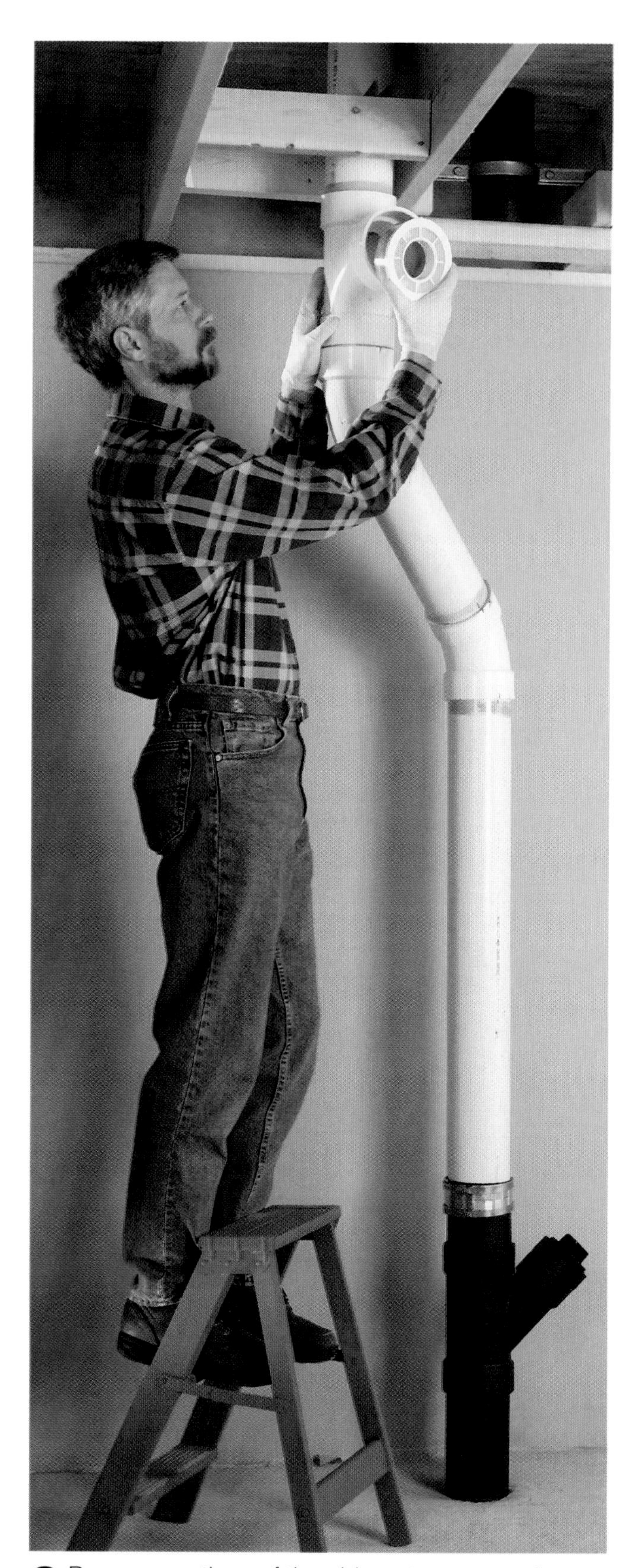

2 Remove sections of the old waste-vent stack, as needed, then install a new main waste-vent stack running from the main drain in the basement to the roof. Include the fittings necessary to connect branch drains and vent lines to the stack.

3 Install new branch drains from the waste-vent stack to the stub-outs for the individual fixtures. If the fixture locations have not changed, you may need to remove sections of the old drain pipes in order to run the new pipes.

4 Remove the old toilet bend and replace it with a new bend running to the new waste-vent stack. This task usually requires that you remove areas of flooring. Framing work may also be required to create a path for the toilet drain.

5 Replace the vent pipes running from the fixtures up to the attic, then connect them to the new waste-vent stack.

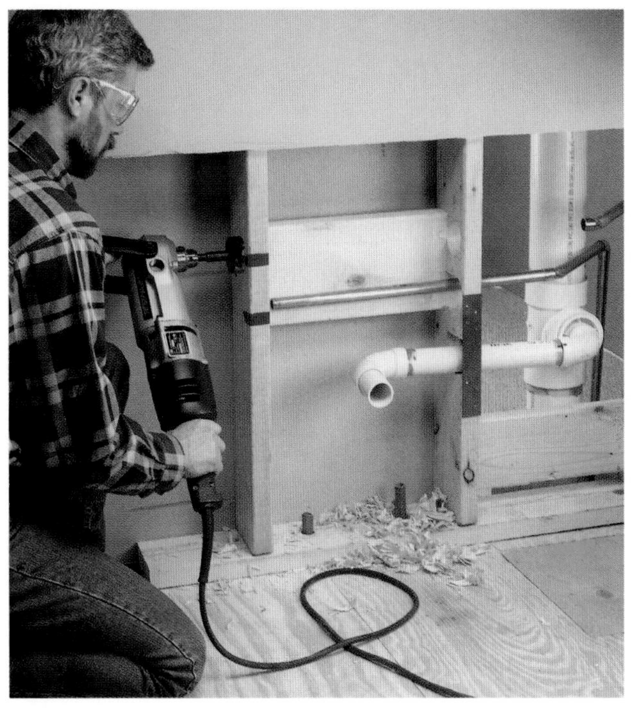

6 Install new copper supply lines running from the water meter to all fixture locations. Test the DWV and water supply pipes and have your work inspected before closing up walls and installing the fixtures.

Build a framed chase. A chase is a false wall created to provide space for new plumbing pipes. It is especially effective for installing a new main waste-vent stack. On a two-story house, chases can be stacked one over the other on each floor in order to run plumbing from the basement to the attic. Once plumbing is completed and inspected, the chase is covered with wallboard and finished to match the room.

Planning Pipe Routes

The first, and perhaps most important, step when replacing old plumbing is to decide how and where to run the new pipes. Since the stud cavities and joist spaces are often covered with finished wall surfaces, finding routes for running new pipes can be challenging.

When planning pipe routes, choose straight, easy pathways whenever possible. Rather than running water supply pipes around wall corners and through studs, for example, it may be easiest to run them straight up wall cavities from the basement. Instead of running a bathtub drain across floor joists, run it straight down into the basement, where the branch drain can be easily extended underneath the joists to the main waste-vent stack.

In some situations, it is most practical to route the new pipes in wall and floor cavities that already hold plumbing pipes, since these spaces often are framed to provide long, unobstructed runs. A detailed map of your plumbing system can be very helpful when planning routes for new plumbing pipes (pages 18 to 23).

To maximize their profits, plumbing contractors generally try to avoid opening walls or changing wall framing when installing new plumbing. But the do-it-yourselfer does not have these limitations. Faced with the difficulty of running pipes through enclosed spaces, you may find it easiest to remove wall surfaces or to create a newly framed space for running new pipes.

On these pages, you will see some common methods used to create pathways for replacing old pipes with new plumbing.

Tips for Planning Pipe Routes

Use existing access panels to disconnect fixtures and remove old pipes. Plan the location of new fixtures and pipe runs to make use of existing access panels, minimizing the amount of demolition and repair work you will need to do.

Convert a laundry chute into a channel for running new plumbing pipes. The door of the chute can be used to provide access to control valves, or it can be removed and covered with wall materials, then finished to match the surrounding wall.

Run pipes inside a closet. If they are unobtrusive, pipes can be left exposed at the back of the closet. Or, you can frame a chase to hide the pipes after the installation is complete.

Remove false ceiling panels to route new plumbing pipes in joist cavities. Or, you can route pipes across a standard plaster or wallboard ceiling, then construct a false ceiling to cover the installation, provided there is adequate height. Most Building Codes require a minimum of 7 ft. from floor to finished ceiling.

(continued next page)

Use a drill extension and spade bit or hole saw to drill through wall plates from unfinished attic or basement spaces above or below the wall.

Look for "wet walls." Walls that hold old plumbing pipes can be good choices for running long vertical runs of new pipe. These spaces usually are open, without obstacles such as fireblocks and insulation.

Probe wall and floor cavities with a long piece of plastic pipe to ensure that a clear pathway exists for running new pipe (left). Once you have established a route using the narrow pipe, you can use the pipe as a guide when running larger drain pipes up into the wall (right).

Remove flooring when necessary. Because replacing toilet and bathtub drains usually requires that you remove sections of floor, a full plumbing replacement job is often done in conjunction with a complete bathroom remodeling project.

Remove wall surfaces when access from above or below the wall is not possible. This demolition work can range from cutting narrow channels in plaster or wallboard to removing the entire wall surface. Remove wall surfaces back to the centers of adjoining studs; the exposed studs provide a nailing surface for attaching new wall materials once the plumbing project is completed.

Create a detailed map showing the planned route for your new plumbing pipes. Such a map can help you get your plans approved by the inspector, and it makes work much simpler. If you have already mapped your existing plumbing system (pages 18 to 23), those drawings can be used to plan new pipe routes.

A new main waste-vent stack is best installed near the location of the old stack. In this way, the new stack can be connected to the basement floor cleanout fitting used by the old cast-iron stack.

Replacing a Main Waste-Vent Stack

Although a main waste-vent stack rarely rusts through entirely, it can be nearly impossible to join new branch drains and vents to an old cast-iron stack. For this reason, plumbing contractors sometimes recommend replacing the iron stack with plastic pipe during a plumbing renovation project.

Be aware that replacing a main waste-vent stack is not an easy job (pages 69 to 71). You will be cutting away heavy sections of cast iron, so working with a helper is essential. Before beginning work, make sure you have a complete plan for your plumbing system and have designed a stack that includes all the fittings you will need to connect branch drains and vent pipes. While work is in progress, none of your plumbing fixtures will be usable. To speed up the project and minimize inconvenience, do as much of the demolition and preliminary construction work as you can before starting work on the stack.

Because main waste-vent stacks may be as large as 4" in diameter, running a new stack through existing walls can be troublesome. To solve this problem, our project employs a common solution: framing a chase in the corner of a room to provide the necessary space for running the new stack from the basement to the attic. When the installation is complete, the chase will be finished with wallboard to match the room.

How to Replace a Main Waste-Vent Stack

1 Secure the cast-iron waste-vent stack near the ceiling of your basement, using a riser clamp installed between the floor joists. Use wood blocks attached to the joists with 3" wallboard screws to support the clamp. Also clamp the stack in the attic, at the point where the stack passes down into the wall cavity. WARNING: A cast-iron stack running from basement to attic can weigh several hundred pounds. Never cut into a cast-iron stack before securing it with riser clamps above the cut.

2 Use a cast iron snap cutter to sever the stack near the floor of the basement, about 8" above the cleanout, and near the ceiling, flush with the bottom of the joists. Have a helper hold the stack while you are cutting out the section. NOTE: After cutting into the main waste-vent stack, plug the open end of the pipe with a cloth to prevent sewer gases from rising into your home.

3 Nail blocking against the bottom of the joists across the severed stack. Then, cut a 6"-diameter hole in the basement ceiling where the new waste-vent stack will run, using a reciprocating saw. Suspend a plumb bob at the centerpoint of the opening as a guide for aligning the new stack.

(continued next page)

4 Attach a 5-ft. segment of PVC plastic pipe the same diameter as the old waste-vent stack to the exposed end of the cast-iron cleanout fitting, using a banded coupling with neoprene sleeve (page 70).

5 Dry-fit 45° elbows and straight lengths of plastic pipe to offset the new stack, lining it up with the plumb bob centered on the ceiling opening.

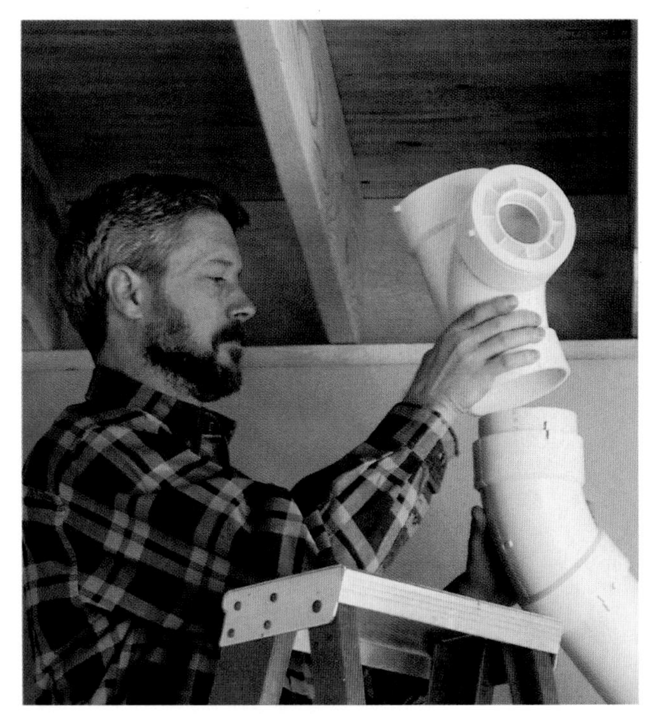

6 Dry-fit a waste T-fitting on the stack, with the inlets necessary for any branch drains that will be connected in the basement. Make sure the fitting is positioned at a height that will allow the branch drains to have the correct ¼" per foot downward slope toward the stack.

7 Determine the length for the next piece of waste-vent pipe by measuring from the basement T-fitting to the next planned fitting in the vertical run. In our project, we will be installing a T-fitting between floor joists, where the toilet drain will be connected.

8 Cut a PVC plastic pipe to length, raise it into the opening, and dry-fit it to the T-fitting. NOTE: For very long pipe runs, you may need to construct this vertical run by solvent-gluing two or more segments of pipe together with couplings.

9 Check the length of the stack, then solvent-glue all fittings together. Support the new stack with a riser clamp resting on blocks attached between basement ceiling joists.

10 Attach the next waste T-fitting to the stack. In our demonstration project, the waste T lies between floor joists and will be used to connect the toilet drain. Make sure the waste T is positioned at a height which will allow for the correct ⅛" per foot downward slope for the toilet drain.

11 Add additional lengths of pipe, with waste T-fittings installed where other fixtures will drain into the stack. In our example, a waste T with a 1½" bushing insert is installed where the vanity sink drain will be attached to the stack. Make sure the T-fittings are positioned to allow for the correct downward pitch of the branch drains.

(continued next page)

12 Cut a hole in the ceiling where the waste-vent stack will extend into the attic, then measure, cut, and solvent-glue the next length of pipe in place. The pipe should extend at least 1 ft. up into the attic.

13 Remove the roof flashing from around the old waste-vent stack. You may need to remove shingles in order to accomplish this. NOTE: Always use caution when working on a roof. If you are unsure of your ability to do this work, hire a roof repair specialist to remove the old flashing and install new flashing around the new vent pipe.

14 In the attic, remove old vent pipes, where necessary, then sever the cast-iron soil stack with a cast iron cutter and lower the stack down from the roof opening with the aid of a helper. Support the old stack with a riser clamp installed between joists.

15 Solvent-glue a vent T with a 1½" bushing in the side inlet to the top of the new waste-vent stack. The side inlet should point toward the nearest auxiliary vent pipe extending up from below.

16 Finish the waste-vent stack installation by using 45° elbows and straight lengths of pipe to extend the stack through the same roof opening used by the old vent stack. The new stack should extend at least 1 ft. through the roof, but no more than 2 ft.

How to Flash a Waste-Vent Stack

1 Loosen the shingles directly above the new vent stack, and remove any nails, using a flat pry bar. When installed, the metal vent flashing will lie flat on the shingles surrounding the vent pipe. Apply roofing cement to the underside of the flashing.

2 Slide the flashing over the vent pipe, and carefully tuck the base of the flashing up under the shingle. Press the flange firmly against the roof deck to spread the roofing cement, then anchor it with rubber gasket flashing nails. Reattach loose shingles as necessary.

Replacing Branch Drains & Vent Pipes

Remove old pipes only where they obstruct the planned route for the new pipes. You will probably need to remove drain and water supply pipes at each fixture location, but the remaining pipes usually can be left in place. A reciprocating saw with metal-cutting blade works well for this job.

In our demonstration project, we are replacing branch drains for a bathtub and vanity sink. The tub drain will run down into the basement before connecting to the main waste-vent stack, while the vanity drain will run horizontally to connect directly to the stack.

A vent pipe for the bathtub runs up into the attic, where it will join the main waste-vent stack. The vanity sink, however, requires no secondary vent pipe, since its location falls within the critical distance (page 29) of the new waste-vent stack.

How to Replace Branch Drains

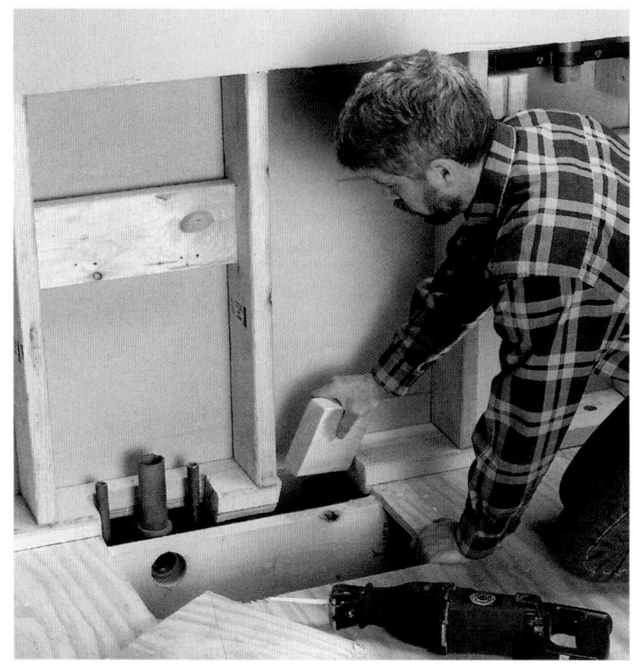

1 Establish a route for vertical drain pipes running through wall cavities down into the basement. For our project, we are cutting away a section of the wall sole plate in order to run a 1½" bathtub drain pipe from the basement up to the bathroom.

2 From the basement, cut a hole in the bottom of the wall, below the opening cut (step 1). Measure, cut, and insert a length of vertical drain pipe up into the wall to the bathroom. A length of flexible CPVC pipe can be useful for guiding the drain pipe up into the wall. For very long pipe runs, you may need to join two or more lengths of pipe with couplings as you insert the run.

3 Secure the vertical drain pipe with a riser clamp supported on 2 × 4 blocks nailed between joists. Take care not to overtighten the clamps.

4 Install a horizontal pipe from the waste T-fitting on the waste-vent stack to the vertical drain pipe. Maintain a downward slope toward the stack of ¼" per foot, and use a Y-fitting with 45° elbow to form a cleanout where the horizontal and vertical drain pipes meet.

5 Solvent-glue a waste T-fitting to the top of the vertical drain pipe. For a bathtub drain, as shown here, the T-fitting must be well below floor level to allow for the bathtub drain trap (page 108). You may need to notch or cut a hole in floor joists to connect the drain trap to the waste T (page 76).

6 From the attic, cut a hole into the top of the bathroom wet wall, directly above the bathtub drain pipe. Run a 1½" vent pipe down to the bathtub location, and solvent-glue it to the waste T. Make sure the pipe extends at least 1 ft. into the attic.

(continued next page)

7 Remove wall surfaces as necessary to provide access for running horizontal drain pipes from fixtures to the new waste-vent stack. In our project, we are running 1½" drain pipe from a vanity sink to the stack. Mark the drain route on the exposed studs, maintaining a ¼" per foot downward slope toward the stack. Use a reciprocating saw or jig saw to notch out the studs (page 76).

8 Secure the old drain and vent pipes with riser clamps supported by blocking attached between the studs.

9 Remove the old drain and water supply pipes, where necessary, to provide space for running the new drain pipes.

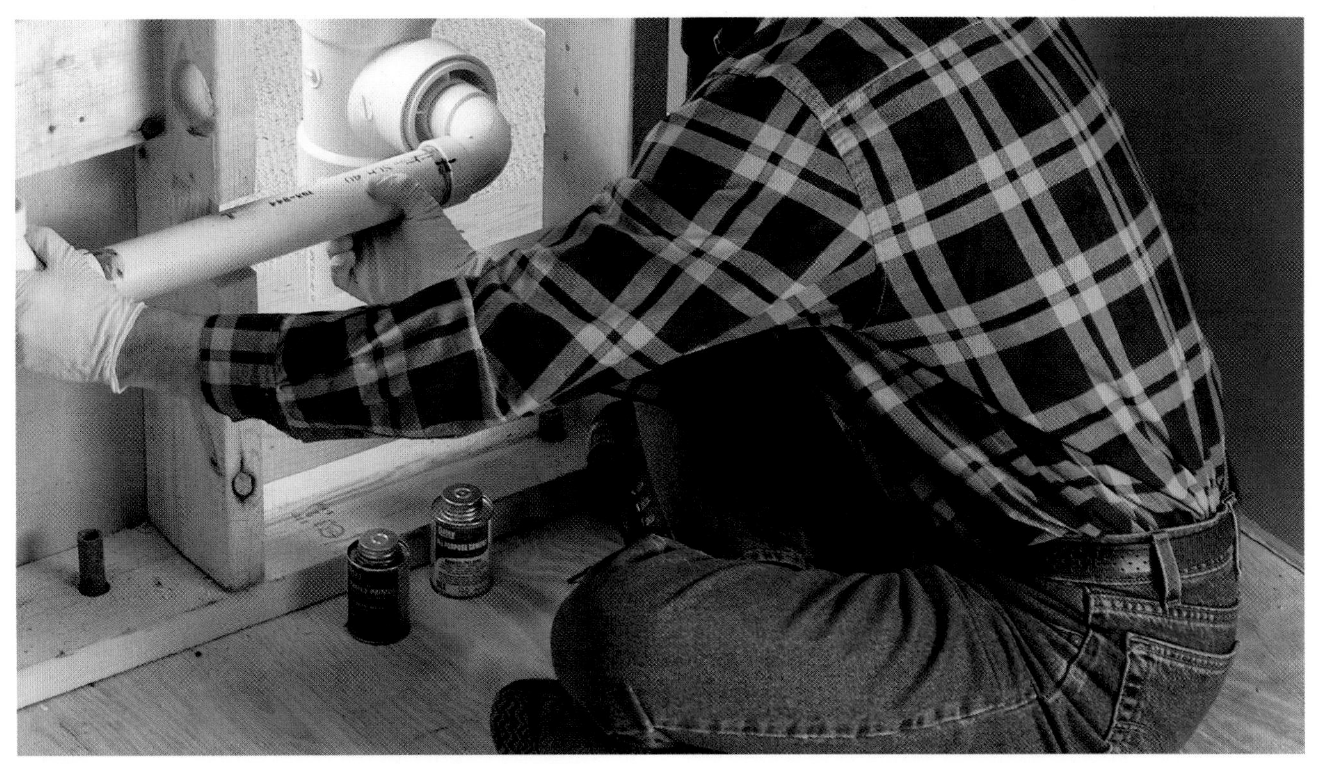

10 Using a sweep elbow and straight length of pipe, assemble a drain pipe to run from the drain stub-out location to the waste T-fitting on the new waste-vent stack. Use a 90° elbow and a short length of pipe to create a stub-out extending at least 2" out from the wall. Secure the stub-out to a ¾" backer board attached between studs.

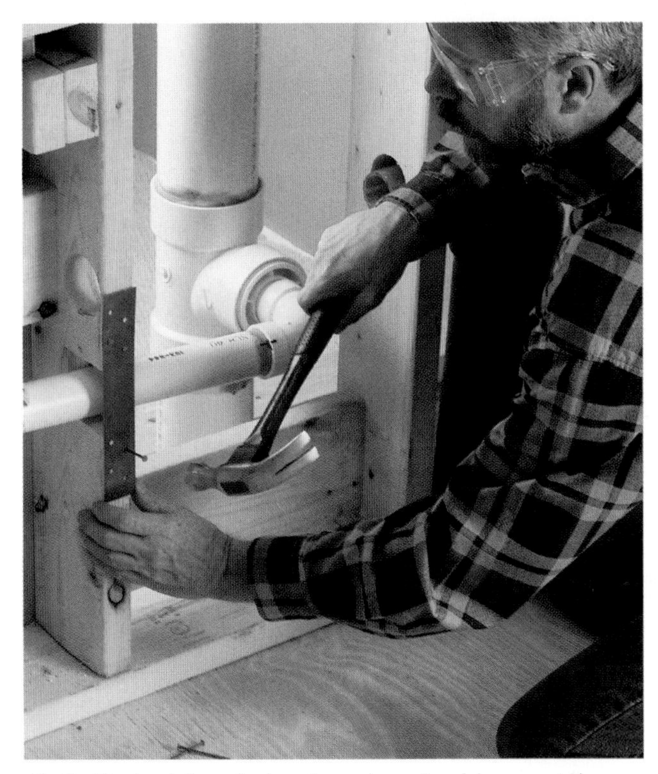

11 Protect the drain pipes by attaching metal protector plates over the notches in the studs. Protector plates prevent drain pipes from being punctured when wall surfaces are replaced.

12 In the attic, use a vent elbow and straight length of pipe to connect the vertical vent pipe from the tub to the new waste-vent stack.

Replacing a Toilet Drain

Replacing a toilet drain usually requires that you remove flooring and wall surface to gain access to the pipes.

Replacing a toilet drain is sometimes a troublesome task, mostly because the cramped space makes it difficult to route the large, 3" or 4" pipe. You likely will need to remove flooring around the toilet and wall surface behind the toilet.

Replacing a toilet drain may require framing work, as well, if you find it necessary to cut into joists in order to route the new pipes. When possible, plan your project to avoid changes to the framing members.

How to Replace a Toilet Drain

1 Remove the toilet, then unscrew the toilet flange from the floor and remove it from the drain pipe. NOTE: If the existing toilet flange is cast iron or bronze, it may be joined to the toilet bend with poured lead or solder; in this case, it is easiest to break up the flange with a masonry hammer (make sure to wear eye protection) and remove it in pieces.

2 Cut away the flooring around the toilet drain along the center of the floor joists, using a circular saw with the blade set to a depth ⅛" more than the thickness of the subfloor. The exposed joist will serve as a nailing surface when the subfloor is replaced.

3 Cut away the old toilet bend as close as possible to the old waste-vent stack, using a reciprocating saw with metal-cutting blade, or a snap cutter.

4 If a joist obstructs the route to the new waste-vent stack, cut away a section of the floor joist. Install double headers and metal joist hangers to support the ends of the cut joist.

5 Create a new toilet drain running to the new waste-vent stack, using a toilet bend and a straight length of pipe. Position the drain so there will be at least 15" of space between the center of the bowl and side wall surfaces when the toilet is installed. Make sure the drain slopes at least ⅛" per foot toward the stack, then support the pipe with plastic pipe strapping attached to framing members. Insert a 6" length of pipe in the top inlet of the closet bend; once the new drain pipes have been tested, this pipe will be cut off with a handsaw and fitted with a toilet flange.

6 Cut a piece of exterior-grade plywood to fit the cutout floor area, and use a jig saw to cut an opening for the toilet drain stub-out. Position the plywood, and attach it to joists and blocking with 2" screws.

Support copper supply pipes every 6 ft. along vertical runs and 10 ft. along horizontal runs. Always use copper or plastic support materials with copper; never use steel straps, which can interact with copper and cause corrosion.

Replacing Supply Pipes

When replacing old galvanized water supply pipes, we recommend that you use type-M rigid copper. Use ¾" pipe for the main distribution pipes and ½" pipes for the branch lines running to individual fixtures.

For convenience, run hot and cold water pipes parallel to one another, between 3" and 6" apart. Use the straightest, most direct routes possible when planning the layout, because too many bends in the pipe runs can cause significant friction and reduce water pressure.

It is a good idea to removed old supply pipes that are exposed, but pipes hidden in walls can be left in place unless they interfere with the installation of the new supply pipes.

How to Replace Water Supply Pipes

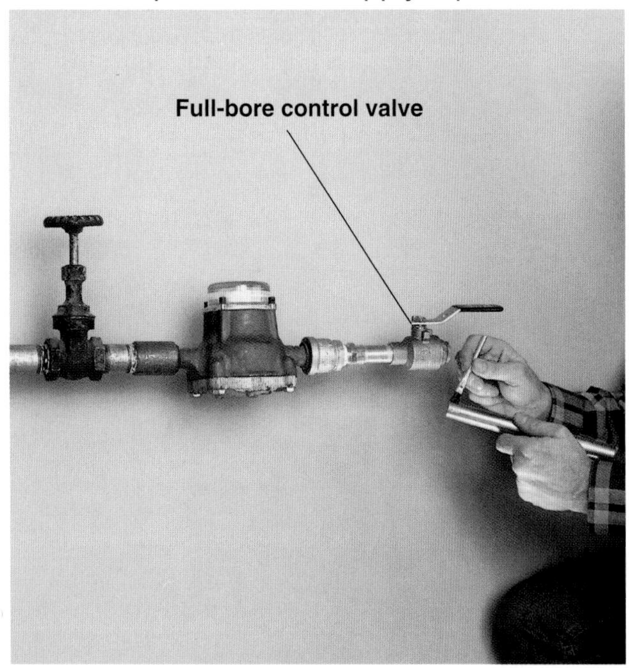

1 Shut off the water on the street side of the water meter, then disconnect and remove the old water pipes from the house side. Solder a ¾" male threaded adapter and full-bore control valve to a short length of ¾" copper pipe, then attach this assembly to the house side of the water meter. Extend the ¾" cold-water distribution pipe toward the nearest fixture, which is usually the water heater.

2 At the water heater, install a ¾" T-fitting in the cold-water distribution pipe. Use two lengths of ¾" copper pipe and a full-bore control valve to run a branch pipe to the water heater. From the outlet opening on the water heater, extend a ¾" hot-water distribution pipe, also with a full-bore control valve (page 245). Continue the hot and cold supply lines on parallel routes toward the next group of fixtures in your house.

3 Establish routes for branch supply lines by drilling holes into stud cavities. Install T-fittings, then begin the branch lines by installing brass control valves. Branch lines should be made with ¾" pipe if they are supplying more than one fixture; ½" if they are supplying only one fixture.

4 Extend the branch lines to the fixtures. In our project, we are running ¾" vertical branch lines up through the framed chase to the bathroom. Route pipes around obstacles, such as a main waste-vent stack, by using 45° and 90° elbows and short lengths of pipe.

5 Where branch lines run through studs or floor joists, drill holes or cut notches in the framing members (page 76), then insert the pipes. For long runs of pipe, you may need to join two or more shorter lengths of pipe, using couplings as you create the runs.

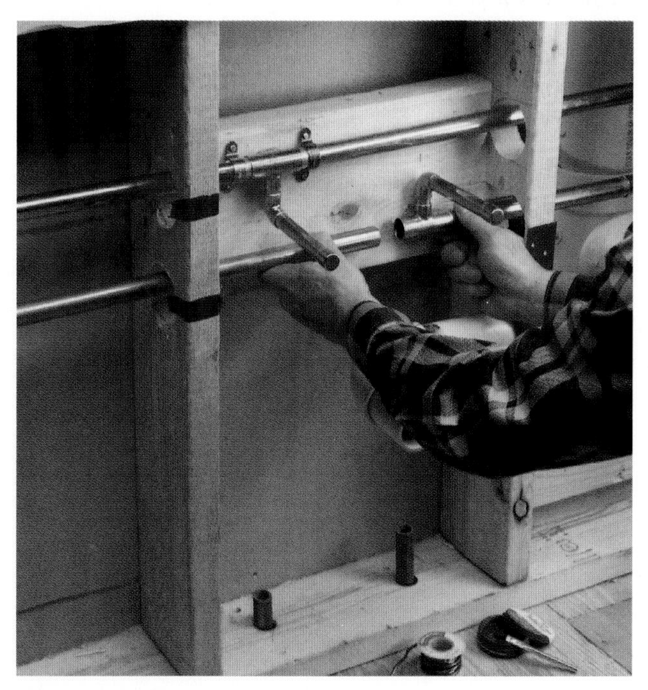

6 Install ¾" to ½" reducing T-fittings and elbows to extend the branch lines to individual fixtures. In our bathroom, we are installing hot and cold stub-outs for the bathtub and sink, and a cold-water stub-out for the toilet. Cap each stub-out until your work has been inspected and the wall surfaces have been completed.

PRICE
PFISTER

Repairing Old Plumbing

Every homeowner eventually deals with plumbing problems. Most can be repaired with basic skills, tools, and instructions and require no professional help. This section will help you recognize and fix problems with all types of common household plumbing fixtures.

On the following pages you'll find directions for repairing faucets, including shutoff valves, supply tubes, diverter valves, sprayers, and aerators. The information on tub and shower plumbing includes directions for repairing tub and shower faucets, plus shower heads. For problems with an outdoor water supply, see the directions for repairing valves, hose bibs, and sillcocks.

Toilet problems, too, are easily remedied. We'll show you how to adjust and replace toilet parts and how to stop a toilet from running or leaking.

The section on clogs includes detailed instructions for clearing all types of drains and for fixing separate drain system parts.

Because water heaters are usually out of sight, they are easy to ignore—until something goes wrong. This section shows how water heaters work, and how to repair or replace them. The section ends with advice on fixing burst, frozen, or noisy plumbing pipes.

Faucet Problems & Repairs

Most faucet problems are easy to fix. You can save money and time by making these simple repairs yourself. Replacement parts for faucet repairs usually are inexpensive and readily available at hardware stores and home centers. Techniques for repair vary, depending on the faucet design (pages 176 to 177).

If a badly worn faucet continues to leak, even after repairs are made, the faucet should be replaced. In less than an hour, you can replace an old, problem faucet with a new model that is designed to provide years of trouble-free service.

Problems	Repairs
Faucet drips from the end of the spout, or leaks around the base.	Identify the faucet design (page 177), then install replacement parts, using directions on following pages.
Old, worn-out faucet continues to leak after repairs are made.	Replace the old faucet (pages 188 to 191).
Water pressure at spout seems low, or water flow is partially blocked.	1. Clean faucet aerator (page 194). 2. Replace corroded galvanized pipes with copper (pages 46 to 55).
Water pressure from sprayer seems low, or sprayer leaks from handle.	1. Clean sprayer head (page 194). 2. Fix diverter valve (page 195).
Water leaks onto floor underneath faucet.	1. Replace cracked sprayer hose (page 195). 2. Tighten water connections, or replace supply tubes and shutoff valves (pages 192 to 193). 3. Fix leaky sink strainer (page 225).
Hose bib or valve drips from spout or leaks around handle.	Take valve apart and replace washers and O-rings (pages 206 to 207).

Fixing Leaky Faucets

Cartridge

Spout

Aerator

Mixing chamber

Hot water supply tube

Cold water supply tube

Typical faucet has a single handle attached to a hollow cartridge. The cartridge controls hot and cold water flowing from the supply tubes into the mixing chamber. Water is forced out the spout and through the aerator. When repairs are needed, replace the entire cartridge.

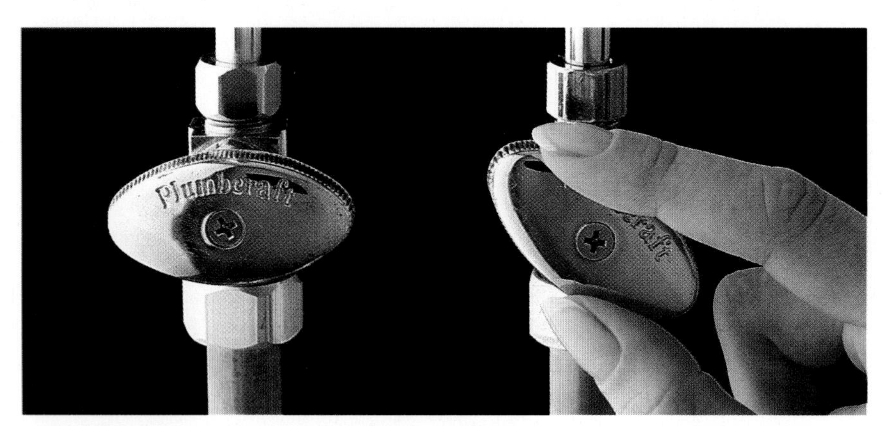

Turn off water before starting any faucet repair, using shutoff valves underneath faucet, or main service valve found near water meter (page 6). When opening shutoff valves after finishing repairs, keep faucet handle in open position to release trapped air. When water runs steadily, close faucet.

A leaky faucet is the most common home plumbing problem. Leaks occur when washers, O-rings, or seals inside the faucet are dirty or worn. Fixing leaks is easy, but the techniques for making repairs will vary, depending on the design of the faucet. Before beginning work, you must first identify your faucet design and determine what replacement parts are needed.

There are four basic faucet designs: ball-type, cartridge, disc, or compression. Many faucets can be identified easily by outer appearance, but others must be taken apart before the design can be recognized.

The compression design is used in many double-handle faucets. Compression faucets all have washers or seals that must be replaced from time to time. These repairs are easy to make, and replacement parts are inexpensive.

Ball-type, cartridge, and disc faucets are all known as washerless faucets. Many washerless faucets are controlled with a single handle, although some cartridge models use two handles. Washerless faucets are more trouble-free than compression faucets, and are designed for quick repair.

When installing new faucet parts, make sure the replacements match the original parts. Replacement parts for popular washerless faucets are identified by brand name and model number. To ensure a correct selection, you may want to bring the worn parts to the store for comparison.

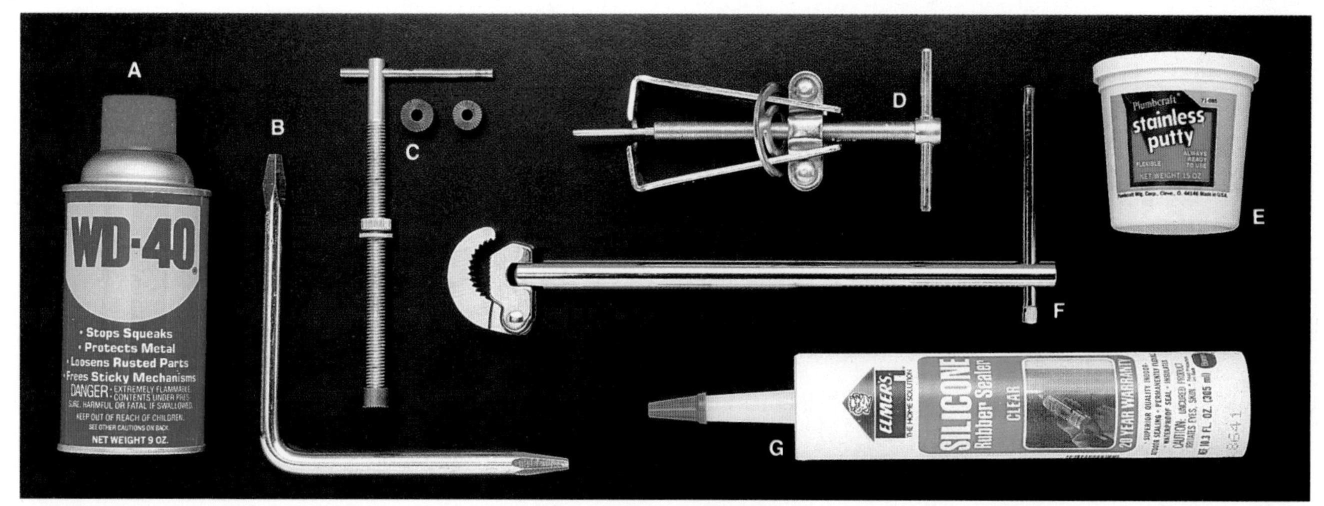

Specialty tools and materials for faucet repairs include: penetrating oil (A), seat wrench (B), seat-dressing (reamer) tool (C), handle puller (D), plumber's putty (E), basin wrench (F), silicone caulk (G).

How to Identify Faucet Designs

Ball-type faucet has a single handle over a dome-shaped cap. If your single-handle faucet is made by Delta or Peerless, it is probably a ball-type faucet. See pages 178 to 179 to fix a ball-type faucet.

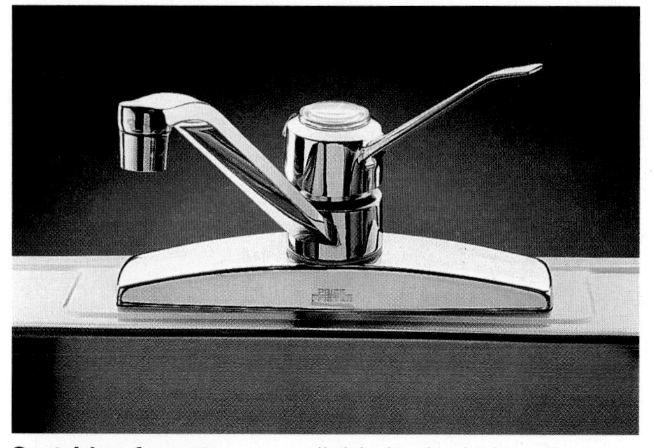

Cartridge faucets are available in single-handle or double-handle models. Popular cartridge faucet brands include Price Pfister, Moen, Valley, and Aqualine. See pages 180 to 181 to fix a cartridge faucet.

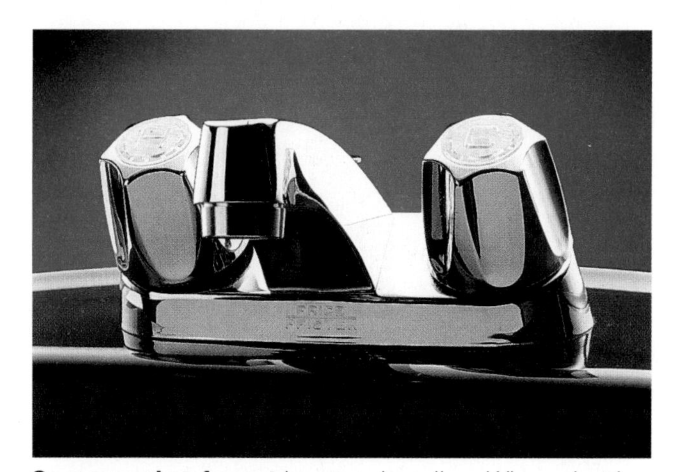

Compression faucet has two handles. When shutting the faucet off, you usually can feel a rubber washer being squeezed inside the faucet. Compression faucets are sold under many brand names. See pages 182 to 185 to fix a compression faucet.

Disc faucet has a single handle and a solid, chromed-brass body. If your faucet is made by American Standard or Reliant, it may be a disc faucet. See pages 186 to 187 to fix a disc faucet.

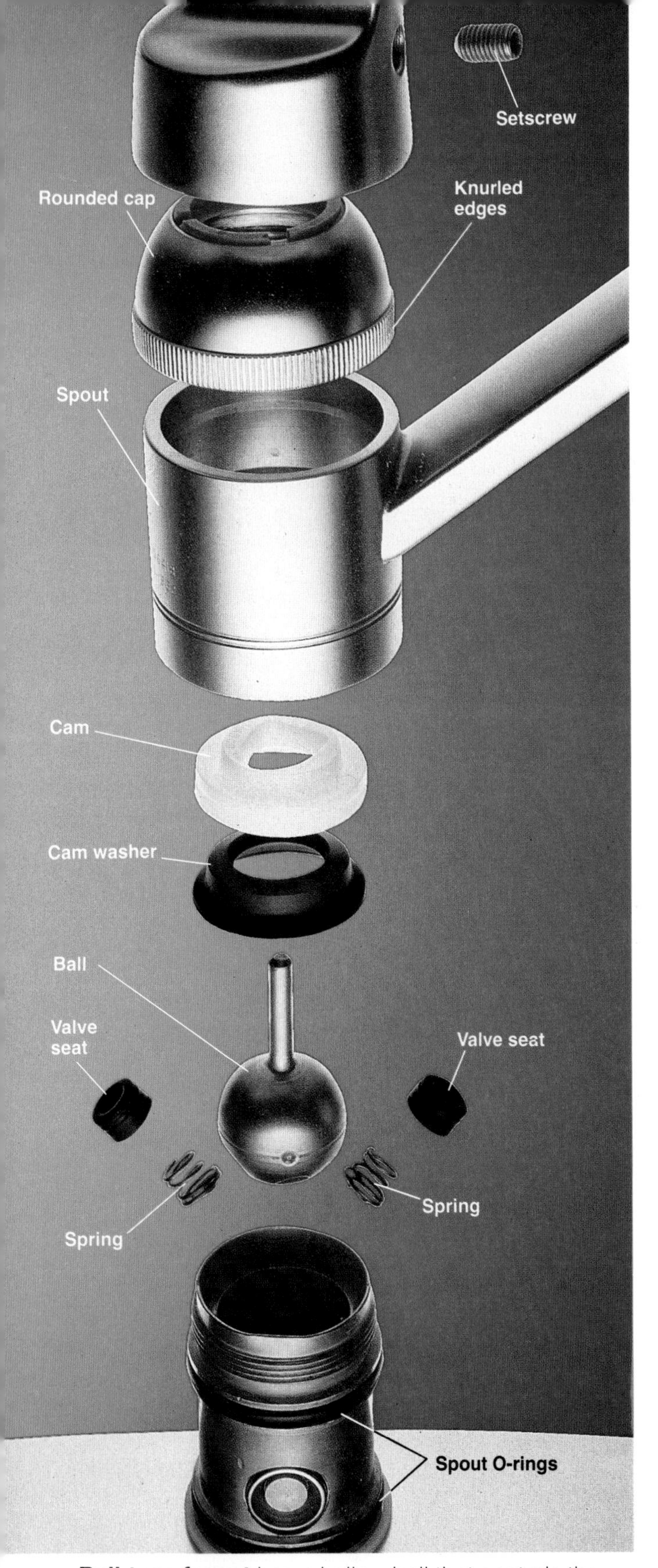

Ball-type faucet has a hollow ball that controls the temperature and flow of water. Dripping at the faucet spout is caused by worn-out valve seats, springs, or a damaged ball. Leaks around the base of the faucet are caused by worn O-rings.

Fixing Ball-type Faucets

A ball-type faucet has a single handle, and is identified by the hollow metal or plastic ball inside the faucet body. Many ball-type faucets have a rounded cap with knurled edges located under the handle. If your faucet leaks from the spout and has this type of cap, first try tightening the cap with channel-type pliers. If tightening does not fix the leak, disassemble the faucet and install replacement parts.

Faucet manufacturers offer several types of replacement kits for ball-type faucets. Some kits contain only the springs and neoprene valve seats, while better kits also include the cam and cam washer.

Replace the rotating ball only if it is obviously worn or scratched. Replacement balls are either metal or plastic. Metal balls are slightly more expensive than plastic, but are more durable.

Remember to turn off the water before beginning work (page 176).

Everything You Need:

Tools: channel-type pliers, Allen wrench, screwdriver, utility knife.

Materials: ball-type faucet repair kit, new rotating ball (if needed), masking tape, O-rings, heatproof grease.

Repair kit for a ball-type faucet includes rubber valve seats, springs, cam, cam washer, and spout O-rings. Kit may also include small allen wrench tool used to remove faucet handle. Make sure kit is made for your faucet model. Replacement ball can be purchased separately, but is not needed unless old ball is obviously worn.

How to Fix a Ball-type Faucet

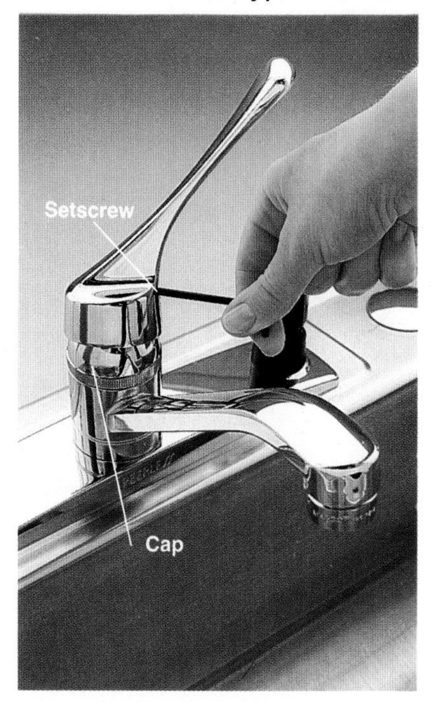

1 Loosen handle setscrew with an Allen wrench. Remove handle to expose faucet cap.

2 Remove the cap with channel-type pliers. To prevent scratches to the shiny chromed finish, wrap masking tape around the jaws of the pliers.

3 Lift out the faucet cam, cam washer, and the rotating ball. Check the ball for signs of wear.

4 Reach into the faucet with a screwdriver and remove the old springs and neoprene valve seats.

5 Remove spout by twisting it upward, then cut off old O-rings. Coat new O-rings with heatproof grease, and install. Reattach spout, pressing downward until the collar rests on plastic slip ring. Install new springs and valve seats.

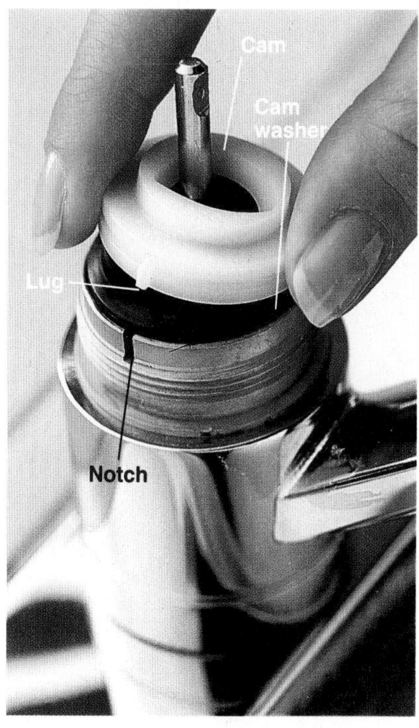

6 Insert ball, new cam washer, and cam. Small lug on cam should fit into notch on faucet body. Screw cap onto faucet and attach handle.

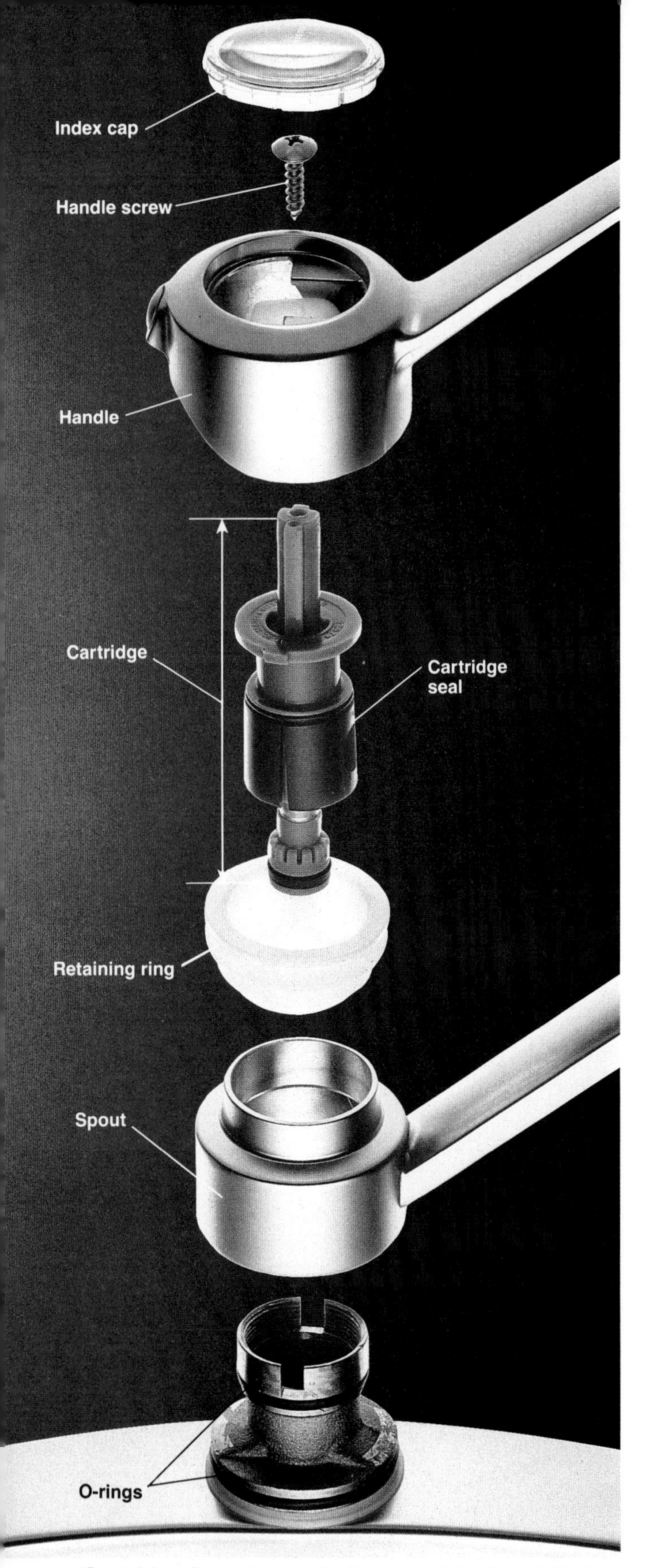

Cartridge faucet has a hollow cartridge insert that lifts and rotates to control the flow and temperature of water. Dripping at the spout occurs when the cartridge seals become worn. Leaks around the base of the faucet are caused by worn O-rings.

Fixing Cartridge Faucets

A cartridge faucet is identified by the narrow metal or plastic cartridge inside the faucet body. Many single-handle faucets and some double-handle models use cartridge designs.

Replacing a cartridge is an easy repair that will fix most faucet leaks. Faucet cartridges come in many styles, so you may want to bring the old cartridge along for comparison when shopping for a replacement.

Make sure to insert the new cartridge so it is aligned in the same way as the old cartridge. If the hot and cold water controls are reversed, take the faucet apart and rotate the cartridge 180°.

Remember to turn off the water before beginning work (page 176).

Everything You Need:

Tools: screwdriver, channel-type pliers, utility knife.

Materials: replacement cartridge, O-rings, heat-proof grease.

Replacement cartridges come in dozens of styles. Cartridges are available for popular faucet brands, including (from left): Price-Pfister, Moen, Kohler. O-ring kits may be sold separately.

How to Fix a Cartridge Faucet

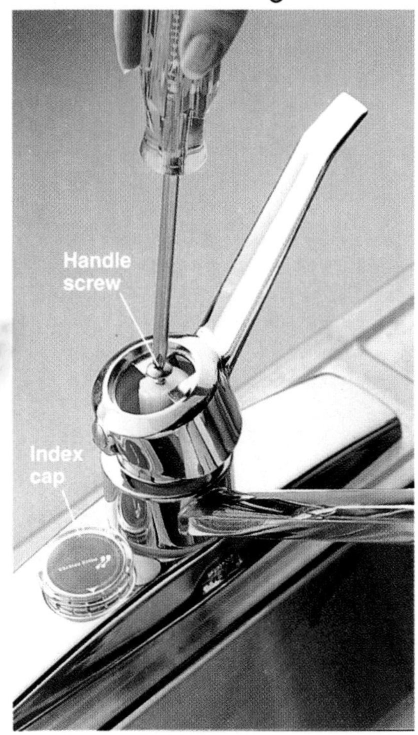

1 Pry off the index cap on top of faucet, and remove the handle screw underneath the cap.

2 Remove faucet handle by lifting it up and tilting it backwards.

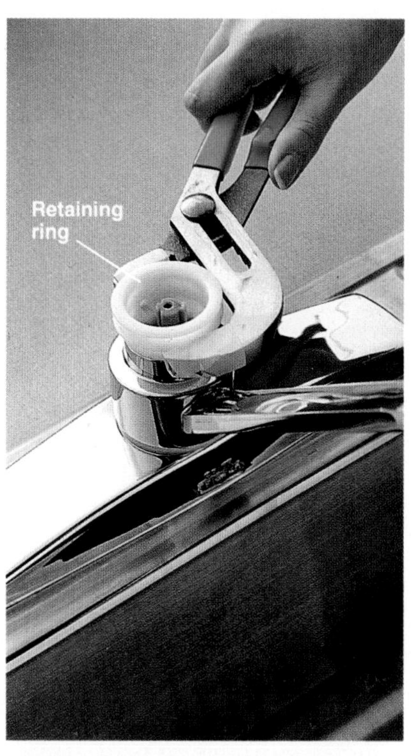

3 Remove the threaded retaining ring with channel-type pliers. Remove any retaining clip holding cartridge in place.

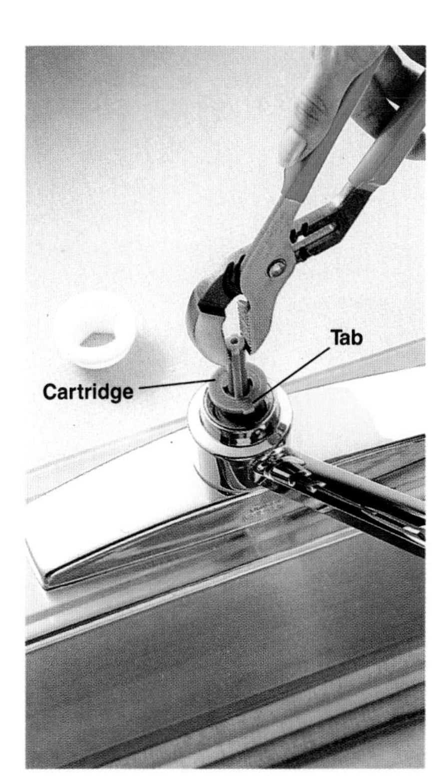

4 Grip top of the cartridge with channel-type pliers. Pull straight up to remove cartridge. Install replacement cartridge so that tab on cartridge faces forward.

5 Remove the spout by pulling up and twisting, then cut off old O-rings with a utility knife. Coat new O-rings with heatproof grease, and install.

6 Reattach the spout. Screw the retaining ring onto the faucet, and tighten with channel-type pliers. Attach the handle, handle screw, and index cap.

Fixing Compression Faucets

Compression faucets have separate controls for hot and cold water, and are identified by the threaded stem assemblies inside the faucet body. Compression stems come in many different styles, but all have some type of neoprene washer or seal to control water flow. Compression faucets leak when stem washers and seals become worn.

Older compression faucets often have corroded handles that are difficult to remove. A specialty tool called a handle puller makes this job easier. Handle pullers may be available at rental centers.

When replacing washers, also check the condition of the metal valve seats inside the faucet body. If the valve seats feel rough, they should be replaced or resurfaced.

Remember to turn off the water before beginning work (page 176).

Everything You Need:

Tools: screwdriver, handle puller (if needed), channel-type pliers, utility knife, seat wrench or seat-dressing tool (if needed).

Materials: universal washer kit, packing string, heatproof grease, replacement valve seats (if needed).

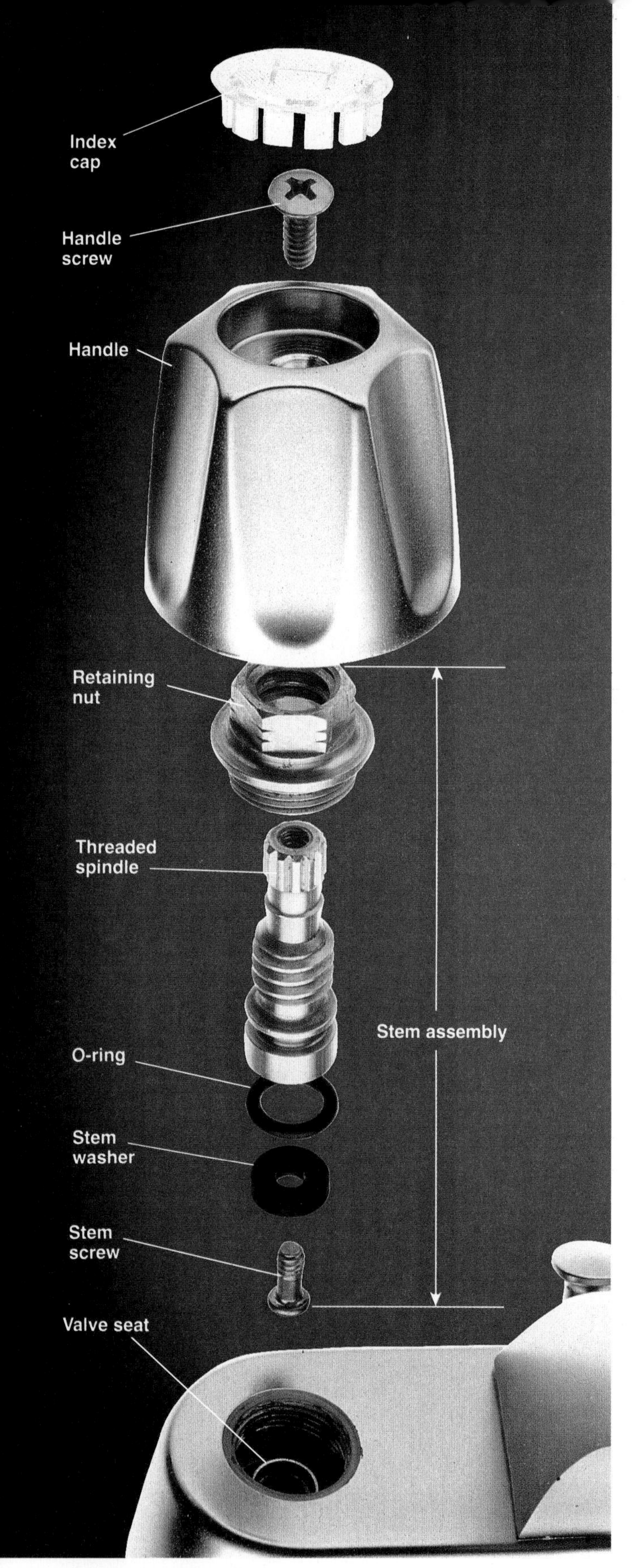

A compression faucet has a stem assembly that includes a retaining nut, threaded spindle, O-ring, stem washer, and stem screw. Dripping at the spout occurs when the washer becomes worn. Leaks around the handle are caused by a worn O-ring.

Universal washer kit contains parts needed to fix most types of compression faucets. Choose a kit that has an assortment of neoprene washers, O-rings, packing washers, and brass stem screws.

Tips for Fixing a Compression Faucet

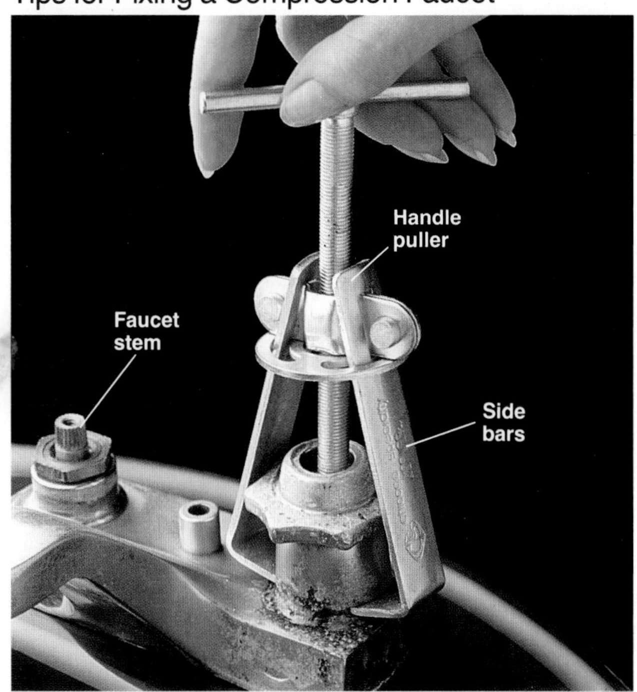

Remove stubborn handles with a handle puller. Remove the faucet index cap and handle screw, and clamp the side bars of the puller under the handle. Thread the puller into the faucet stem, and tighten until the handle comes free.

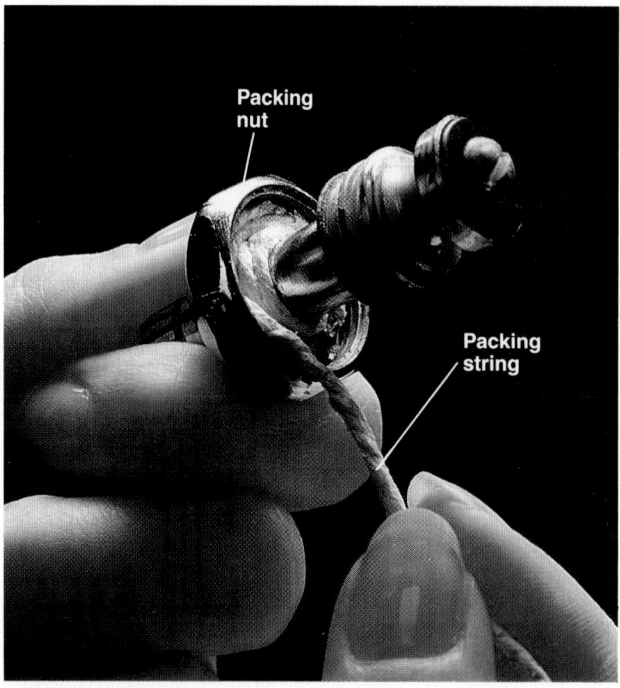

Packing string is used instead of an O-ring on some faucets. To fix leaks around the faucet handle, wrap new packing string around the stem, just underneath the packing nut or retaining nut.

Three Common Types of Compression Stems

Standard stem has a brass stem screw that holds either a flat or beveled neoprene washer to the end of the spindle. If stem screw is worn, it should be replaced.

Tophat stem has a snap-on neoprene diaphragm instead of a standard washer. Fix leaks by replacing the diaphragm.

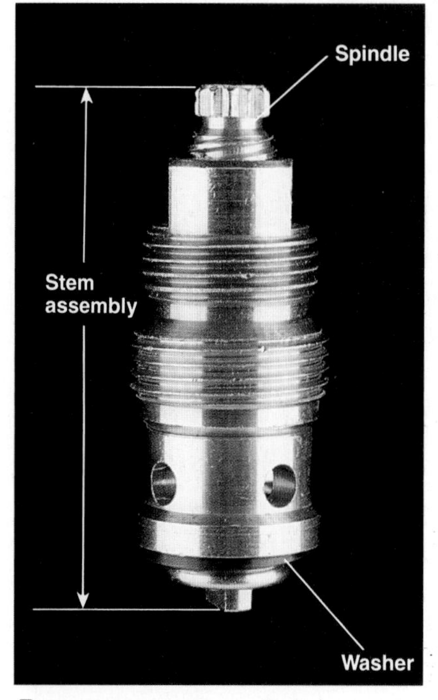

Reverse-pressure stem has a beveled washer at the end of the spindle. To replace washer, unscrew spindle from rest of the stem assembly. Some stems have a small nut that holds washer.

How to Fix a Compression Faucet

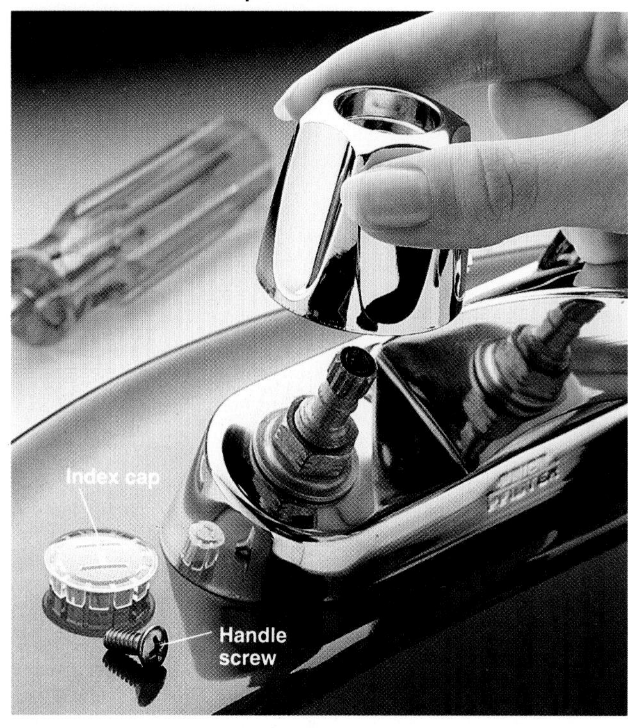

1 Remove index cap from top of faucet handle, and remove handle screw. Remove handle by pulling straight up. If necessary, use a handle puller to remove handle (page 183).

2 Unscrew the stem assembly from body of faucet, using channel-type pliers. Inspect valve seat for wear, and replace or resurface as needed (page opposite). If faucet body or stems are badly worn, it usually is best to replace the faucet (pages 188 to 191).

3 Remove the brass stem screw from the stem assembly. Remove worn stem washer.

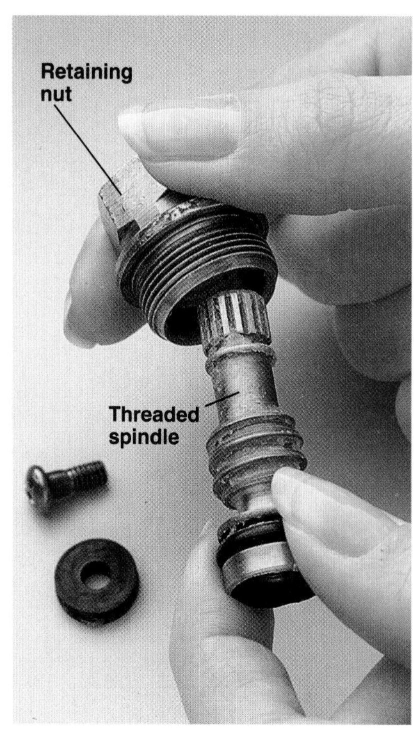

4 Unscrew the threaded spindle from the retaining nut.

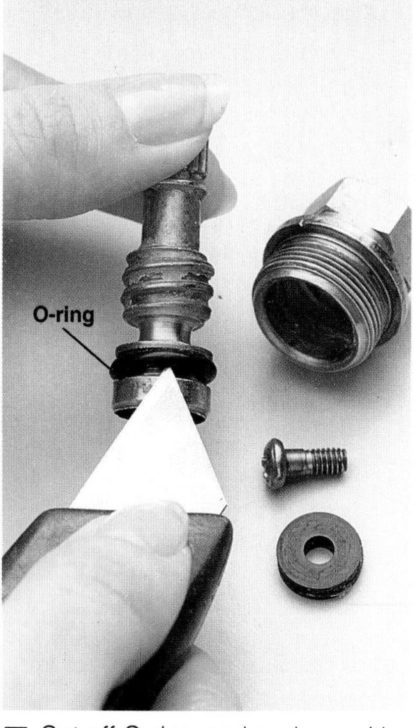

5 Cut off O-ring and replace with an exact duplicate. Install new washer and stem screw. Coat all parts with heatproof grease, then reassemble the faucet.

How to Replace Worn Valve Seats

1 Check valve seat for damage by running a fingertip around the rim of the seat. If the valve seat feels rough, replace the seat, or resurface it with a seat-dressing (reamer) tool (below).

2 Remove valve seat, using a seat wrench. Select end of wrench that fits seat, and insert into faucet. Turn counterclockwise to remove seat, then install an exact duplicate. If seat cannot be removed, resurface with a seat-dressing tool (below).

How to Resurface Valve Seats

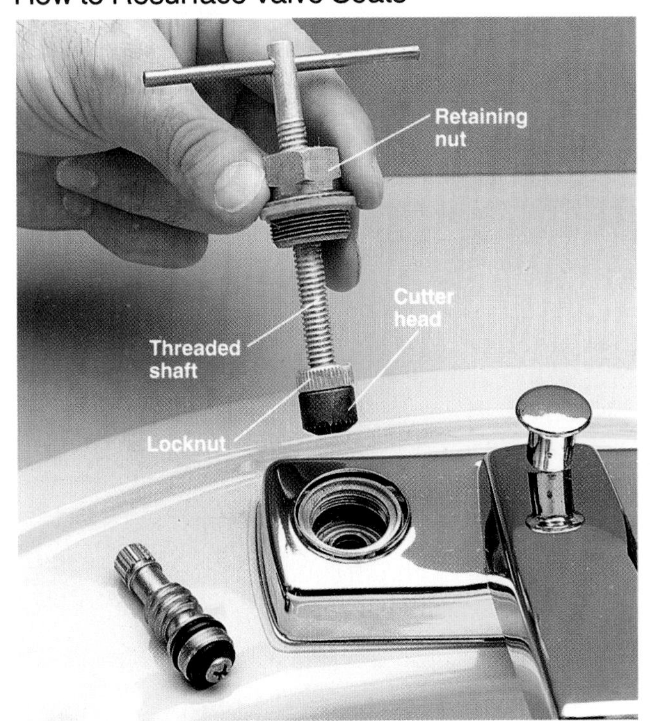

1 Select cutter head to fit the inside diameter of retaining nut. Slide retaining nut over threaded shaft of seat-dressing tool, then attach the locknut and cutter head to the shaft.

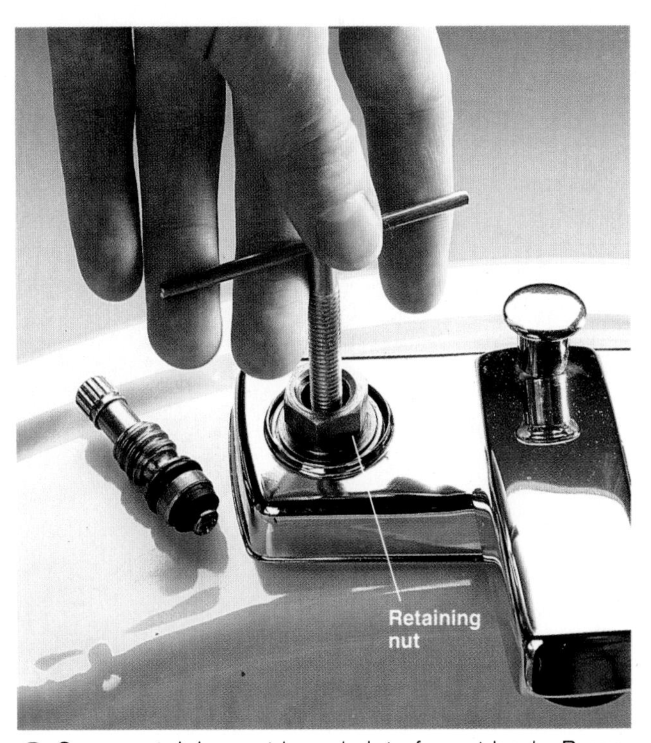

2 Screw retaining nut loosely into faucet body. Press the tool down lightly and turn tool handle clockwise two or three rotations. Reassemble faucet.

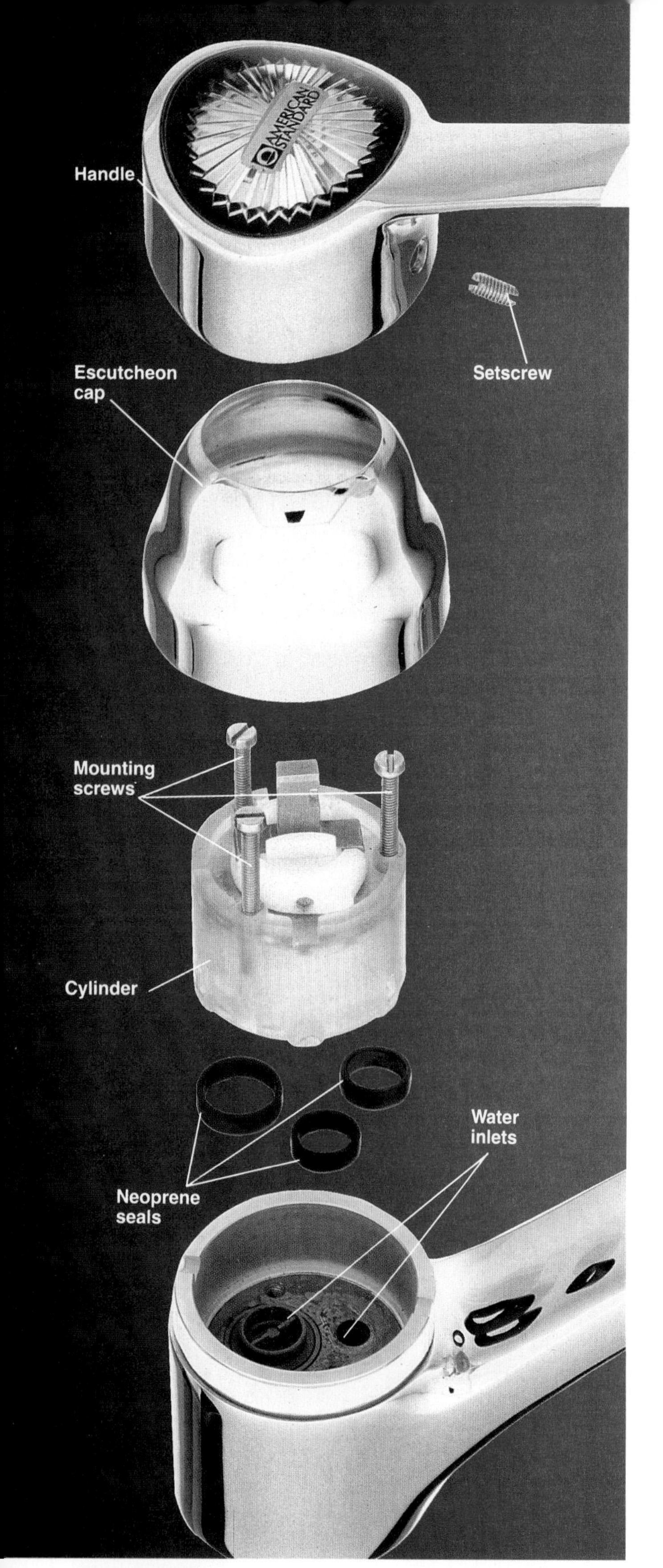

Fixing Disc Faucets

A disc faucet has a single handle and is identified by the wide cylinder inside the faucet body. The cylinder contains a pair of closely fitting ceramic discs that control the flow of water.

A ceramic disc faucet is a top-quality fixture that is easy to repair. Leaks usually can be fixed by lifting out the cylinder and cleaning the neoprene seals and the cylinder openings. Install a new cylinder only if the faucet continues to leak after cleaning.

After making repairs to a disc faucet, make sure handle is in the ON position, then open the shutoff valves slowly. Otherwise, ceramic discs can be cracked by the sudden release of air from the faucet. When water runs steadily, close the faucet.

Remember to turn off the water before beginning work (page 176).

Everything You Need:

Tools: screwdriver.

Materials: Scotch Brite pad, replacement cylinder (if needed).

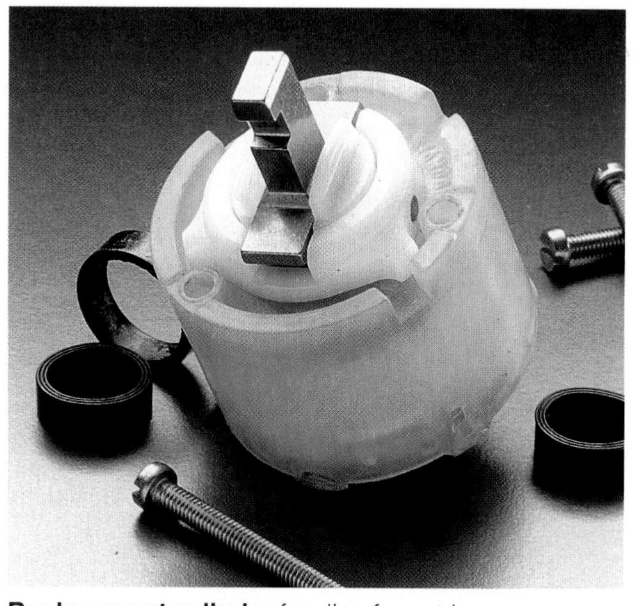

Disc faucet has a sealed cylinder containing two closely fitting ceramic discs. Faucet handle controls water by sliding the discs into alignment. Dripping at the spout occurs when the neoprene seals or cylinder openings are dirty.

Replacement cylinder for disc faucet is necessary only if faucet continues to leak after cleaning. Continuous leaking is caused by cracked or scratched ceramic discs. Replacement cylinders come with neoprene seals and mounting screws.

How to Fix a Ceramic Disc Faucet

1 Rotate faucet spout to the side, and raise the handle. Remove the setscrew and lift off the handle.

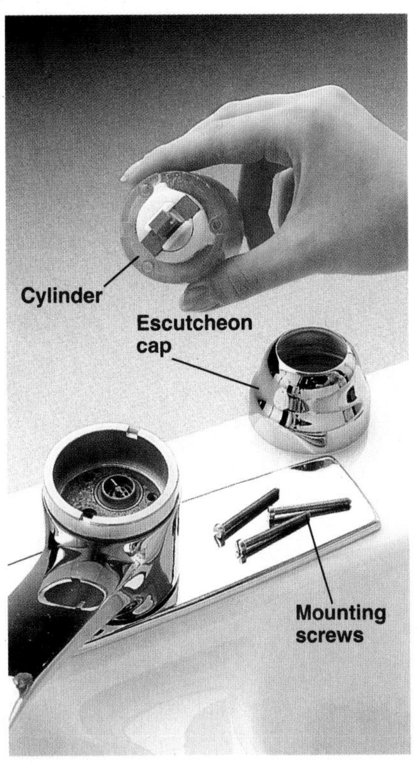

2 Remove the escutcheon cap. Remove cartridge mounting screws, and lift out the cylinder.

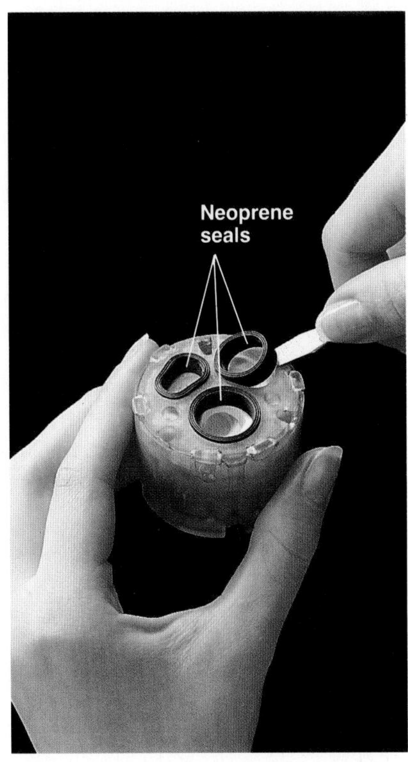

3 Remove the neoprene seals from the cylinder openings.

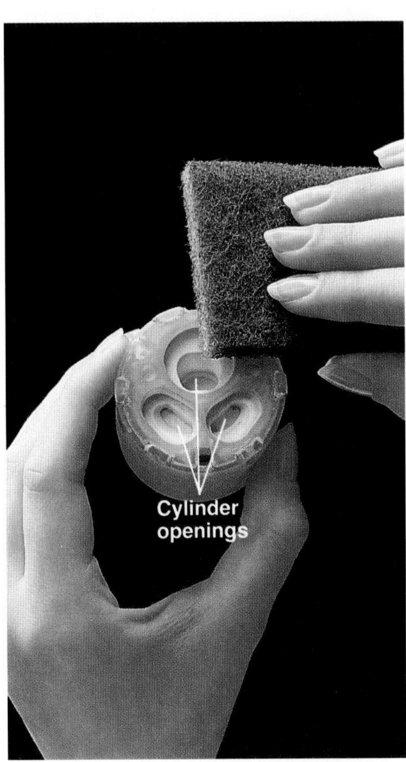

4 Clean the cylinder openings and the neoprene seals with a Scotch Brite pad. Rinse cylinder with clear water.

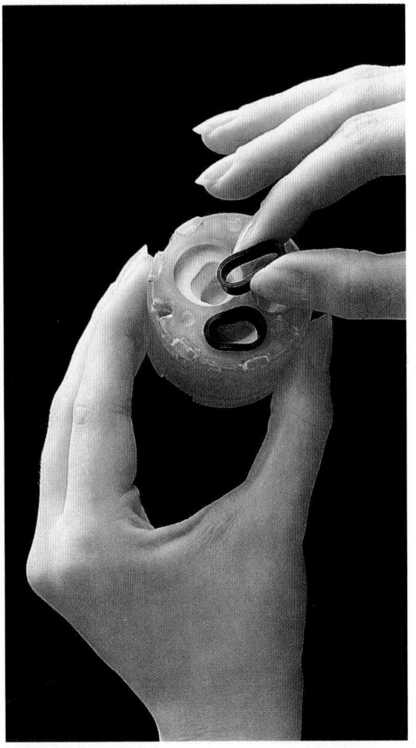

5 Return seals to the cylinder openings, and reassemble faucet. Move handle to ON position, then slowly open shutoff valves. When water runs steadily, close faucet.

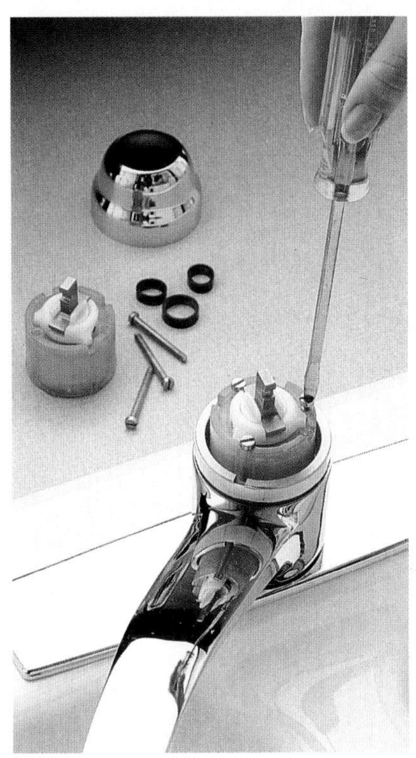

Install a new cylinder only if the faucet continues to leak after cleaning.

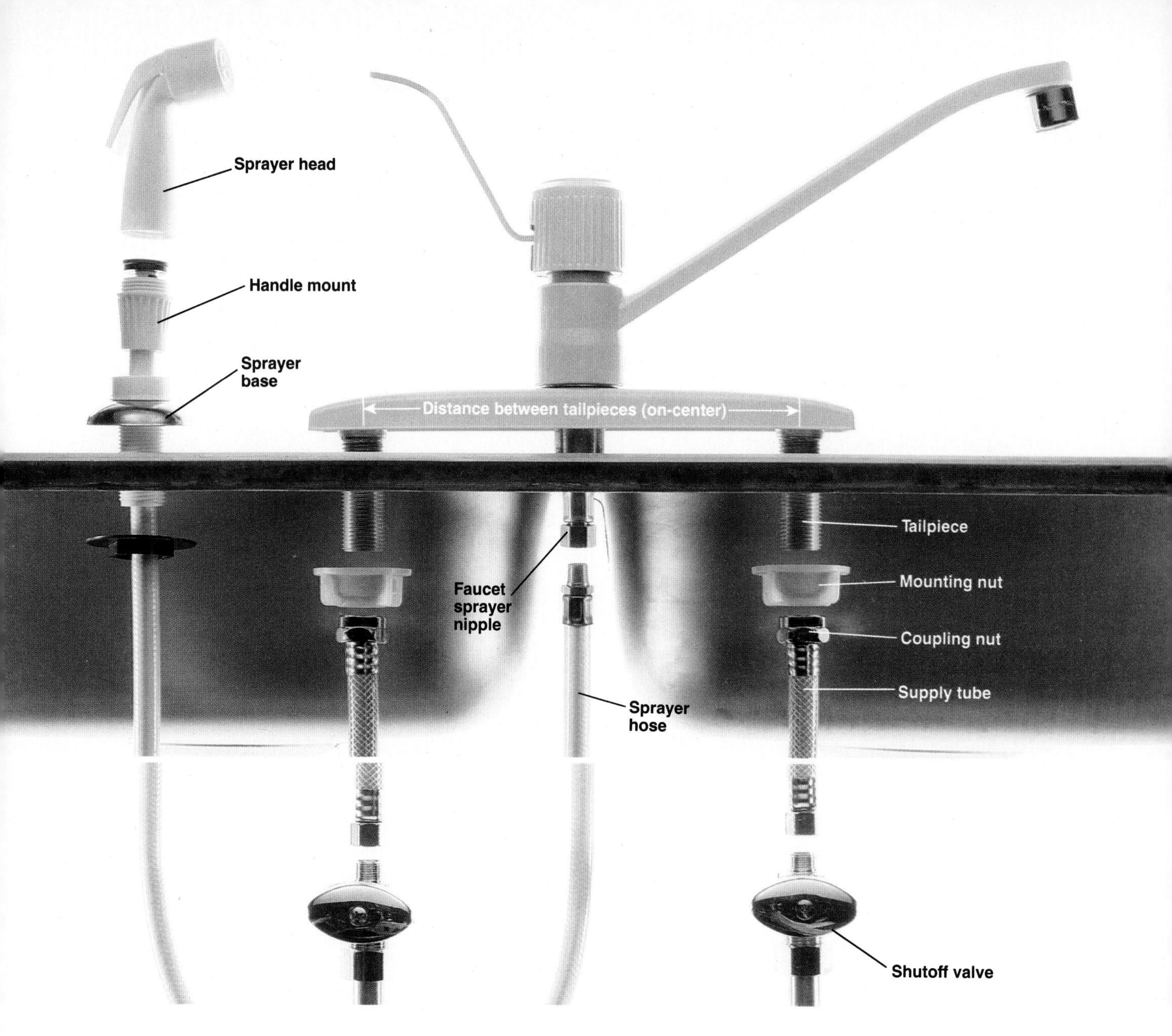

Replacing a Sink Faucet

Installing a new faucet is an easy project that takes about one hour. Before buying a new faucet, first find the diameter of the sink openings, and measure the distance between the tailpieces (measured on-center). Make sure the tailpieces of the new faucet match the sink openings.

When shopping for a new faucet, choose a model made by a reputable manufacturer. Replacement parts for a well-known brand will be easy to find if the faucet ever needs repairs. Better faucets have solid brass bodies. They are easy to install and provide years of trouble-free service. Some washerless models have lifetime warranties.

Always install new supply tubes when replacing a faucet. Old supply tubes should not be reused. If water pipes underneath the sink do not have shutoff valves, you may choose to install the valves while replacing the faucet (pages 192 to 193).

Remember to turn off the water before beginning work (page 176).

Everything You Need:

Tools: basin wrench or channel-type pliers, putty knife, caulk gun, adjustable wrenches.

Materials: penetrating oil, silicone caulk or plumber's putty, two flexible supply tubes.

How to Remove an Old Sink Faucet

1 Spray penetrating oil on tailpiece mounting nuts and supply tube coupling nuts. Remove the coupling nuts with a basin wrench or channel-type pliers.

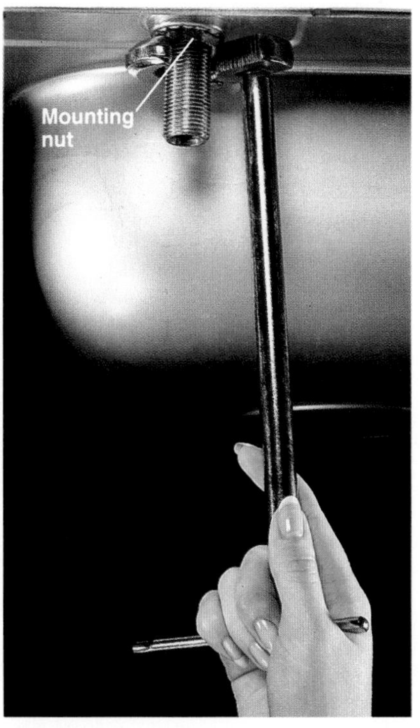

2 Remove the tailpiece mounting nuts with a basin wrench or channel-type pliers. Basin wrench has a long handle that makes it easy to work in tight areas.

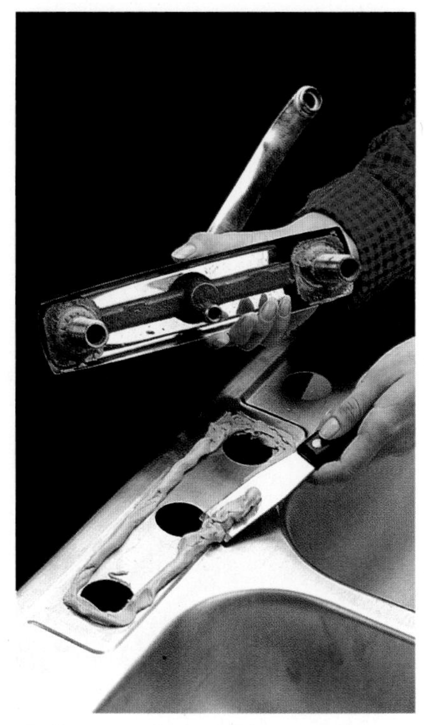

3 Remove faucet. Use a putty knife to clean away old putty from surface of sink.

Faucet Hookup Variations

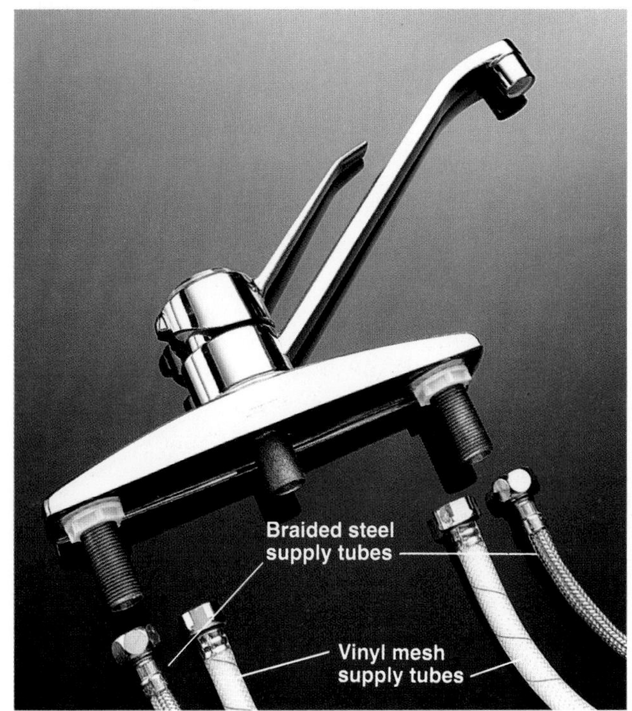

New faucet without supply tubes: Buy two supply tubes. Supply tubes are available in braided steel or vinyl mesh (shown above), PB plastic, or chromed copper (page 192).

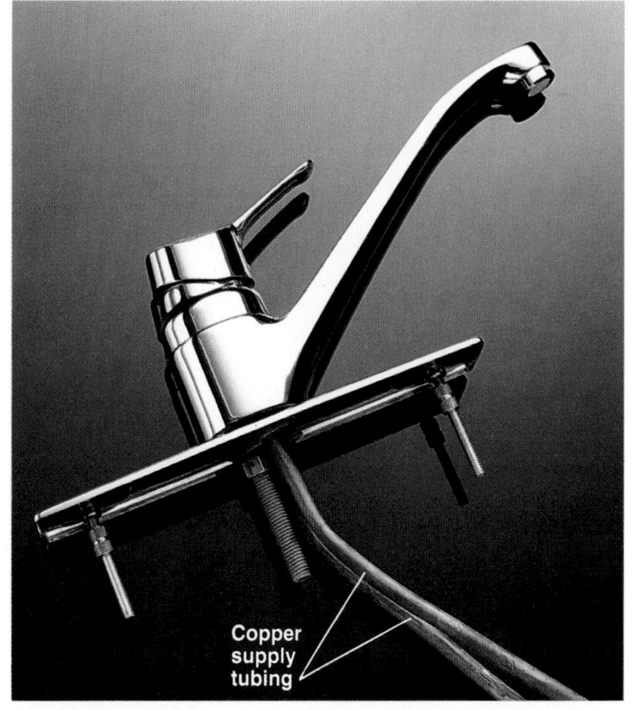

New faucet with preattached copper supply tubing: Make water connections by attaching the supply tubing directly to the shutoff valves with compression fittings (page 191).

How to Install a New Sink Faucet

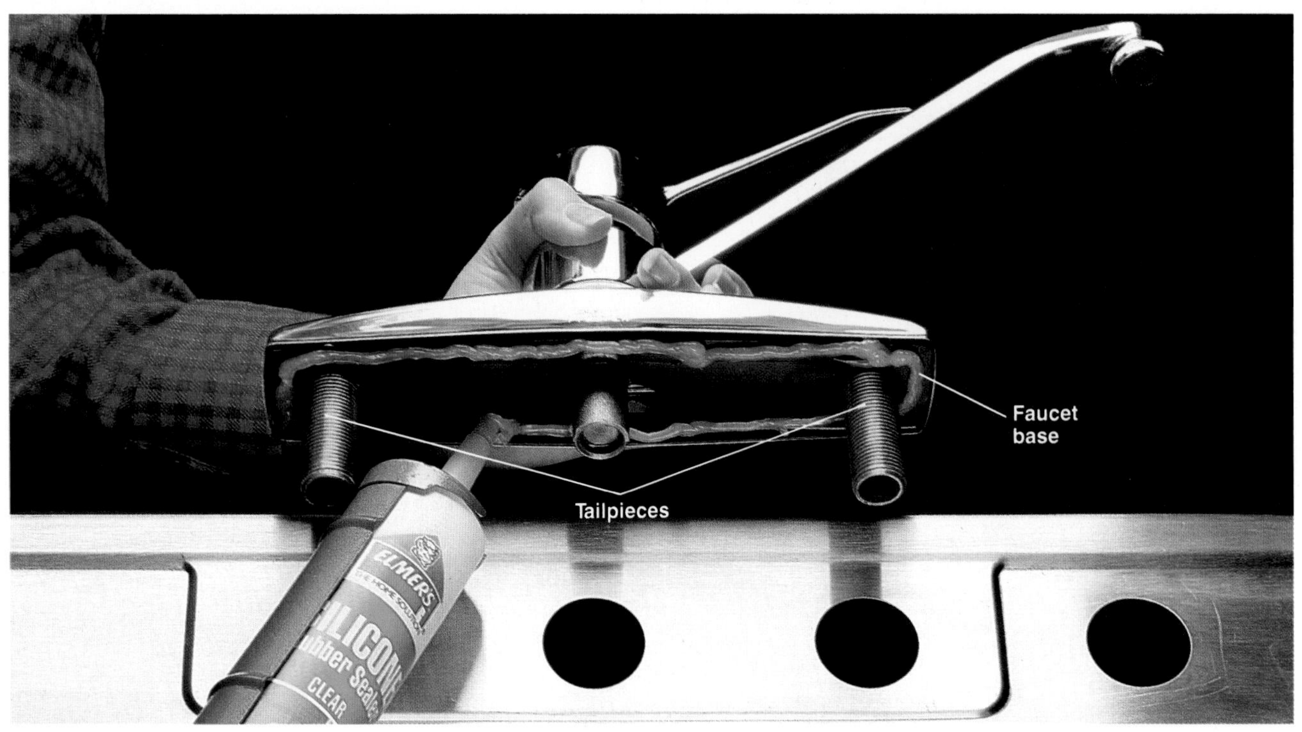

1 Apply a 1/4" bead of silicone caulk or plumber's putty around the base of the faucet. Insert the faucet tailpieces into the sink openings. Position the faucet so base is parallel to back of sink, and press the faucet down to make sure caulk forms a good seal.

2 Screw the metal friction washers and the mounting nuts onto the tailpieces, then tighten with a basin wrench or channel-type pliers. Wipe away excess caulk around base of faucet.

3 Connect flexible supply tubes to faucet tailpieces. Tighten coupling nuts with a basin wrench or channel-type pliers.

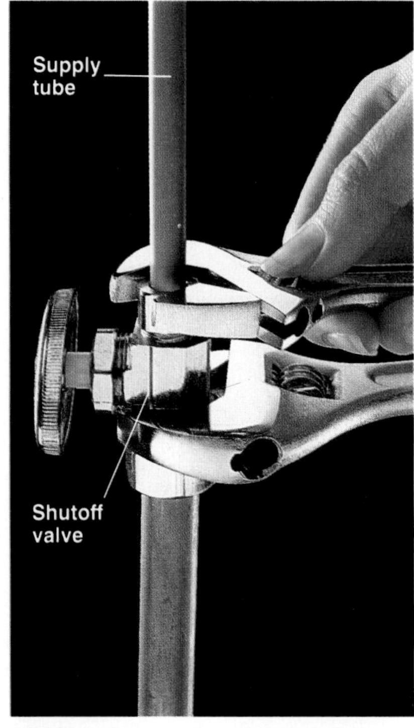

4 Attach supply tubes to shutoff valves, using compression fittings (pages 52 to 53). Hand-tighten nuts, then use an adjustable wrench to tighten nuts 1/4 turn. If necessary, hold valve with another wrench while tightening.

How to Connect a Faucet with Preattached Supply Tubing

1 Attach faucet to sink by placing rubber gasket, retainer ring, and locknut onto threaded tailpiece. Tighten locknut with a basin wrench or channel-type pliers. Some center-mounted faucets have a decorative coverplate. Secure coverplate from underneath with washers and locknuts screwed onto coverplate bolts.

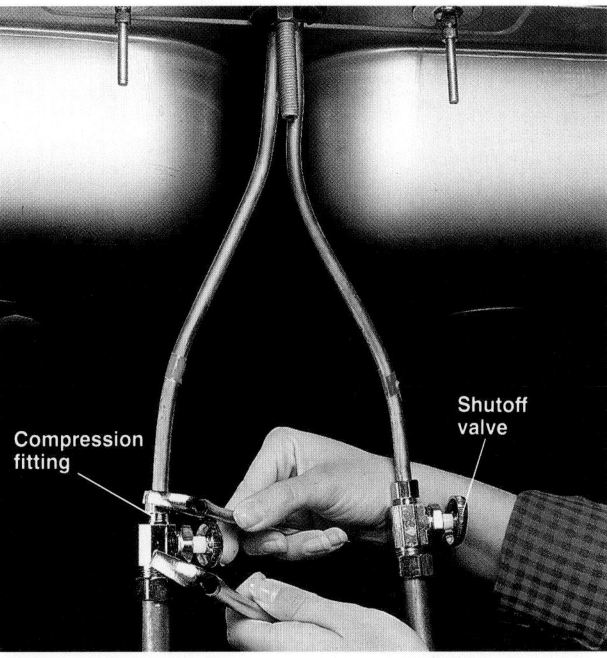

2 Connect preattached supply tubing to shutoff valves with compression fittings (pages 52 to 53). Red-coded tube should be attached to the hot water pipe, blue-coded tube to the cold water pipe.

How to Attach a Sink Sprayer

1 Apply a 1/4" bead of plumber's putty or silicone caulk to bottom edge of sprayer base. Insert tailpiece of sprayer base into sink opening.

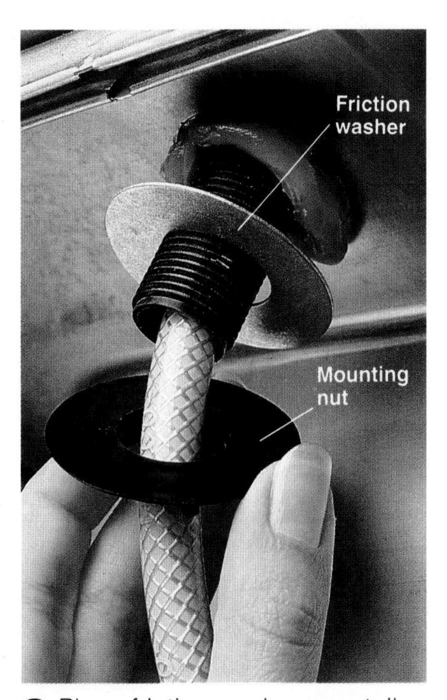

2 Place friction washer over tailpiece. Screw the mounting nut onto tailpiece and tighten with a basin wrench or channel-type pliers. Wipe away excess putty around base of sprayer.

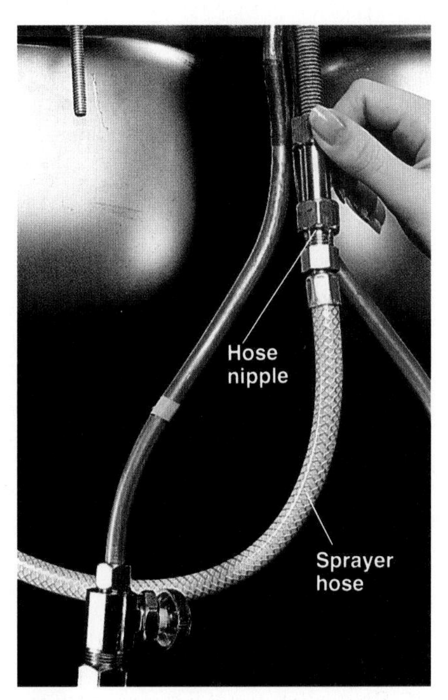

3 Screw sprayer hose onto the hose nipple on the bottom of the faucet. Tighten 1/4 turn, using a basin wrench or channel-type pliers.

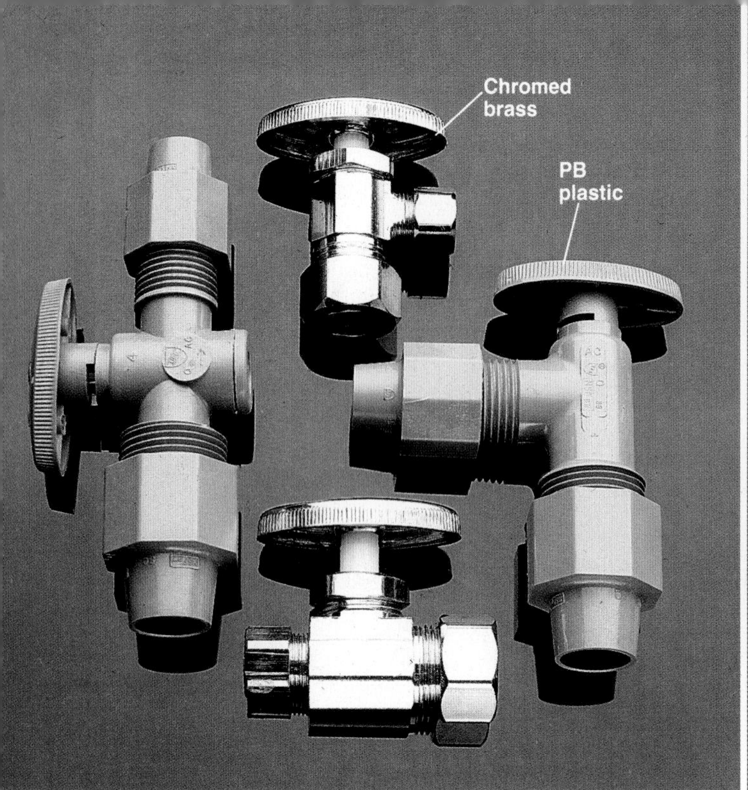

Shutoff valves allow you to shut off the water to an individual fixture so it can be repaired. They can be made from durable chromed brass or lightweight plastic. Shutoff valves come in ½" and ¾" diameters to match common water pipe sizes.

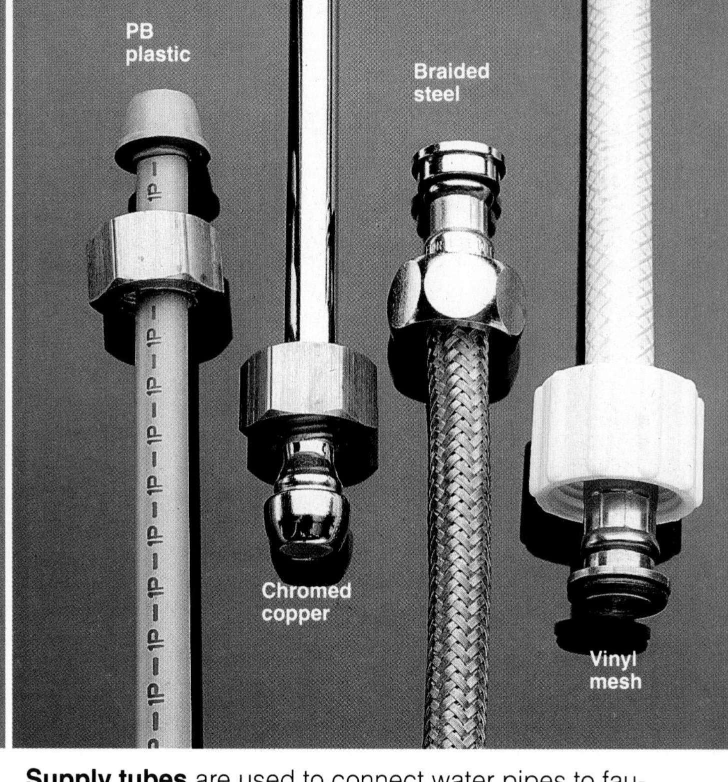

Supply tubes are used to connect water pipes to faucets, toilets, and other fixtures. They come in 12", 20", and 30" lengths. PB plastic and chromed copper tubes are inexpensive. Braided steel and vinyl mesh supply tubes are easy to install.

Installing Shutoff Valves & Supply Tubes

Worn-out shutoff valves or supply tubes can cause water to leak underneath a sink or other fixture. First, try tightening the fittings with an adjustable wrench. If this does not fix the leak, replace the shutoff valves and supply tubes.

Shutoff valves are available in several fitting types. For copper pipes, valves with compression-type fittings (pages 52 to 53) are easiest to install. For plastic pipes (pages 56 to 63), use grip-type valves. For galvanized iron pipes (pages 64 to 67), use valves with female threads.

Older plumbing systems often were installed without fixture shutoff valves. When repairing or replacing plumbing fixtures, you may want to install shutoff valves if they are not already present.

Everything You Need:

Tools: hacksaw, tubing cutter, adjustable wrench, tubing bender, felt-tipped pen.

Materials: shutoff valves, supply tubes, pipe joint compound.

How to Install Shutoff Valves & Supply Tubes

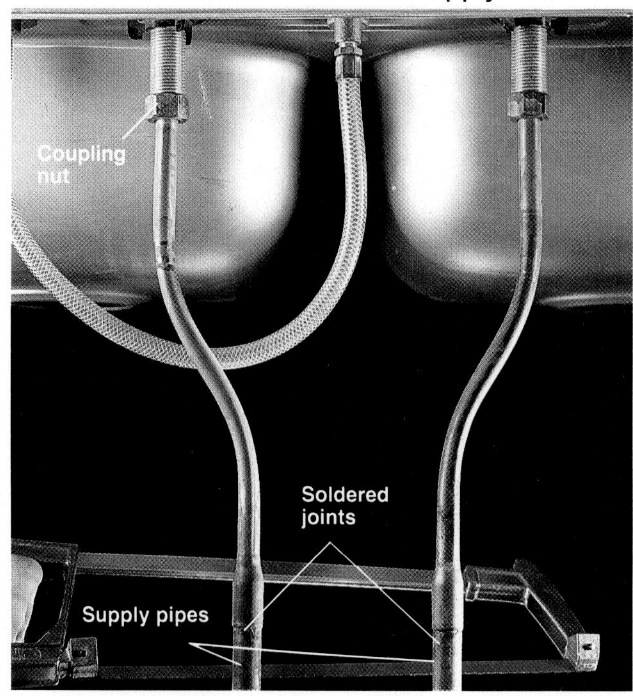

1 Turn off water at the main shutoff valve (page 6). Remove old supply pipes. If pipes are soldered copper, cut them off just below the soldered joint, using a hacksaw or tubing cutter. Make sure the cuts are straight. Unscrew the coupling nuts, and discard the old pipes.

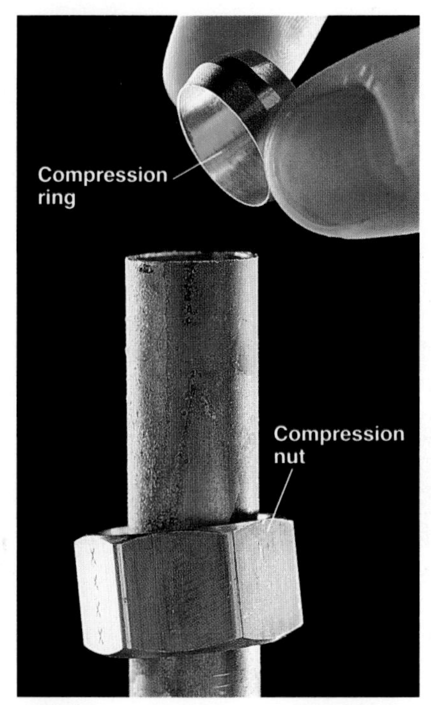

2 Slide a compression nut and compression ring over copper water pipe. Threads of nut should face end of pipe.

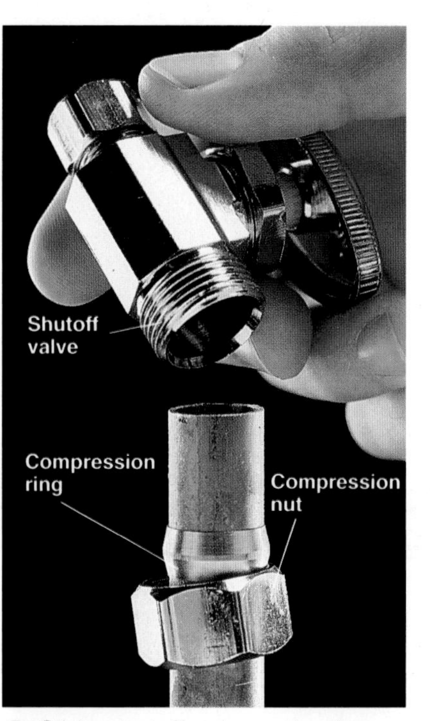

3 Slide shutoff valve onto pipe. Apply a layer of pipe joint compound to compression ring. Screw the compression nut onto the shutoff valve and tighten with an adjustable wrench.

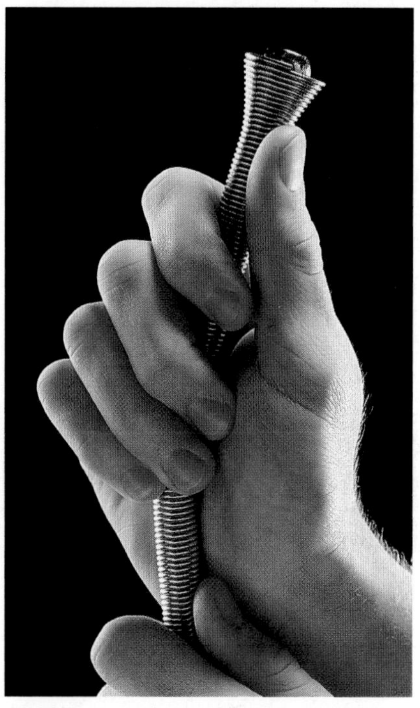

4 Bend chromed copper supply tube to reach from the tailpiece of the fixture to the shutoff valve, using a tubing bender (page 45). Bend the tube slowly to avoid crimping the metal.

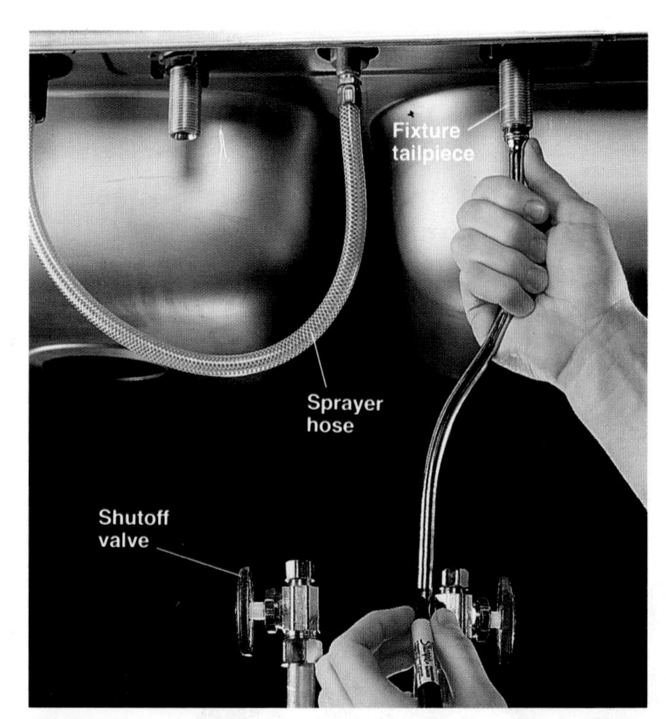

5 Position the supply tube between fixture tailpiece and shutoff valve, and mark tube to length. Cut supply tube with a tubing cutter (page 47).

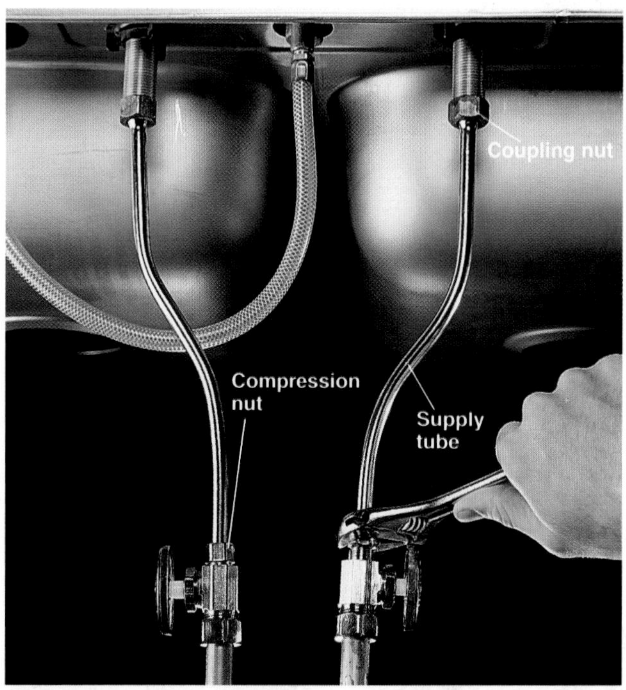

6 Attach bell-shaped end of supply tube to fixture tailpiece with coupling nut, then attach other end to shutoff valve with compression ring and nut (pages 52 to 53). Tighten all fittings with adjustable wrench.

Fixing Sprayers & Aerators

If water pressure from a sink sprayer seems low, or if water leaks from the handle, it is usually because lime buildup and sediment have blocked small openings inside the sprayer head. To fix the problem, first take the sprayer head apart and clean the parts. If cleaning the sprayer head does not help, the problem may be caused by a faulty diverter valve. The diverter valve inside the faucet body shifts water flow from the faucet spout to the sprayer when the sprayer handle is pressed. Cleaning or replacing the diverter valve may fix water pressure problems.

Whenever making repairs to a sink sprayer, check the sprayer hose for kinks or cracks. A damaged hose should be replaced.

If water pressure from a faucet spout seems low, or if the flow is partially blocked, take the spout aerator apart and clean the parts. The aerator is a screw-on attachment with a small wire screen that mixes tiny air bubbles into the water flow. Make sure the wire screen is not clogged with sediment and lime buildup. If water pressure is low throughout the house, it may be because galvanized iron water pipes are corroded. Corroded pipes should be replaced with copper (pages 46 to 55).

Everything You Need:

Tools: screwdriver, channel-type pliers, needle-nose pliers, small brush.

Materials: vinegar, universal washer kit, heat-proof grease, replacement sprayer hose.

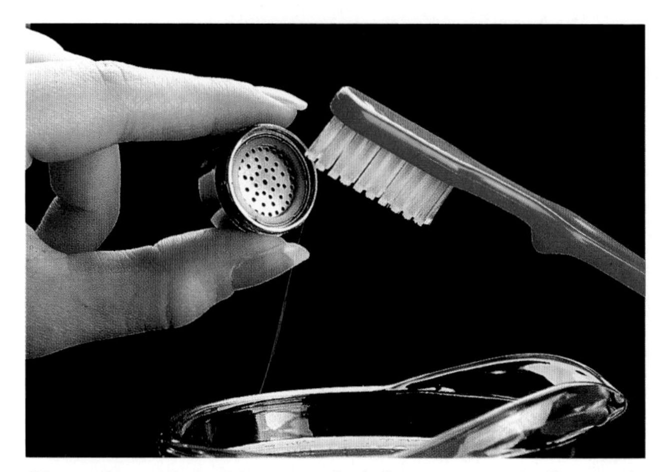

Clean faucet aerators and sink sprayers to fix most low water pressure problems. Take aerator or sprayer head apart, then use a small brush dipped in vinegar to remove sediment.

How to Fix a Diverter Valve

1 Shut off the water (page 176). Remove the faucet handle and the spout (see directions for your faucet type, pages 178 to 187).

2 Pull diverter valve from faucet body with needlenose pliers. Use a small brush dipped in vinegar to clean lime buildup and debris from valve.

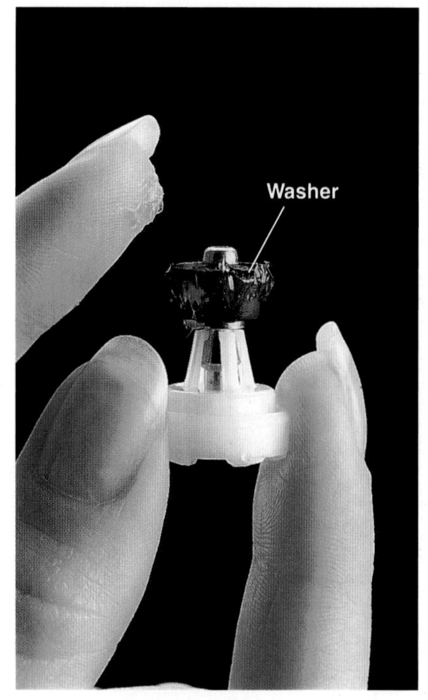

3 Replace any worn O-rings or washers, if possible. Coat the new parts with heatproof grease, then reinstall the diverter valve and reassemble the faucet.

How to Replace a Sprayer Hose

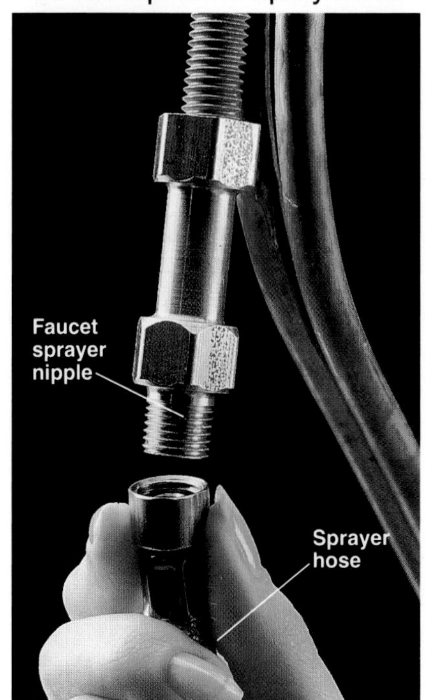

1 Unscrew sprayer hose from faucet sprayer nipple, using channel-type pliers. Pull sprayer hose through sink opening.

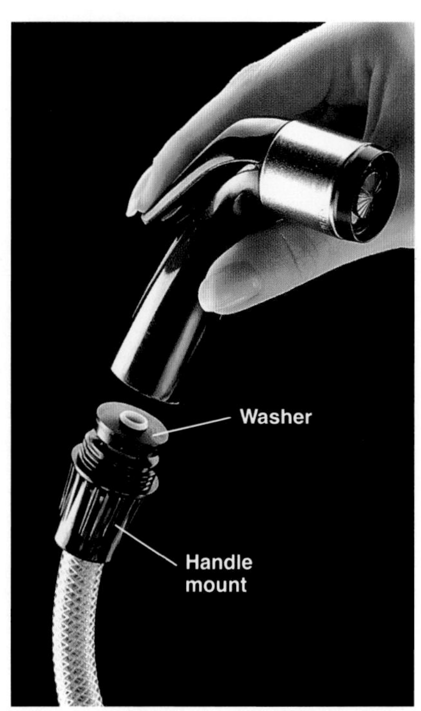

2 Unscrew the sprayer head from the handle mount. Remove washer.

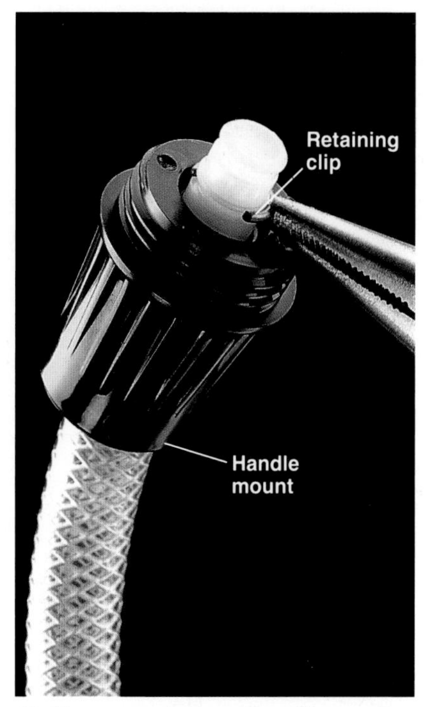

3 Remove retainer clip with needlenose pliers, and discard old hose. Attach handle mount, retainer clip, washer, and sprayer head to new hose. Attach sprayer hose to faucet sprayer nipple on faucet.

Tub & Shower Plumbing

Tub and shower faucets have the same basic designs as sink faucets, and the techniques for repairing leaks are the same as described in the faucet repair section of this book (pages 176 to 187). To identify your faucet design, you may need to take off the handle and disassemble the faucet.

When a tub and shower are combined, the shower head and the tub spout share the same hot and cold water supply lines and handles. Combination faucets are available as three-handle, two-handle,

Tub & Shower Combination Faucets

Three-handle faucet (pages 198 to 199) has valves that are either compression or cartridge design.

or single-handle types (below). The number of handles gives clues as to the design of the faucets and the kinds of repairs that may be necessary.

With combination faucets, a diverter valve or gate diverter is used to direct water flow to the tub spout or the shower head. On three-handle faucet types, the middle handle controls a diverter valve. If water does not shift easily from tub to spout to shower head, or if water continues to run out the spout when the shower is on, the diverter valve probably needs to be cleaned and repaired (pages 198 to 199).

Two-handle and single-handle types use a gate diverter that is operated by a pull lever or knob on the tub spout. Although gate diverters rarely need repair, the lever occasionally may break, come loose, or refuse to stay in the UP position. To repair a gate diverter set in a tub spout, replace the entire spout (page 201).

Tub and shower faucets and diverter valves may be set inside wall cavities. Removing them may require a deep-set ratchet wrench (pages 199, 201).

If spray from the shower head is uneven, clean the spray holes. If the shower head does not stay in an upright position, remove the head and replace the O-ring (page 204).

To add a shower to an existing tub, install a flexible shower adapter (page 205). Several manufacturers make complete conversion kits that allow a shower to be installed in less than one hour.

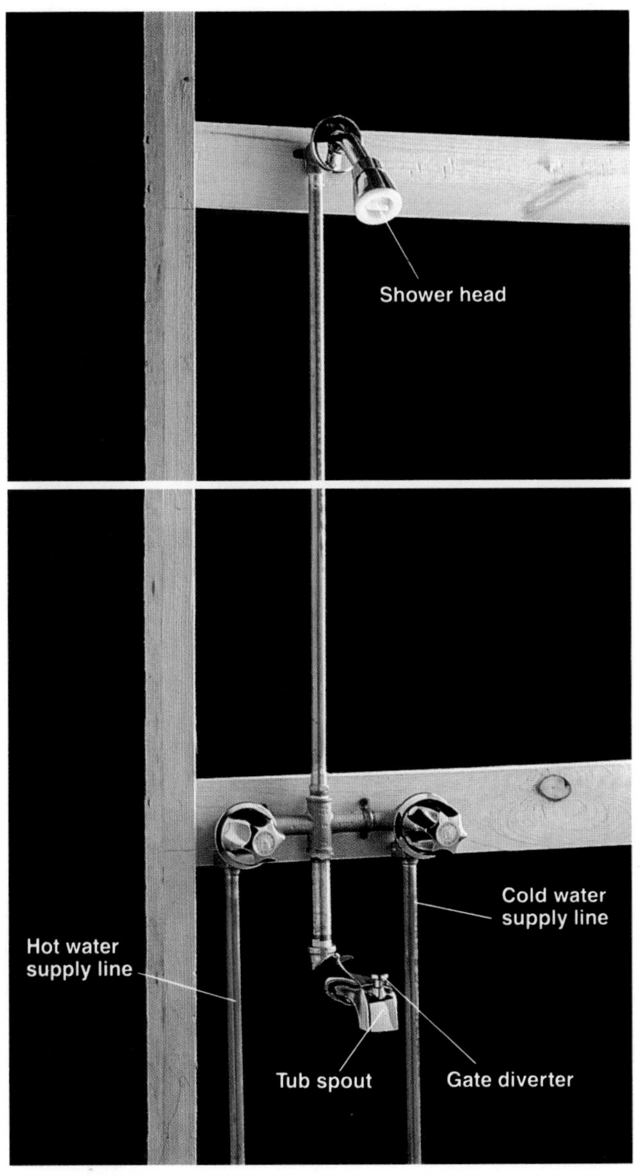

Two-handle faucet (pages 200 to 201) has valves that are either compression or cartridge design.

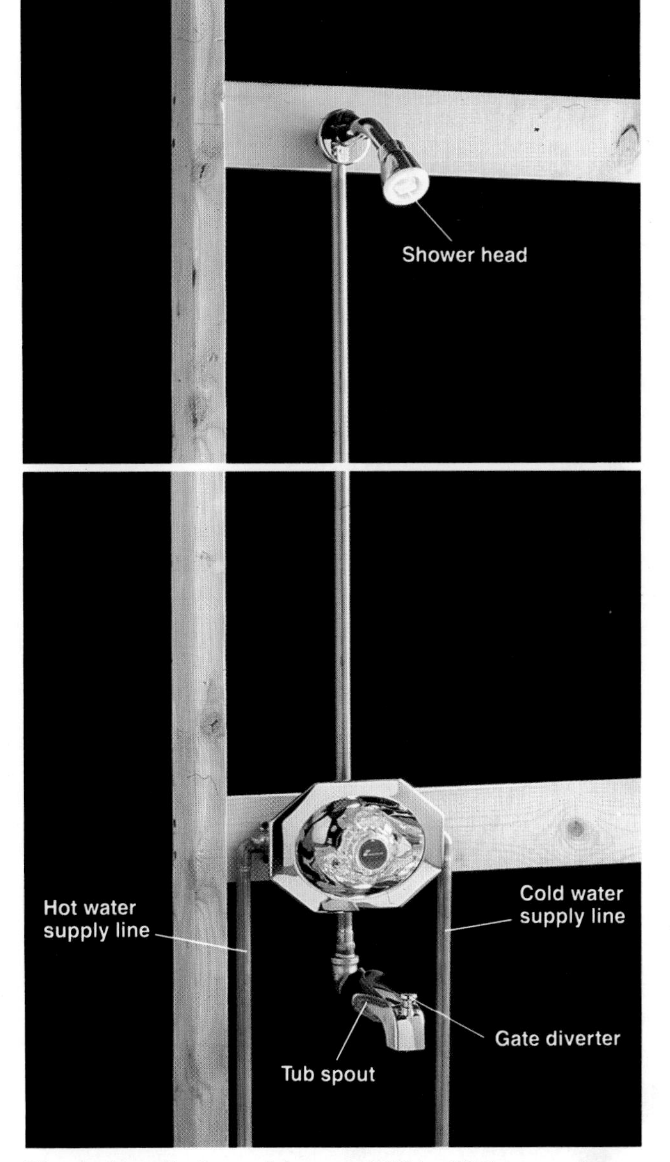

Single-handle faucet (pages 202 to 203) has valves that are cartridge, ball-type, or disc design.

Fixing Three-handle Tub & Shower Faucets

A three-handle faucet type has handles to control hot and cold water, and a third handle that controls the diverter valve and directs water to either a tub spout or a shower head. The separate hot and cold handles indicate cartridge or compression faucet designs. To repair them, see pages 180 to 181 for cartridge, and 182 to 185 for compression.

If a diverter valve sticks, if water flow is weak, or if water runs out of the tub spout when the flow is directed to the shower head, the diverter needs to be repaired or replaced. Most diverter valves are similar to either compression or cartridge faucet valves. Compression type diverters can be repaired, but cartridge types should be replaced.

Remember to turn off the water (page 12) before beginning work.

Everything You Need:

Tools: screwdriver, adjustable wrench or channel-type pliers, deep-set ratchet wrench, small wire brush.

Materials: replacement diverter cartridge or universal washer kit, heatproof grease, vinegar.

How to Repair a Compression Diverter Valve

1 Remove the diverter valve handle with a screwdriver. Unscrew or pry off the escutcheon.

2 Remove bonnet nut with an adjustable wrench or channel-type pliers.

3 Unscrew the stem assembly, using a deep-set ratchet wrench. If necessary, chip away any mortar surrounding the bonnet nut (page 201, step 2).

4 Remove brass stem screw. Replace stem washer with an exact duplicate. If stem screw is worn, replace it.

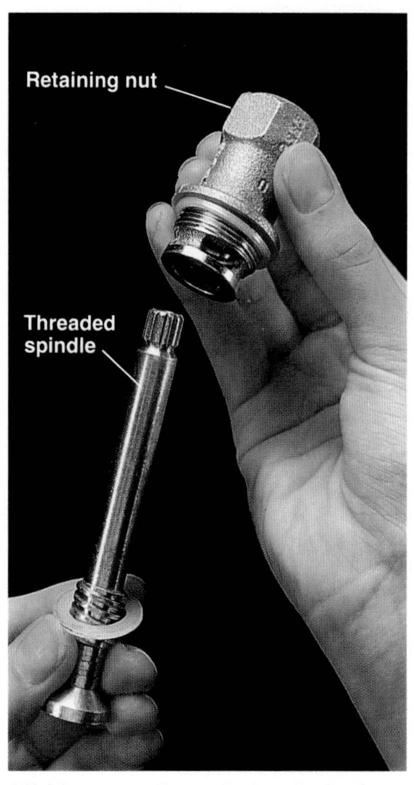

5 Unscrew threaded spindle from retaining nut.

6 Clean sediment and lime build-up from nut, using a small wire brush dipped in vinegar. Coat all parts with heatproof grease and reassemble diverter valve.

Fixing Two-handle Tub & Shower Faucets

Two-handle tub and shower faucets are either cartridge or compression design. They may be repaired following the directions on pages 180 to 181 for cartridge, or pages 182 to 185 for compression. Because the valves of two-handle tub and shower faucets may be set inside the wall cavity, a deep-set socket wrench may be required to remove the valve stem.

Two-handle tub and shower designs have a gate diverter. A gate diverter is a simple mechanism located in the tub spout. A gate diverter closes the supply of water to the tub spout and redirects the flow to the shower head. Gate diverters seldom need repair. Occasionally, the lever may break, come loose, or refuse to stay in the UP position.

If the diverter fails to work properly, replace the tub spout. Tub spouts are inexpensive and easy to replace.

Remember to turn off the water (page 12) before beginning work.

Everything You Need:

Tools: screwdriver, Allen wrench, pipe wrench, channel-type pliers, small cold chisel, ball peen hammer, deep-set ratchet wrench.

Materials: masking tape or cloth, pipe joint compound, replacement faucet parts as needed.

Tips on Replacing a Tub Spout

Check underneath tub spout for a small access slot. The slot indicates the spout is held in place with an Allen screw. Remove the screw, using an Allen wrench. Spout will slide off.

Unscrew faucet spout. Use a pipe wrench, or insert a large screwdriver or hammer handle into the spout opening and turn spout counterclockwise.

Spread pipe joint compound on threads of spout nipple before replacing spout.

How to Remove a Deep-set Faucet Valve

1 Remove handle, and unscrew the escutcheon with channel-type pliers. Pad the jaws of the pliers with masking tape to prevent scratching the escutcheon.

2 Chip away any mortar surrounding the bonnet nut, using a ball peen hammer and a small cold chisel.

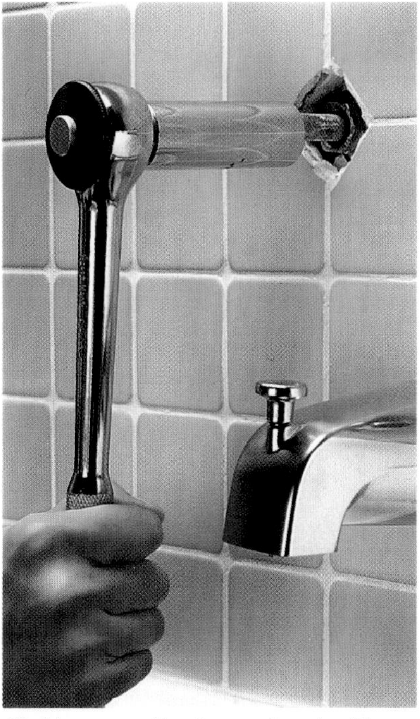

3 Unscrew the bonnet nut with a deep-set ratchet wrench. Remove the bonnet nut and stem from the faucet body.

Fixing Single-handle Tub & Shower Faucets

A single-handle tub and shower faucet has one valve that controls both water flow and temperature. Single-handle faucets may be ball-type, cartridge, or disc designs.

If a single-handle control valve leaks or does not function properly, disassemble the faucet, clean the valve, and replace any worn parts. Use the repair techniques described on pages 178 to 179 for ball-type, or pages 186 to 187 for ceramic disc. Repairing a single-handle cartridge faucet is shown on the opposite page.

Direction of the water flow to either the tub spout or the shower head is controlled by a gate diverter. Gate diverters seldom need repair. Occasionally, the lever may break, come loose, or refuse to stay in the UP position. If the diverter fails to work properly, replace the tub spout (page 201).

Everything You Need:

Tools: screwdriver, adjustable wrench, channel-type pliers.

Materials: replacement parts as needed.

How to Repair a Single-handle Cartridge Tub & Shower Faucet

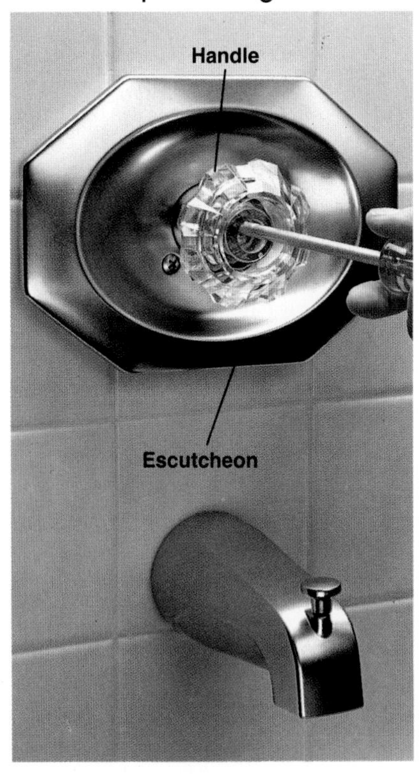

1 Use a screwdriver to remove the handle and escutcheon.

2 Turn off water supply at built-in shutoff valves or main shutoff valve (page 72).

3 Unscrew and remove retaining ring or bonnet nut, using an adjustable wrench.

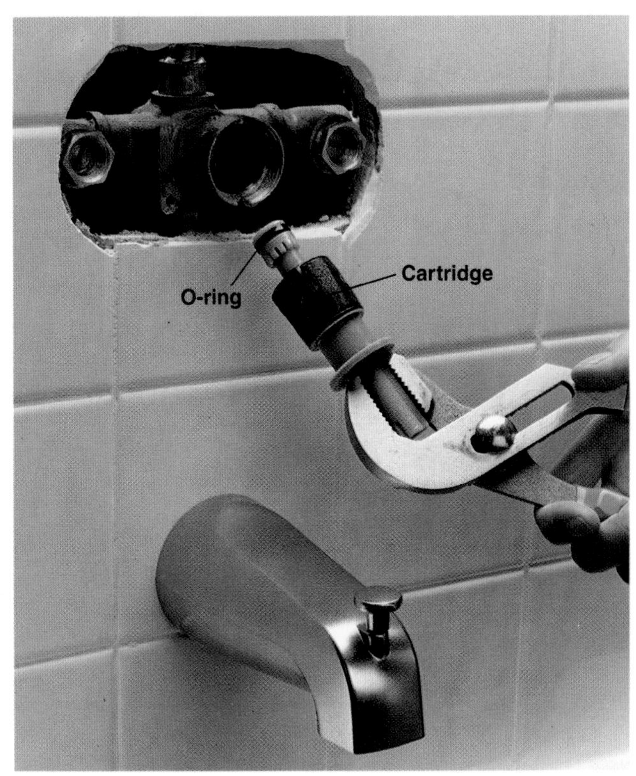

4 Remove cartridge assembly by grasping end of valve with channel-type pliers and pulling gently.

5 Flush valve body with clean water to remove sediment. Replace any worn O-rings. Reinstall cartridge and test valve. If faucet fails to work properly, replace the cartridge.

Shower arm
Collar nut
Swivel ball nut
Spray adjustment cam lever
Swivel ball
O-ring
Spray outlets

A typical shower head can be disassembled easily for cleaning and repair. Some shower heads include a spray adjustment cam lever that is used to change the force of the spray.

Fixing & Replacing Shower Heads

If spray from the shower head is uneven, clean the spray holes. The outlet or inlet holes of the shower head may get clogged with mineral deposits.

Shower heads pivot into different positions. If a shower head does not stay in position, or if it leaks, replace the O-ring that seals against the swivel ball.

A tub can be equipped with a shower by installing a flexible shower adapter kit. Complete kits are available at hardware stores and home centers.

Everything You Need:

Tools: adjustable wrench or channel-type pliers, pipe wrench, drill, glass & tile bit (if needed), mallet, screwdriver.

Materials: masking tape, thin wire (paper clip), heatproof grease, rag, replacement O-rings (if needed), masonry anchors, flexible shower adapter kit (optional).

How to Clean & Repair a Shower Head

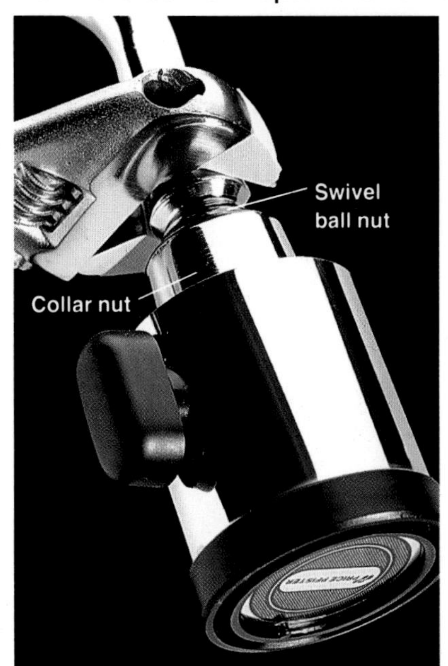

1 Unscrew swivel ball nut, using an adjustable wrench or channel-type pliers. Wrap jaws of the tool with masking tape to prevent marring the finish. Unscrew collar nut from shower head.

2 Clean outlet and inlet holes of shower head with a thin wire. Flush the head with clean water.

3 Replace the O-ring, if necessary. Lubricate the O-ring with heatproof grease before installing.

How to Install a Flexible Shower Adapter

1 Remove old tub spout (page 103). Install new tub spout from kit, using a pipe wrench. New spout will have an adapter hose outlet. Wrap the tub spout with a rag to prevent damage to the chrome finish.

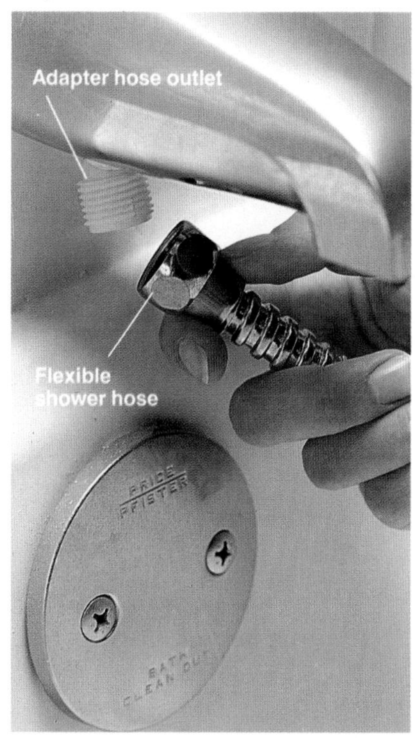

2 Attach flexible shower hose to the adaptor hose outlet. Tighten with an adjustable wrench or channel-type pliers.

3 Determine location of shower head hanger. Use hose length as a guide, and make sure shower head can be easily lifted off hanger.

4 Mark hole locations. Use a glass and tile bit to drill holes in ceramic tile for masonry anchors.

5 Insert anchors into holes, and tap into place with a wooden or rubber mallet.

6 Fasten shower head holder to the wall, and hang shower head.

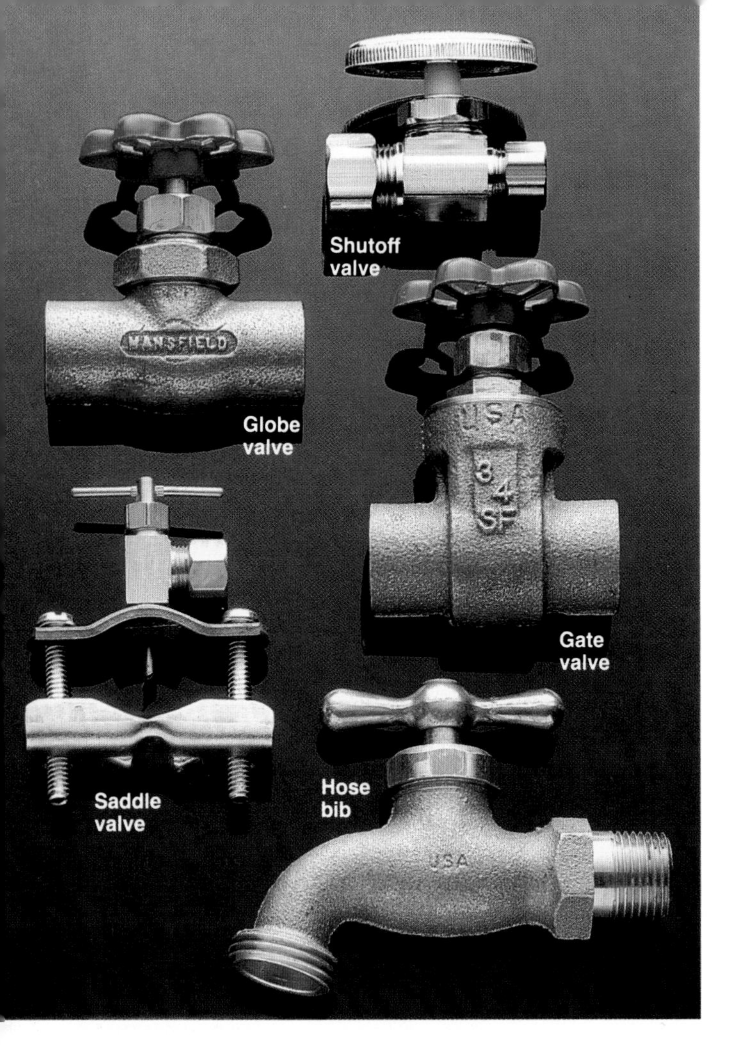

Repairing Valves & Hose Bibs

Valves make it possible to shut off water at any point in the supply system. If a pipe breaks or a plumbing fixture begins to leak, you can shut off water to the damaged area so that it can be repaired. A hose bib is a faucet with a threaded spout, often used to connect rubber utility or appliance hoses.

Valves and hose bibs leak when washers or seals wear out. Replacement parts can be found in the same universal washer kits used to repair compression faucets (page 182). Coat replacement washers with heatproof grease to keep them soft and prevent cracking.

Remember to turn of the water before beginning work (page 12).

Everything You Need:

Tools: screwdriver, adjustable wrench.

Materials: universal washer kit, heatproof grease.

How to Fix a Leaky Hose Bib

1 Remove the handle screw, and lift off the handle. Unscrew the packing nut with an adjustable wrench.

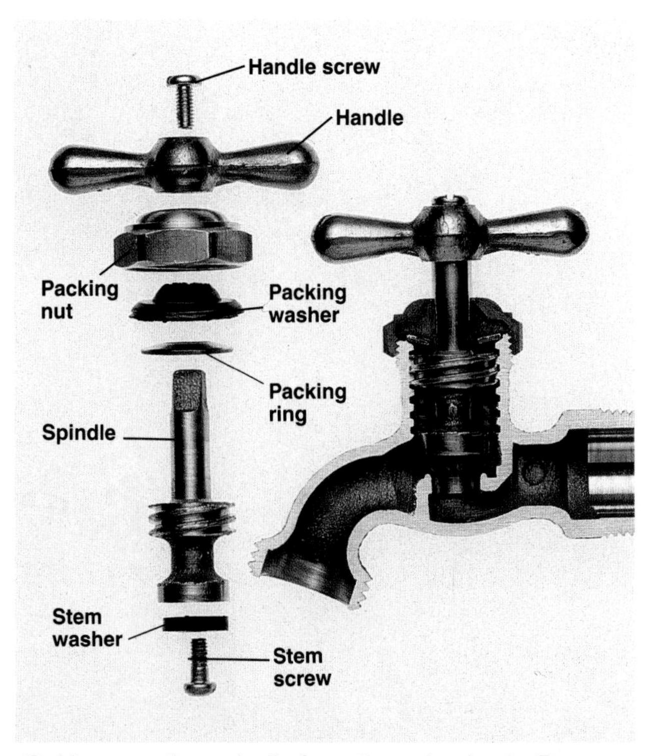

2 Unscrew the spindle from the valve body. Remove the stem screw and replace the stem washer. Replace the packing washer, and reassemble the valve.

Common Types of Valves

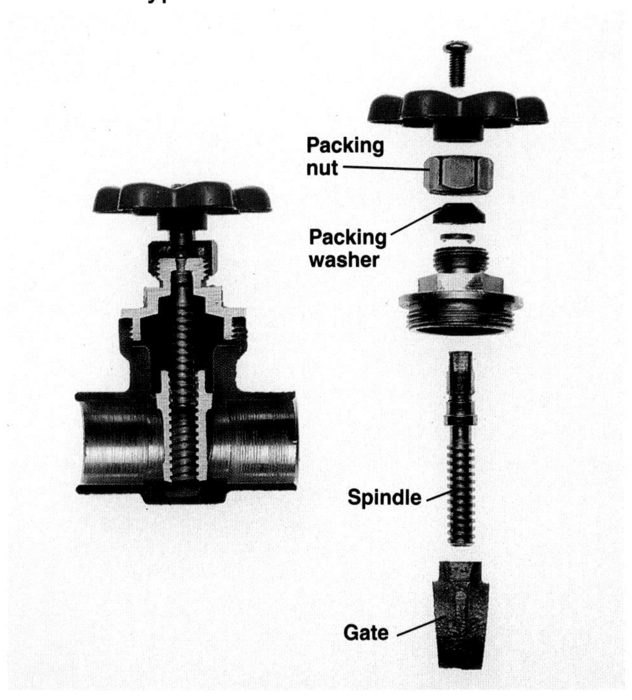

Gate valve has a movable brass wedge, or "gate," that screws up and down to control water flow. Gate valves may develop leaks around the handle. Repair leaks by replacing the packing washer or packing string found underneath the packing nut.

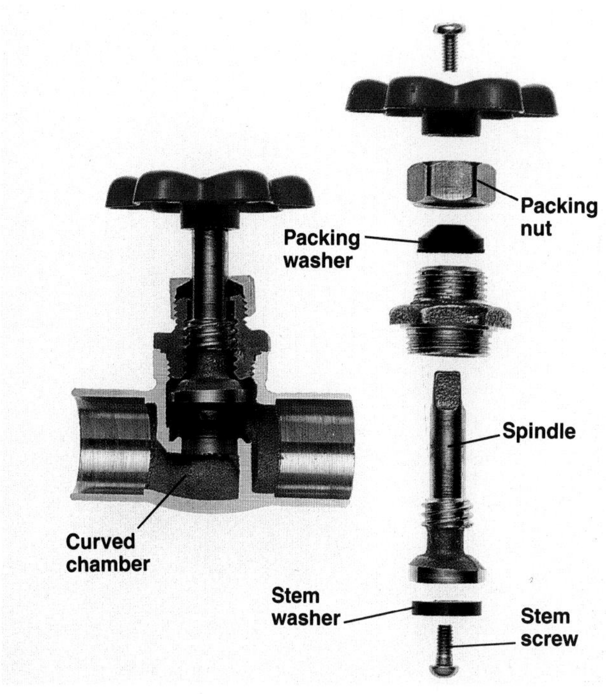

Globe valve has a curved chamber. Repair leaks around the handle by replacing the packing washer. If valve does not fully stop water flow when closed, replace the stem washer.

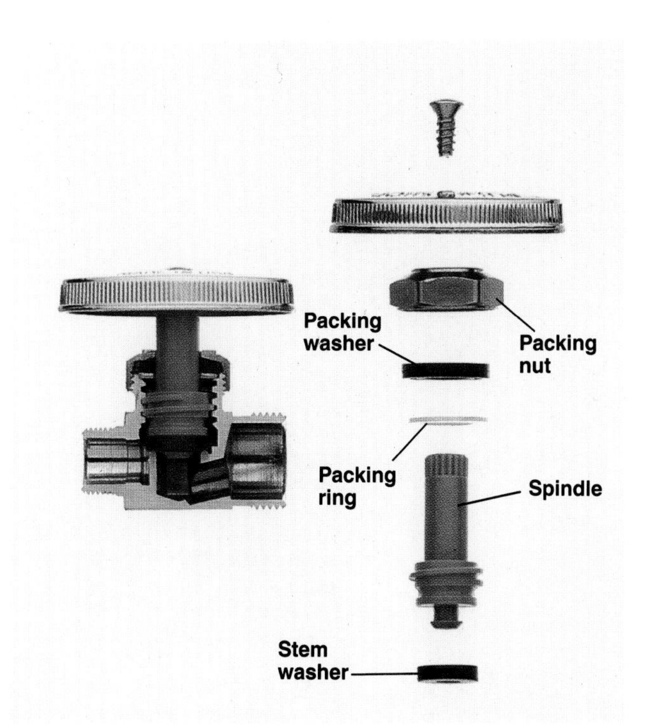

Shutoff valve controls water supply to a single fixture (pages 192 to 193). Shutoff valve has a plastic spindle with a packing washer and a snap-on stem washer. Repair leaks around the handle by replacing the packing washer. If valve does not fully stop water flow when closed, replace the stem washer.

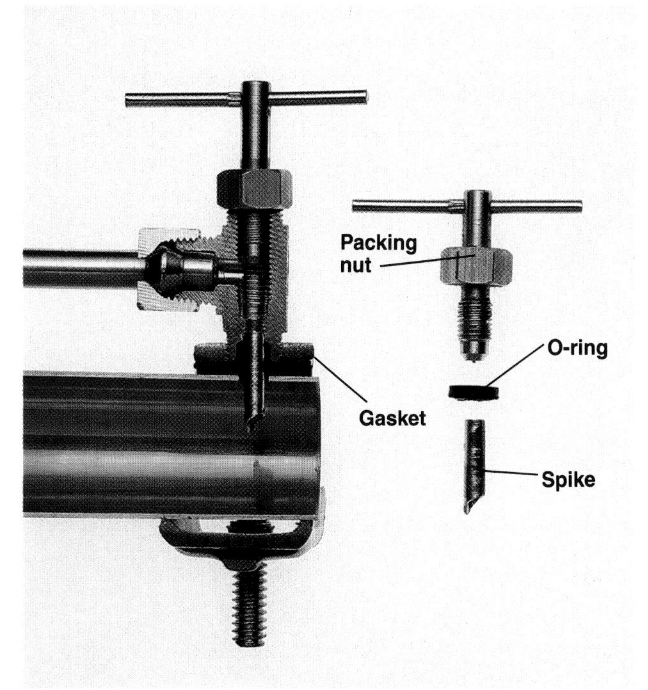

Saddle valve is a small fitting often used to connect a refrigerator icemaker or sink-mounted water filter to a copper water pipe. Saddle valve contains a hollow metal spike that punctures water pipe when valve is first closed. Fitting is sealed with a rubber gasket. Repair leaks around the handle by replacing the O-ring under the packing nut.

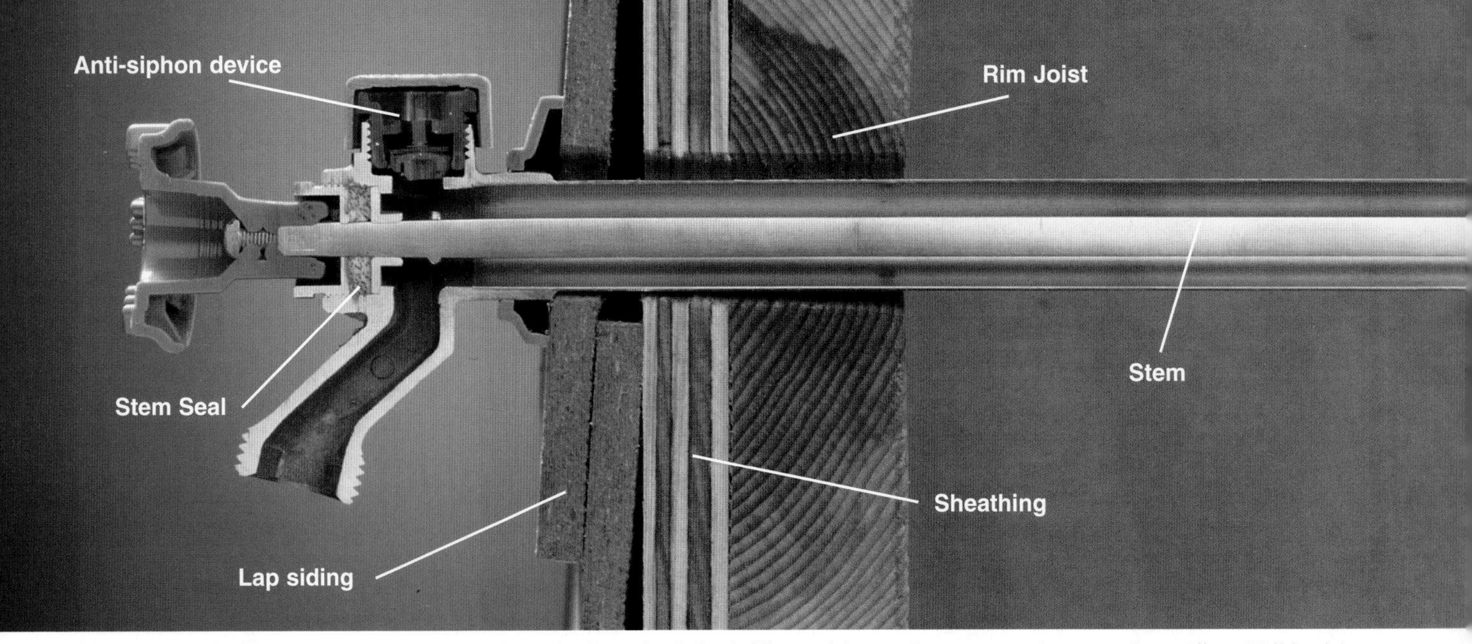

Frost-proof sillcock is mounted against the header joist (sill), and has a long stem that reaches 6" to 30" inside the house to protect the valve from cold. A sillcock should angle downward slightly to provide drainage. The stem washer and O-ring (or packing string) can be replaced if the sillcock begins to leak. In a copper plumbing system,

Installing & Repairing Sillcocks

A sillcock is a compression faucet attached to the outside of the house. Repair a leaky sillcock by replacing the stem washer and the O-ring.

Sillcocks can be damaged by frost. To repair a ruptured pipe, see pages 250 to 251. To prevent pipes from rupturing, close the indoor shutoff valves at the start of the cold weather season, disconnect all garden hoses, and open the sillcock to let trapped water drain out.

A special frost-proof sillcock has a long stem that reaches at least 6" inside the house to protect it from cold. Install a sillcock so the pipe angles downward from the shutoff valve. This allows water to drain away each time the faucet is turned off.

Remember to turn off the water before beginning work (page 12).

Everything You Need:

Tools: screwdriver, channel-type pliers, pencil, right-angle drill or standard drill, 1" spade bit, caulk gun, hacksaw or tubing cutter, propane torch.

Materials: universal washer kit, sillcock, silicone caulk, 2" corrosion-resistant screws, copper pipe, T-fitting, Teflon tape, threaded adapter, shutoff valve, emery cloth, soldering paste (flux), solder.

How to Repair a Sillcock

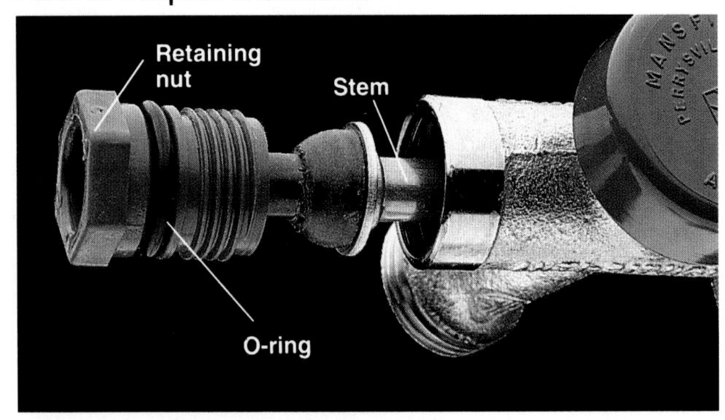

1 Remove sillcock handle, and loosen retaining nut with channel-type pliers. Remove stem. Replace O-ring found on retaining nut or stem.

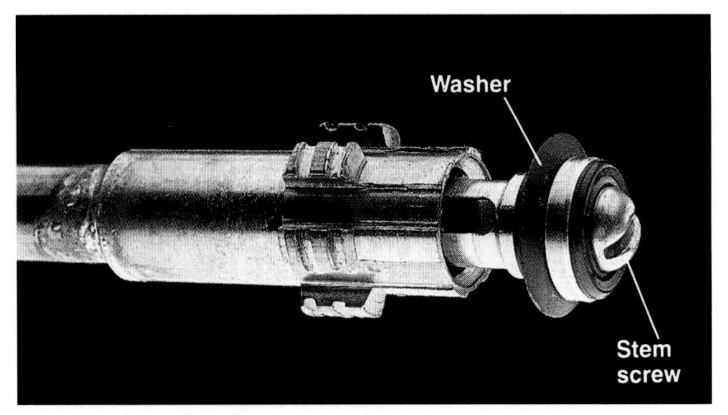

2 Remove the brass stem screw at the end of the stem, and replace the washer. Reassemble the sillcock.

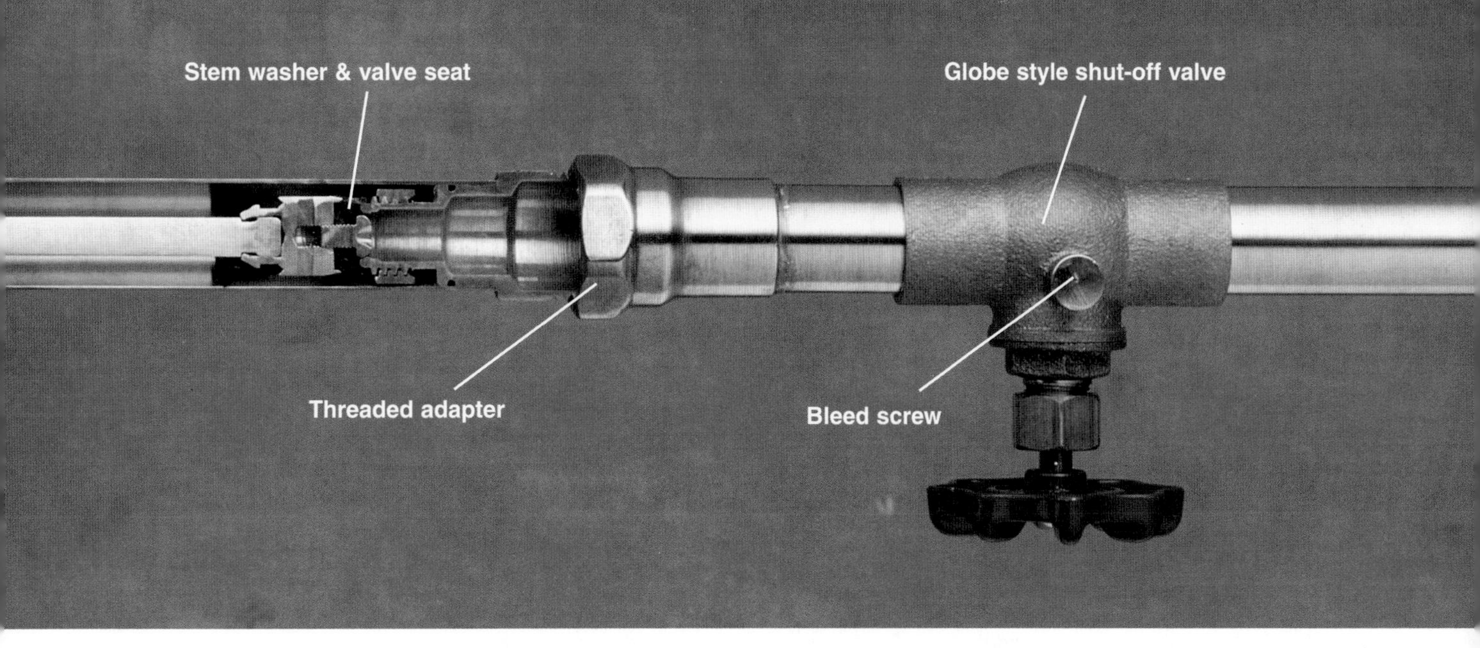

the sillcock is connected to a nearby cold water supply pipe with a threaded adapter, two lengths of soldered copper pipe, and a shutoff valve. A T-fitting (not shown) is used to tap into an existing cold water pipe.

How to Install a Frost-proof Sillcock

1 Locate position of hole for sillcock. From nearest cold water pipe, mark a point on header joist that is slightly lower than water pipe. Drill a hole through header, sheathing, and siding, using a 1" spade bit.

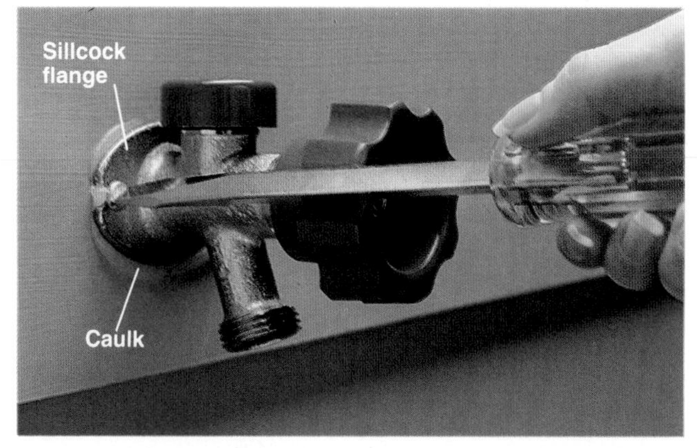

2 Apply a thick bead of silicone caulk to bottom of sillcock flange, then insert sillcock into hole, and attach to siding with 2" corrosion-resistant screws. Turn handle to ON position. Wipe away excess caulk.

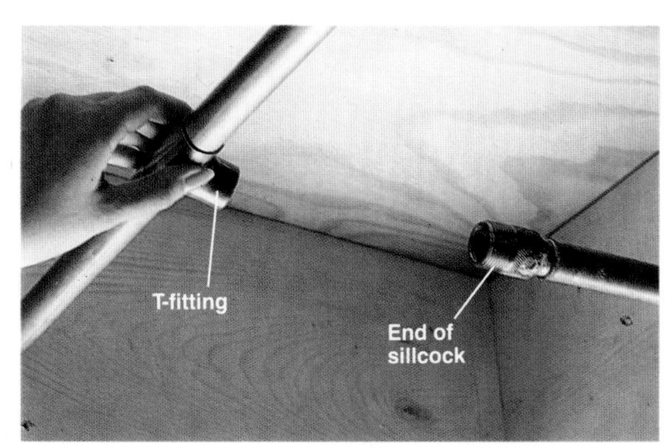

3 Mark cold water pipe, then cut pipe and install a T-fitting (pages 48 to 51). Wrap Teflon tape around threads of sillcock.

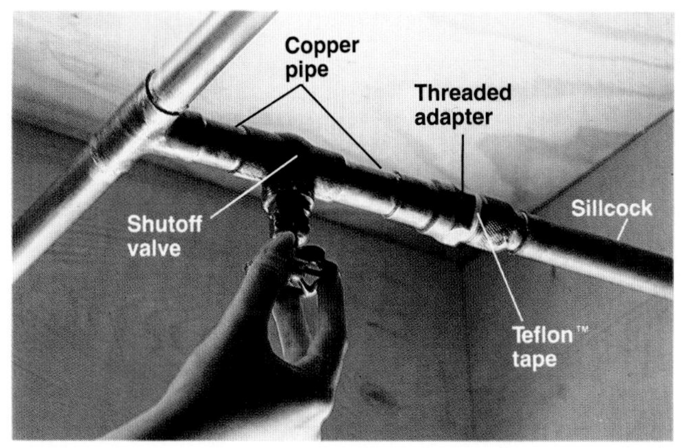

4 Join T-fitting to sillcock with threaded adapter (page 43), a shutoff valve, and two lengths of copper pipe. Prepare pipes and solder the joints. Turn on water, and close sillcock when water runs steadily.

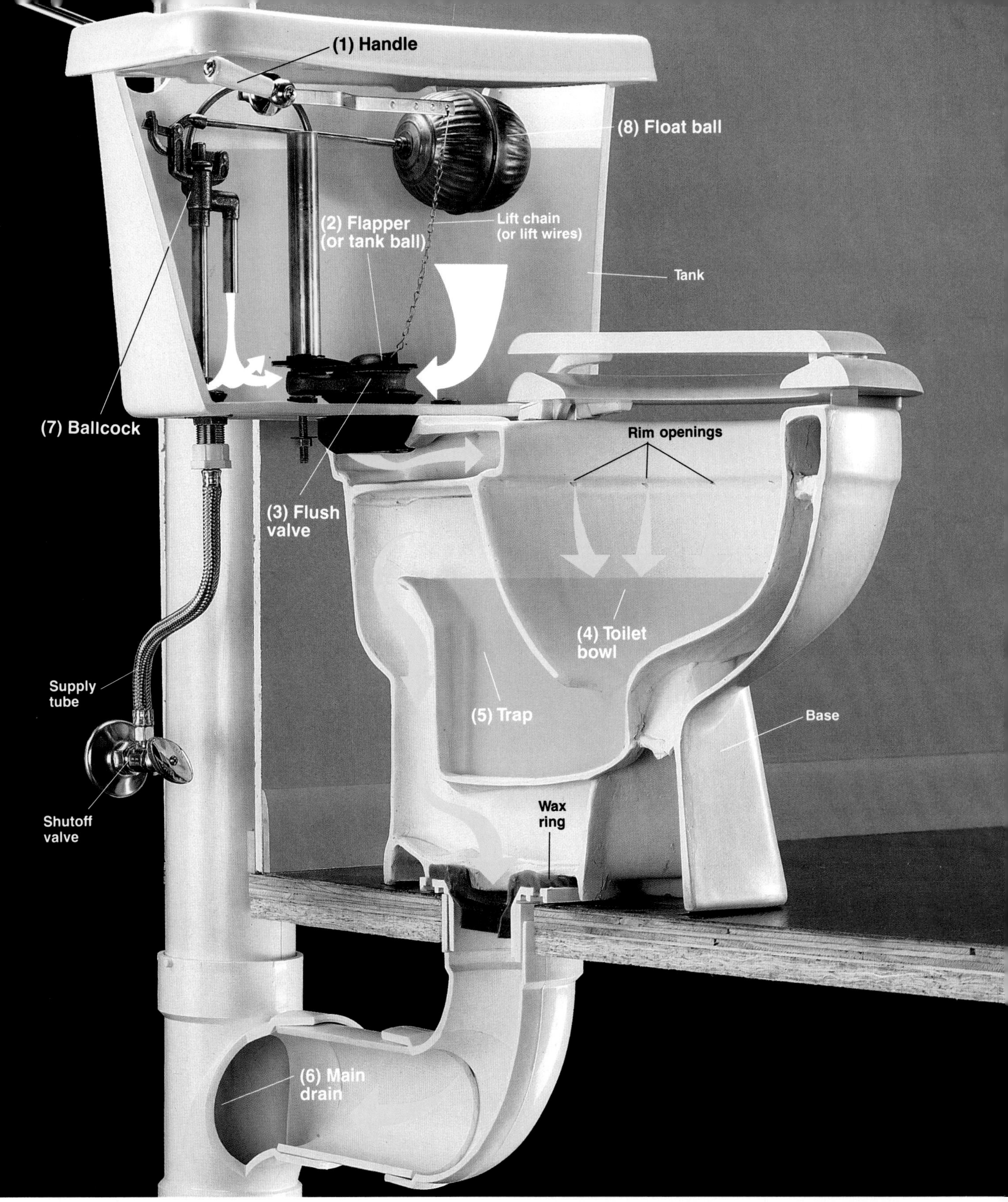

How a toilet works: When the **handle (1)** is pushed, the lift chain raises a rubber seal, called a **flapper or tank ball (2)**. Water in the tank rushes down through the **flush valve opening (3)** in the bottom of the tank, into the **toilet bowl (4)**. Waste water in the bowl is forced through the **trap (5)** into the **main drain (6)**. When the toilet tank is empty, the flapper seals the tank, and a water supply valve, called a **ballcock (7)**, refills the toilet tank. The ballcock is controlled by a **float ball (8)** that rides on the surface of the water. When the tank is full, the float ball automatically shuts off the ballcock.

Common Toilet Problems

A clogged toilet is one of the most common plumbing problems. If a toilet overflows or flushes sluggishly, clear the clog with a plunger or closet auger (page 228). If the problem persists, the clog may be in the main waste and vent stack (page 235).

Most other toilet problems are fixed easily with minor adjustments that require no disassembly or replacement parts. You can make these adjustments in a few minutes, using simple tools (page 212).

If minor adjustments do not fix the problem, further repairs will be needed. The parts of a standard toilet are not difficult to take apart, and most repair projects can be completed in less than an hour.

A recurring puddle of water on the floor around a toilet may be caused by a crack in the toilet base or in the tank. A damaged toilet should be replaced. Installing a new toilet is an easy project that can be finished in three or four hours.

A standard two-piece toilet has an upper tank that is bolted to a base. This type of toilet uses a simple gravity-operated flush system, and can be repaired easily using the directions on the following pages. Some one-piece toilets use a complicated, high-pressure flush valve. Repairing these toilets can be difficult, so this work should be left to a professional.

Problems	Repairs
Toilet handle sticks, or is hard to push.	1. Adjust lift wires (page 212). 2. Clean & adjust handle (page 212).
Handle is loose.	1. Adjust handle (page 212). 2. Reattach lift chain or lift wires to lever (page 212).
Toilet will not flush at all.	1. Make sure water is turned on. 2. Adjust lift chain or lift wires (page 212).
Toilet does not flush completely.	1. Adjust lift chain (page 212). 2. Adjust water level in tank (page 214).
Toilet overflows, or flushes sluggishly.	1. Clear clogged toilet (page 228). 2. Clear clogged main waste and vent stack (page 235).
Toilet runs continuously.	1. Adjust lift wires or lift chain (page 212). 2. Replace leaky float ball (page 213). 3. Adjust water level in tank (page 214). 4. Adjust & clean flush valve (page 217). 5. Replace flush valve (page 217). 6. Repair or replace ballcock (pages 215 to 216).
Water on floor around toilet.	1. Tighten tank bolts and water connections (page 218). 2. Insulate tank to prevent condensation (page 218). 3. Replace wax ring (pages 219 to 220). 4. Replace cracked tank or bowl (pages 218 to 221).

Making Minor Adjustments

Many common toilet problems can be fixed by making minor adjustments to the handle and the attached lift chain (or lift wires).

If the handle sticks or is hard to push, remove the tank cover and clean the handle mounting nut. Make sure the lift wires are straight.

If the toilet will not flush completely unless the handle is held down, you may need to remove excess slack in the lift chain.

If the toilet will not flush at all, the lift chain may be broken or may need to be reattached to the handle lever.

A continuously running toilet (page opposite) can be caused by bent lift wires, kinks in a lift chain, or lime buildup on the handle mounting nut. Clean and adjust the handle and the lift wires or chain to fix the problem.

Everything You Need:

Tools: adjustable wrench, needlenose pliers, screwdriver, small wire brush.

Materials: vinegar.

How to Adjust a Toilet Handle & Lift Chain (or Lift Wires)

Clean and adjust handle mounting nut so handle operates smoothly. Mounting nut has reversed threads. Loosen nut by turning clockwise; tighten by turning counterclockwise. Remove lime buildup by scrubbing handle parts with a brush dipped in vinegar.

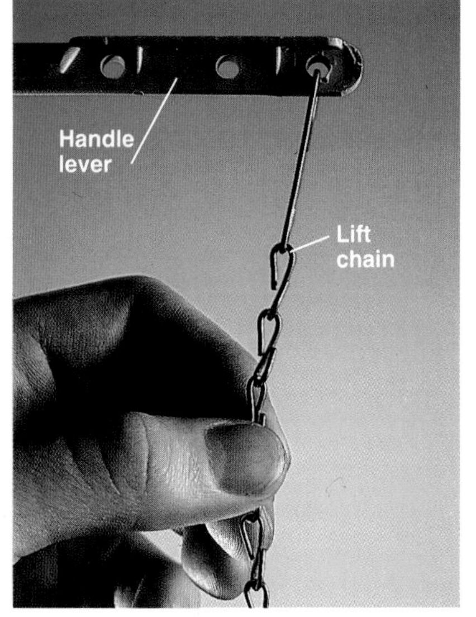

Adjust lift chain so it hangs straight from handle lever, with about ½" of slack. Remove excess slack in chain by hooking the chain in a different hole in the handle lever, or by removing links with needlenose pliers. A broken lift chain must be replaced.

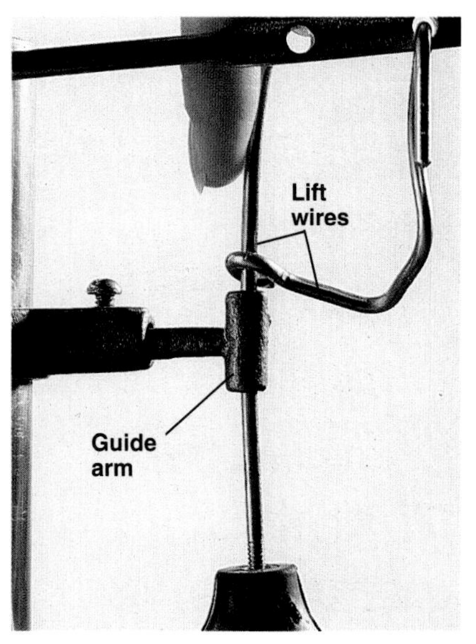

Adjust lift wires (found on toilets without lift chains) so that wires are straight and operate smoothly when handle is pushed. A sticky handle often can be fixed by straightening bent lift wires.

Fixing a Running Toilet

The sound of continuously running water occurs if fresh water continues to enter the toilet tank after the flush cycle is complete. A running toilet can waste 20 or more gallons of fresh water each day.

To fix a running toilet, first jiggle the toilet handle. If the sound of running water stops, then either the handle or the lift wires (or lift chain) need to be adjusted (page opposite).

If the sound of running water does not stop when the handle is jiggled, then remove the tank cover and check to see if the float ball is touching the side of the tank. If necessary, bend the float arm to reposition the float ball away from the side of the tank. Make sure the float ball is not leaking. To check for leaks, unscrew the float ball and shake it gently. If there is water inside the ball, replace it.

If these minor adjustments do not fix the problem, then you will need to adjust or repair the ballcock or the flush valve (photo, right). Follow the directions on the following pages.

Everything You Need:

Tools: screwdriver, small wire brush, sponge, adjustable wrenches, spud wrench or channel-type pliers.

Materials: universal washer kit, ballcock (if needed), ballcock seals, emery cloth, Scotch Brite pad, flapper or tank ball, flush valve (if needed).

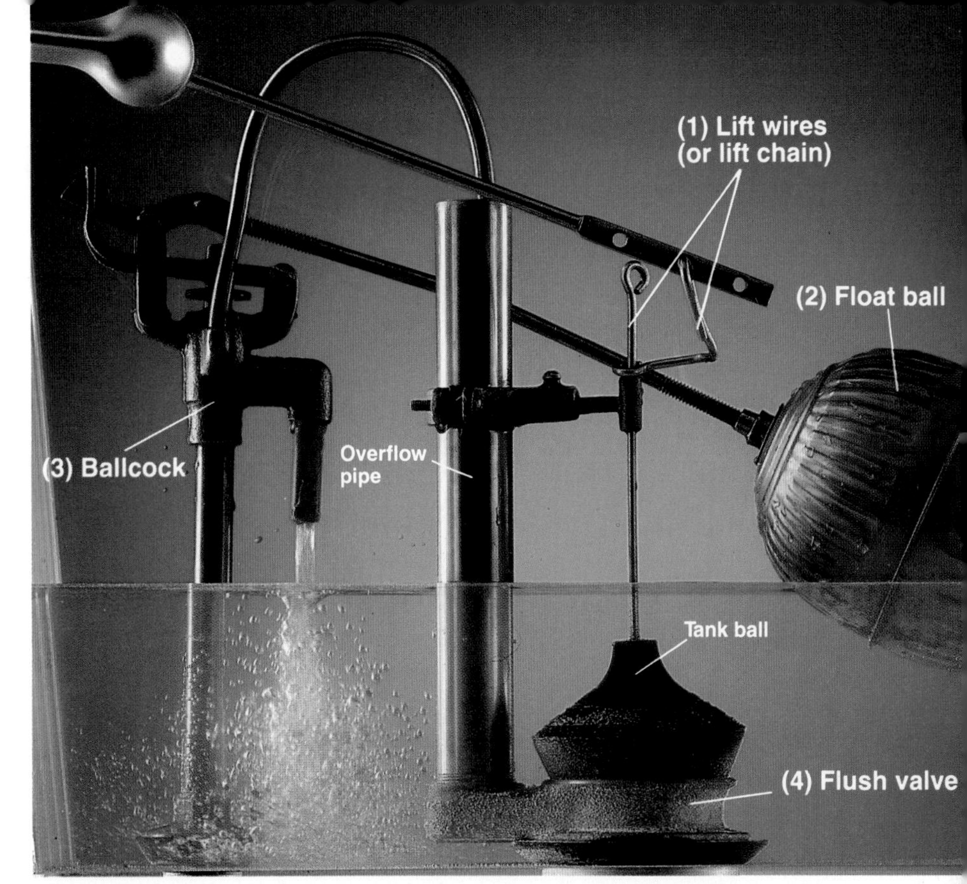

The sound of continuously running water can be caused by several different problems: if the **lift wire (1)** (or lift chain) is bent or kinked; if the **float ball (2)** leaks or rubs against the side of the tank; if a faulty **ballcock (3)** does not shut off the fresh water supply; or if the **flush valve (4)** allows water to leak down into the toilet bowl. First, check the lift wires and float ball. If making simple adjustments and repairs to these parts does not fix the problem, then you will need to repair the ballcock or flush valve (photo, below).

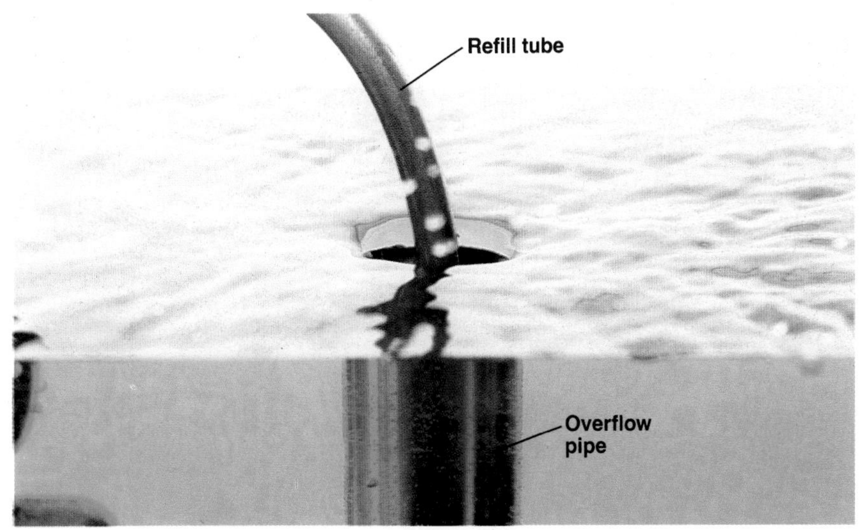

Check the overflow pipe if the sound of running water continues after the float ball and lift wires are adjusted. If you see **water flowing into the overflow pipe,** the ballcock needs to be repaired. First adjust ballcock to lower the water level in the tank (page 214). If problem continues, repair or replace the ballcock (pages 215 to 216). If **water is not flowing into the overflow pipe,** then the flush valve needs to be repaired (page 217). First check the tank ball (or flapper) for wear, and replace if necessary. If problem continues, replace the flush valve.

How to Adjust a Ballcock to Set Water Level

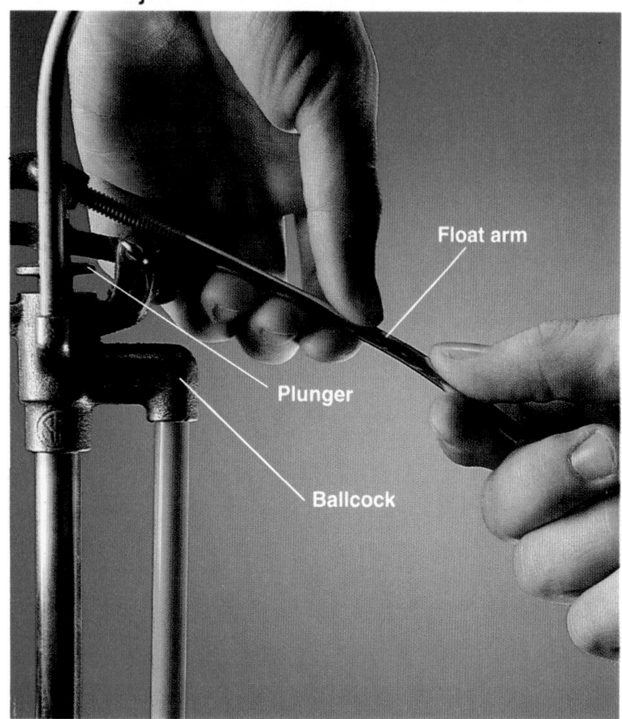

Traditional plunger-valve ballcock is made of brass. Water flow is controlled by a plunger attached to the float arm and ball. Lower the water level by bending the float arm downward slightly. Raise the water level by bending float arm upward.

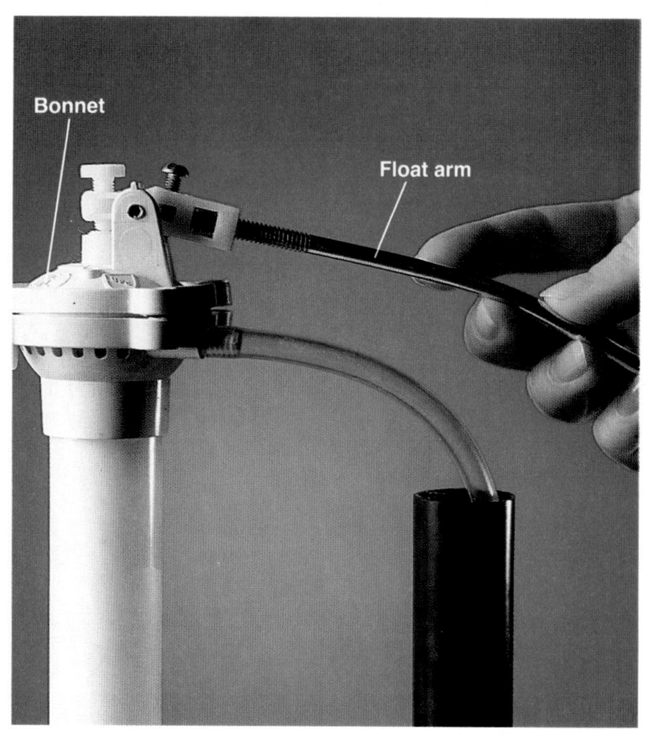

Diaphragm ballcock usually is made of plastic, and has a wide bonnet that contains a rubber diaphragm. Lower the water level by bending the float arm downward slightly. Raise the water level by bending float arm upward.

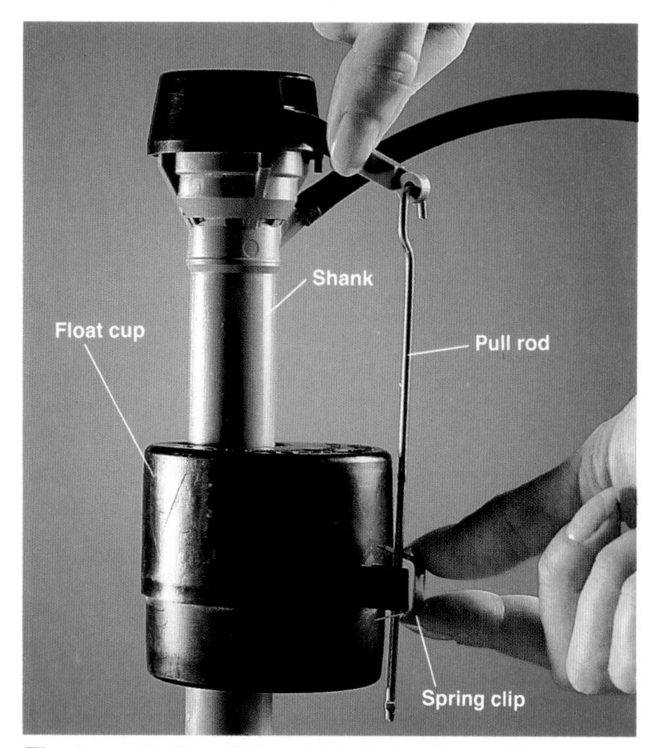

Float cup ballcock is made of plastic, and is easy to adjust. Lower the water level by pinching spring clip on pull rod, and moving float cup downward on the ballcock shank. Raise the water level by moving the cup upward.

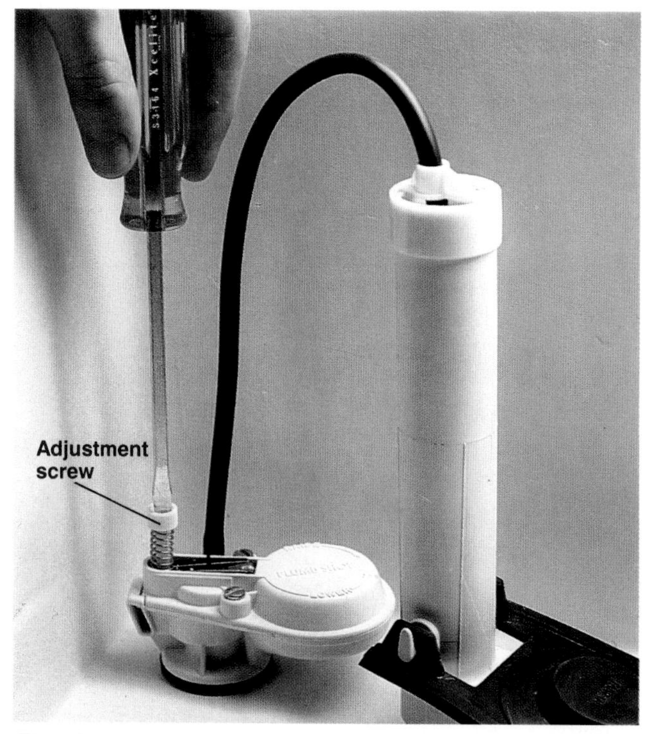

Floatless ballcock controls water level with a pressure-sensing device. Turn the adjustment screw clockwise, 1/2 turn at a time, to raise the water level; counterclockwise to lower it. Note: floatless ballcocks are no longer allowed by Code, and should be replaced.

How to Repair a Plunger-valve Ballcock

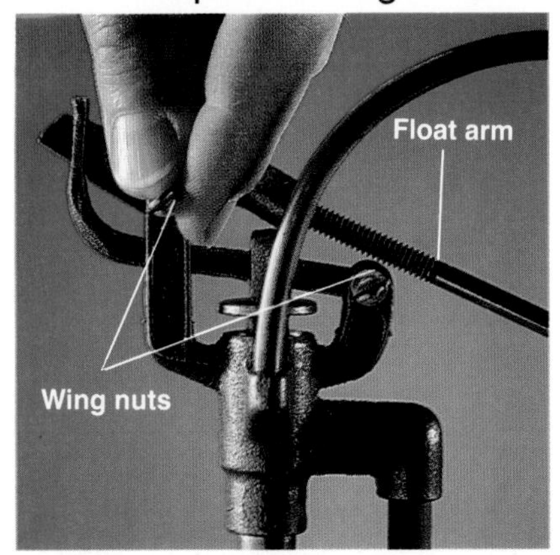

1 Shut off the water, and flush to empty the tank. Remove the wing nuts on the ballcock. Slip out the float arm.

2 Pull up on plunger to remove it. Pry out packing washer or O-ring. Pry out plunger washer. (Remove stem screw, if necessary.)

3 Install replacement washers. Clean sediment from inside of ballcock with a wire brush. Reassemble ballcock.

How to Repair a Diaphragm Ballcock

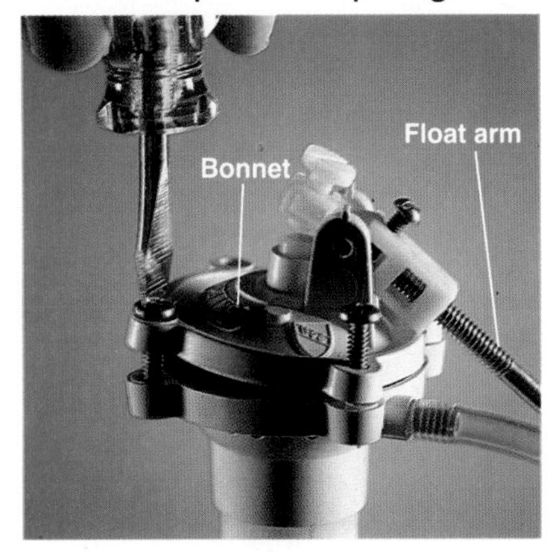

1 Shut off the water, and flush to empty the tank. Remove the screws from the bonnet.

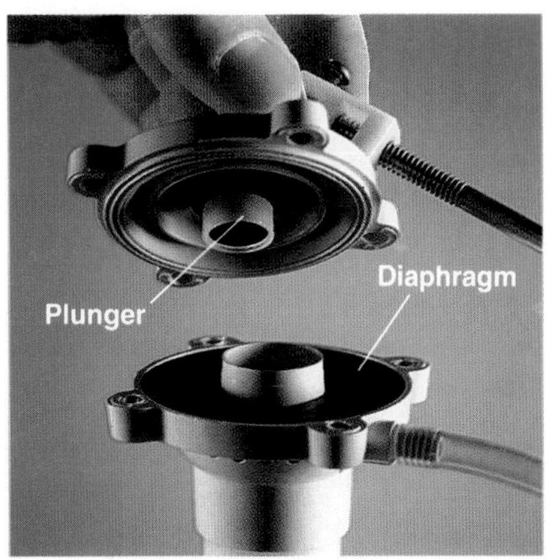

2 Lift off float arm with bonnet attached. Check diaphragm and plunger for wear.

3 Replace any stiff or cracked parts. If assembly is badly worn, replace the entire ballcock (page 216).

How to Repair a Float Cup Ballcock

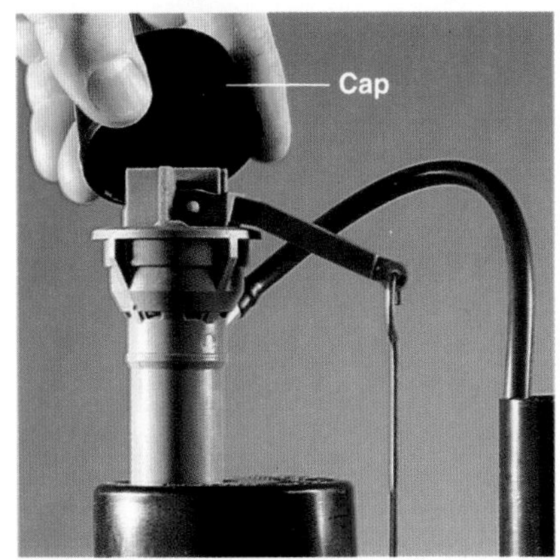

1 Shut off the water, and flush to empty the tank. Remove the ballcock cap.

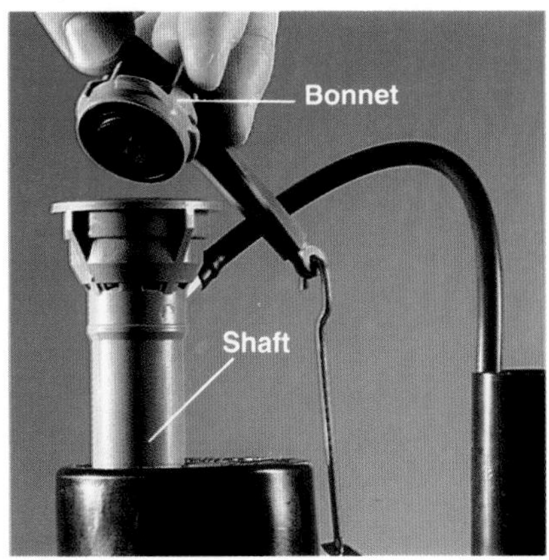

2 Remove bonnet by pushing down on shaft and turning counterclockwise. Clean out sediment inside ballcock with wire brush.

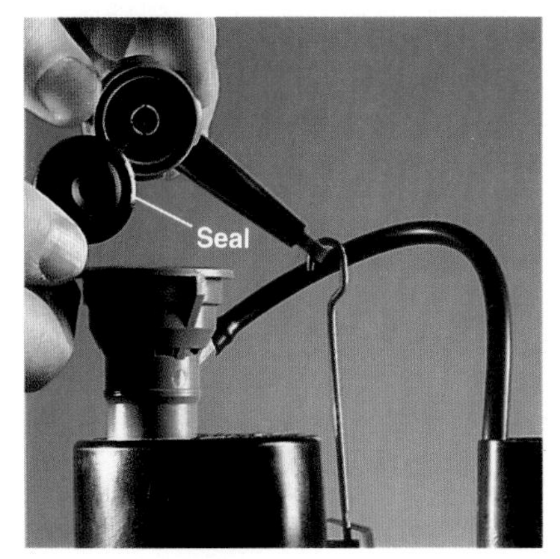

3 Replace the seal. If assembly is badly worn, replace the entire ballcock (page 216).

How to Install a New Ballcock

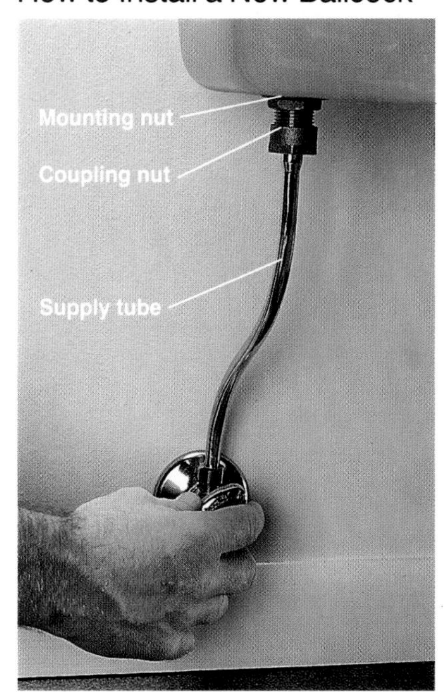

1 Shut off water, and flush toilet to empty tank. Use a sponge to remove remaining water. Disconnect supply tube coupling nut and ballcock mounting nut with adjustable wrench. Remove old ballcock.

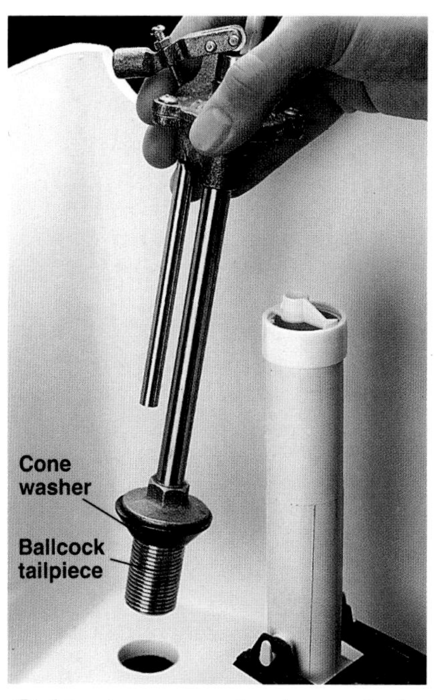

2 Attach cone washer to new ballcock tailpiece and insert tailpiece into tank opening.

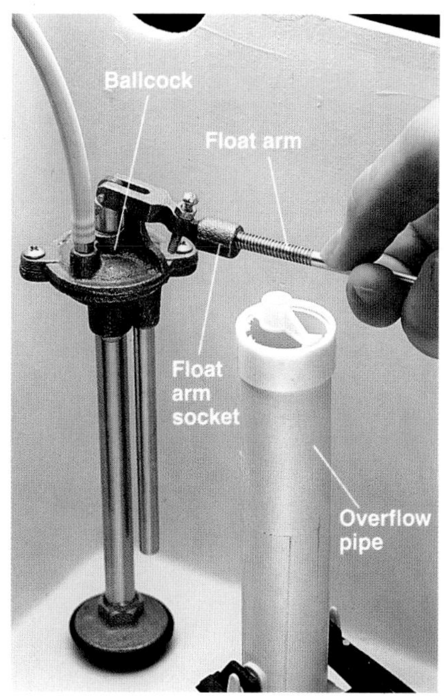

3 Align the float arm socket so that float arm will pass behind overflow pipe. Screw float arm onto ballcock. Screw float ball onto float arm.

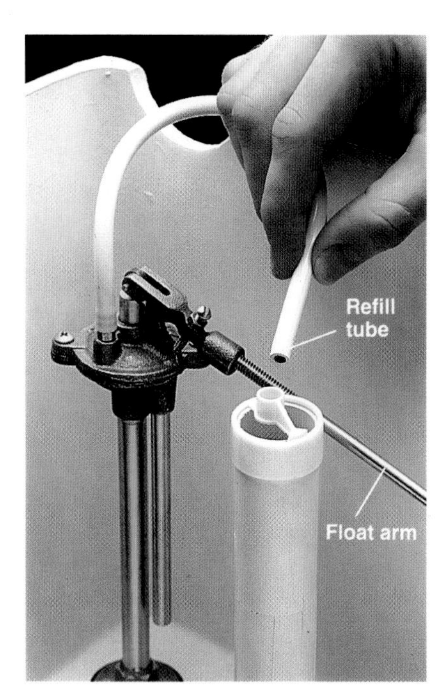

4 Bend or clip refill tube so tip is inside overflow pipe.

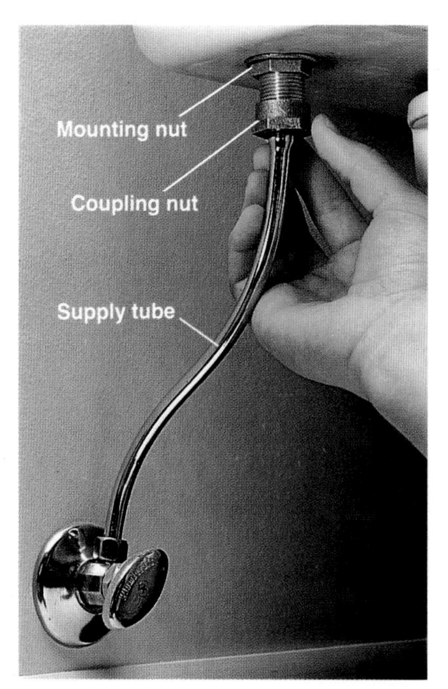

5 Screw mounting nut and supply tube coupling nut onto ballcock tailpiece, and tighten with an adjustable wrench. Turn on the water, and check for leaks.

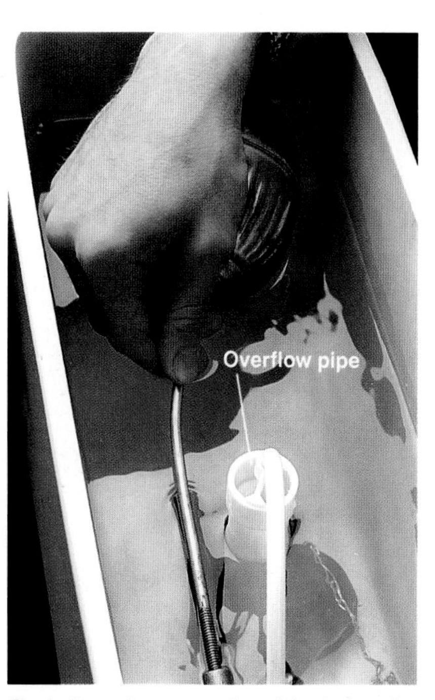

6 Adjust the water level in the tank so it is about 1/2" below top of the overflow pipe (page 214).

How to Adjust & Clean a Flush Valve

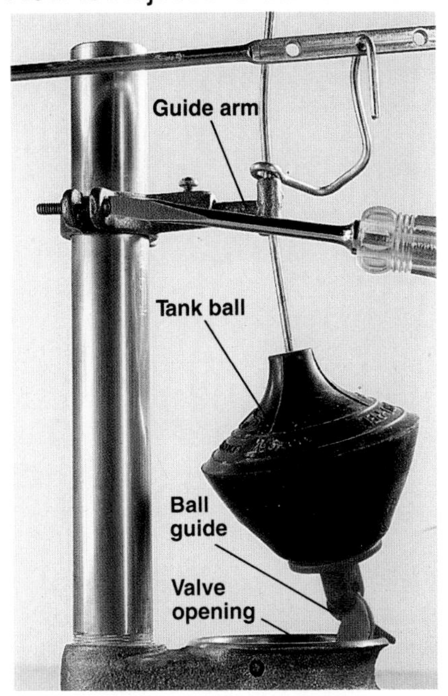

Adjust tank ball (or flapper) so it is directly over flush valve. Tank ball has a guide arm that can be loosened so that tank ball can be repositioned. (Some tank balls have a ball guide that helps seat the tank ball into the flush valve.)

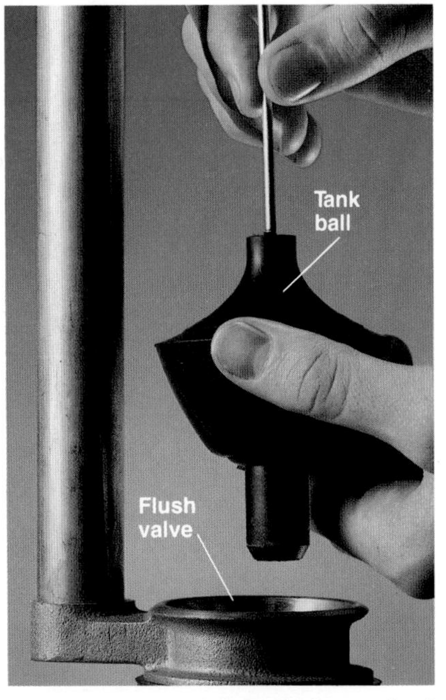

Replace the tank ball if it is cracked or worn. Tank balls have a threaded fitting that screws onto the lift wire. Clean opening of the flush valve, using emery cloth (for brass valves) or a Scotch Brite pad (for plastic valves).

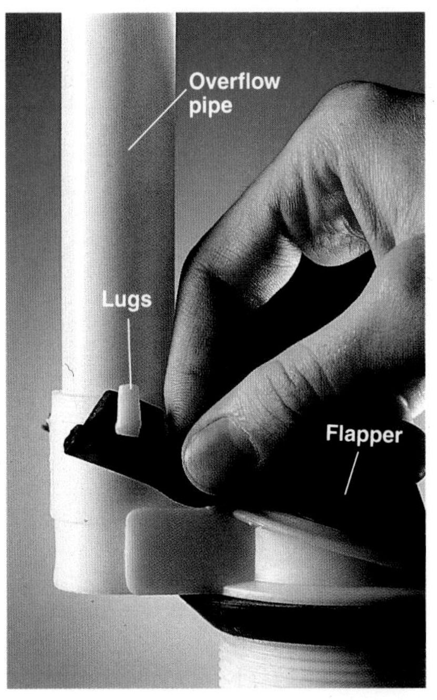

Replace flapper if it is worn. Flappers are attached to small lugs on the sides of overflow pipe.

How to Install a New Flush Valve

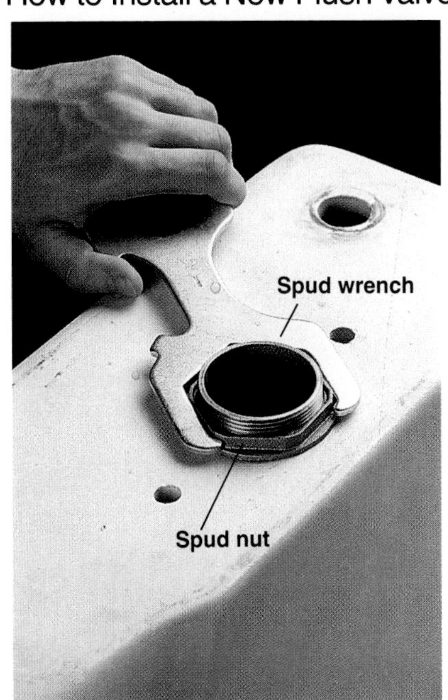

1 Shut off water, disconnect ballcock (page opposite, step 1), and remove toilet tank (page 219, steps 1 and 2). Remove old flush valve by unscrewing spud nut with spud wrench or channel-type pliers.

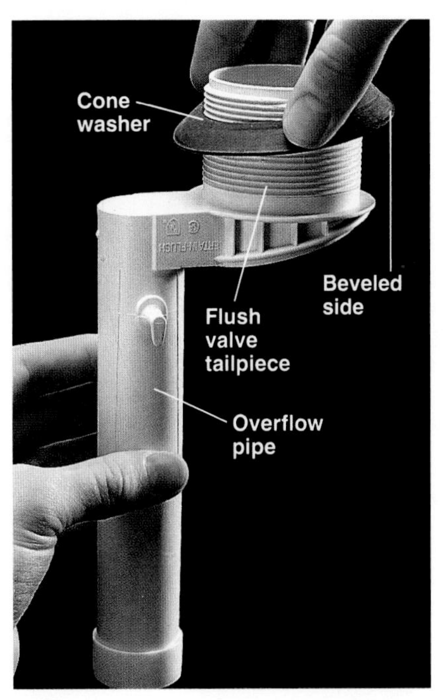

2 Slide cone washer onto tailpiece of new flush valve. Beveled side of cone washer should face end of tailpiece. Insert flush valve into tank opening so that overflow pipe faces ballcock.

3 Screw spud nut onto tailpiece of flush valve, and tighten with a spud wrench or channel-type pliers. Place soft spud washer over tailpiece, and reinstall toilet tank (pages 220 to 221).

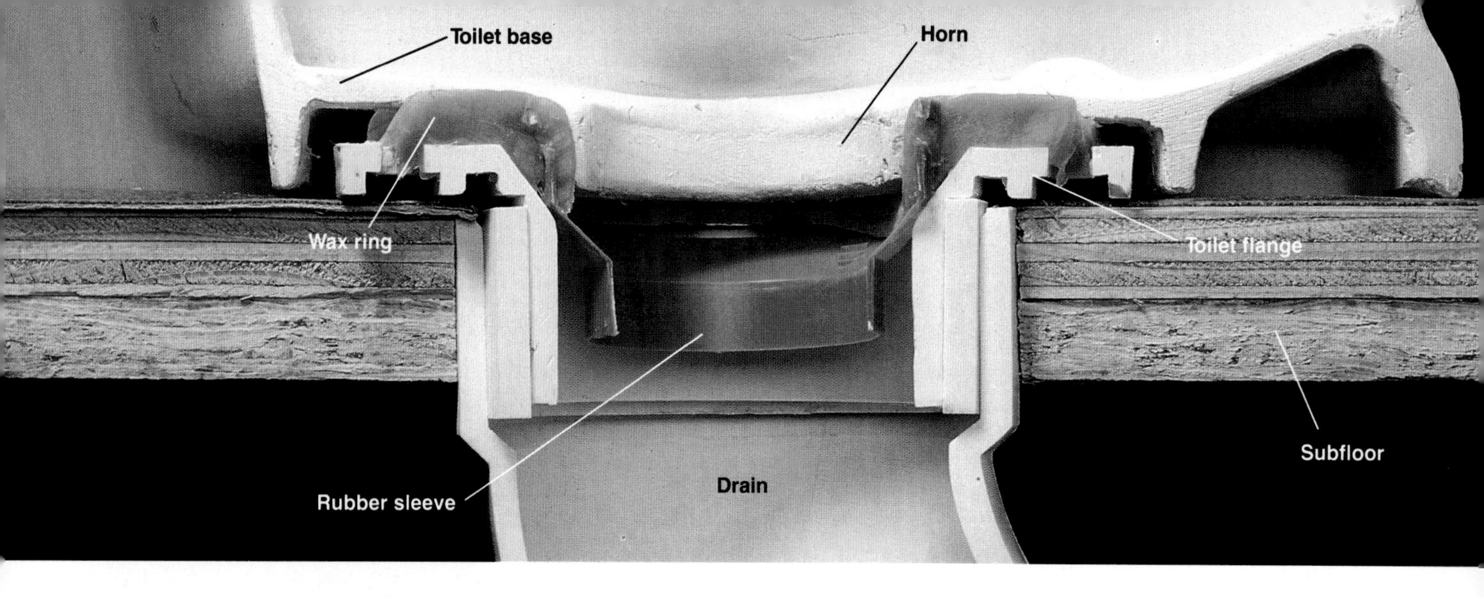

Fixing a Leaking Toilet

Water leaking onto the floor around a toilet may be caused by several different problems. The leaking must be fixed as soon as possible to prevent moisture from damaging the subfloor.

First, make sure all connections are tight. If moisture drips from the tank during humid weather, it is probably condensation. Fix this "sweating" problem by insulating the inside of the tank with foam panels. A crack in a toilet tank also can cause leaks. A cracked tank must be replaced.

Water seeping around the base of a toilet can be caused by an old wax ring that no longer seals against the drain (photo, above), or by a cracked toilet base. If leaking occurs during or just after a flush, replace the wax ring. If leaking is constant, the toilet base is cracked and must be replaced.

New toilets sometimes are sold with flush valves and ballcocks already installed. If these parts are not included, you will need to purchase them. When buying a new toilet, consider a water-saver design. Water-saver toilets use less than half the water needed by a standard toilet.

Everything You Need:

Tools: sponge, adjustable wrench, putty knife, ratchet wrench, screwdriver.

Materials: tank liner kit, abrasive cleanser, rag, wax ring, plumber's putty. *For new installation:* new toilet, toilet handle, ballcock, flush valve, tank bolts, toilet seat.

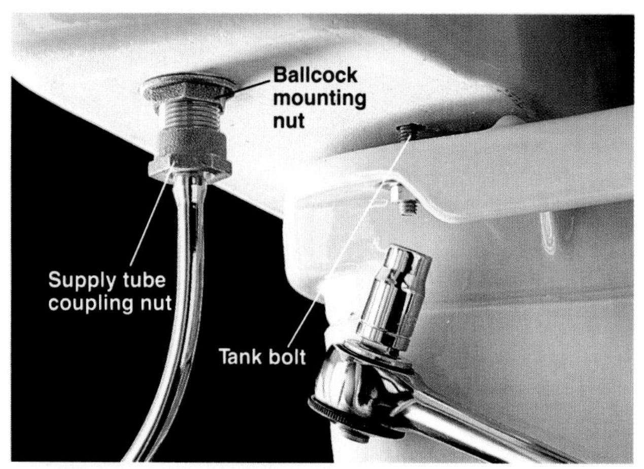

Tighten all connections slightly. Tighten nuts on tank bolts with a ratchet wrench. Tighten ballcock mounting nut and supply tube coupling nut with an adjustable wrench. **Caution: overtightening tank bolts may crack the toilet tank.**

Insulate toilet tank to prevent "sweating," using a toilet liner kit. First, shut off water, drain tank, and clean inside of tank with abrasive cleanser. Cut plastic foam panels to fit bottom, sides, front, and back of tank. Attach panels to tank with adhesive (included in kit). Let adhesive cure as directed.

How to Remove & Replace a Wax Ring & Toilet

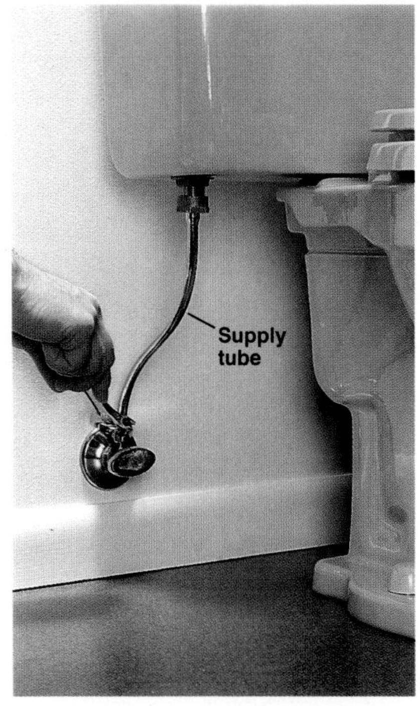

1 Turn off water, and flush to empty toilet tank. Use a sponge to remove remaining water in tank and bowl. Disconnect supply tube with an adjustable wrench.

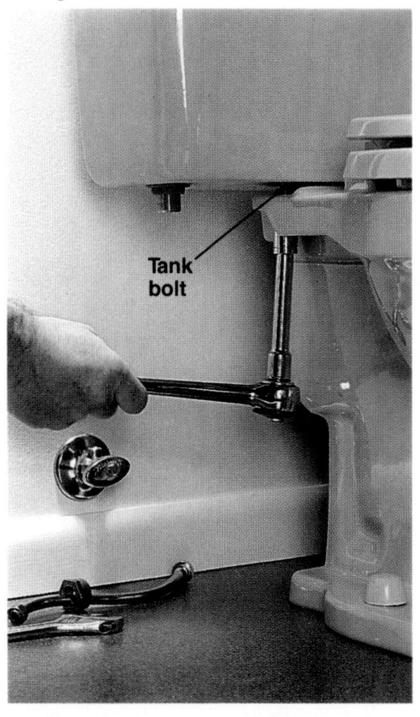

2 Remove the nuts from the tank bolts with a ratchet wrench. Carefully remove the tank and set it aside.

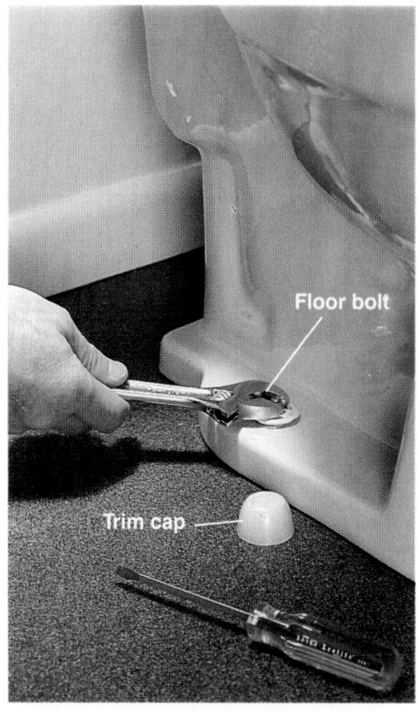

3 Pry off the floor bolt trim caps at the base of the toilet. Remove the floor nuts with an adjustable wrench.

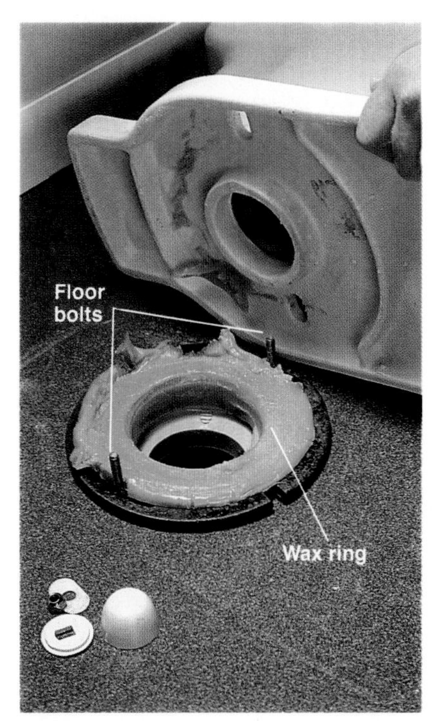

4 Straddle the toilet and rock the bowl from side to side until the seal breaks. Carefully lift the toilet off the floor bolts and set it on its side. Small amount of water may spill from the toilet trap.

5 Remove old wax from the toilet flange in the floor. Plug the drain opening with a damp rag to prevent sewer gases from rising into the house.

6 If old toilet will be reused, clean old wax and putty from the horn and the base of the toilet.

(continued next page)

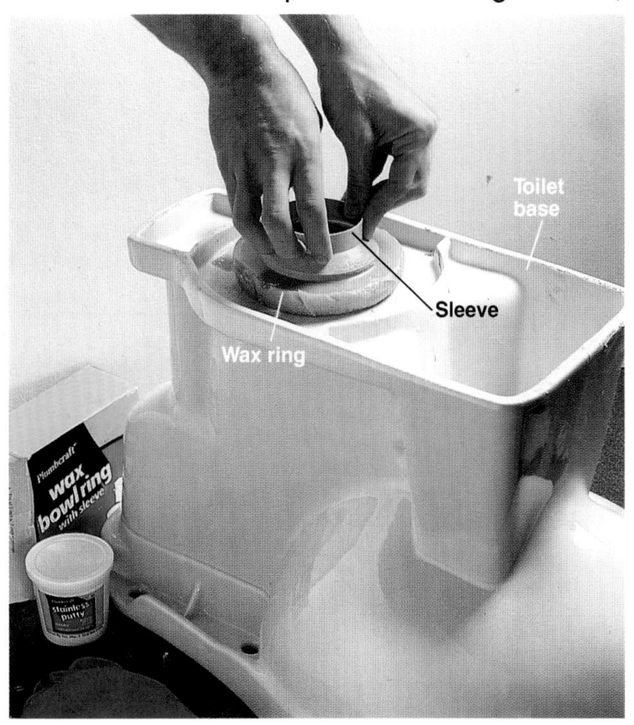

7 Turn stool upside down. Place new wax ring over drain horn. If ring has a rubber or plastic sleeve, sleeve should face away from toilet. Apply a bead of plumber's putty to bottom edge of toilet base.

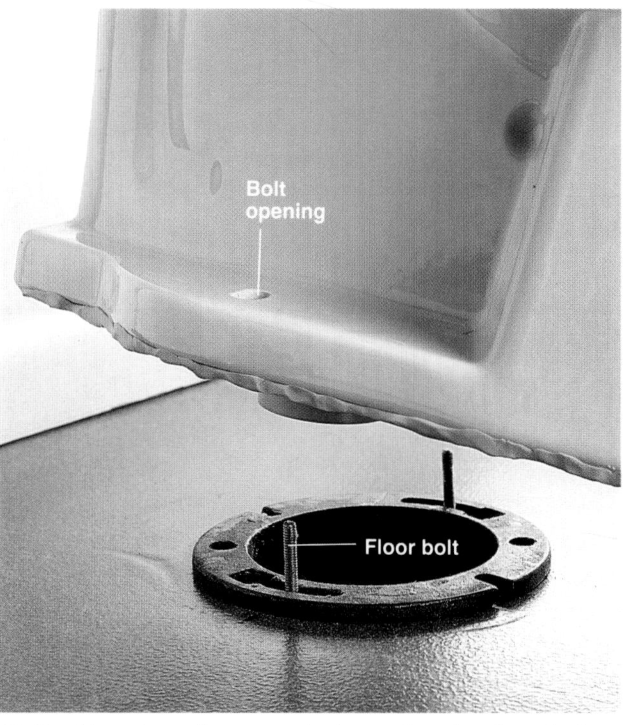

8 Position the toilet over drain so that the floor bolts fit through the openings in the base of the toilet.

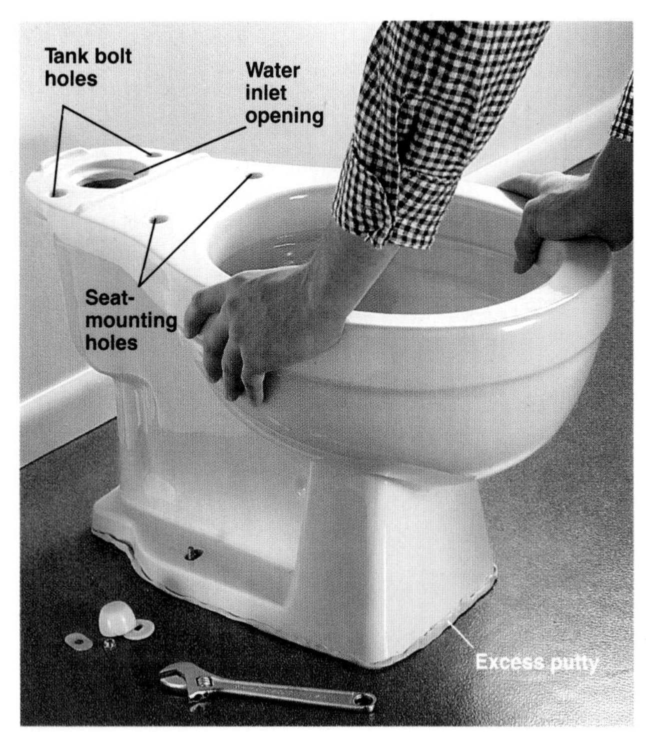

9 Press down on toilet base to compress wax and putty. Thread washers and nuts onto floor bolts, and tighten with adjustable wrench until snug. **Caution: overtightening nuts may crack the base.** Wipe away excess plumber's putty. Cover nuts with trim caps.

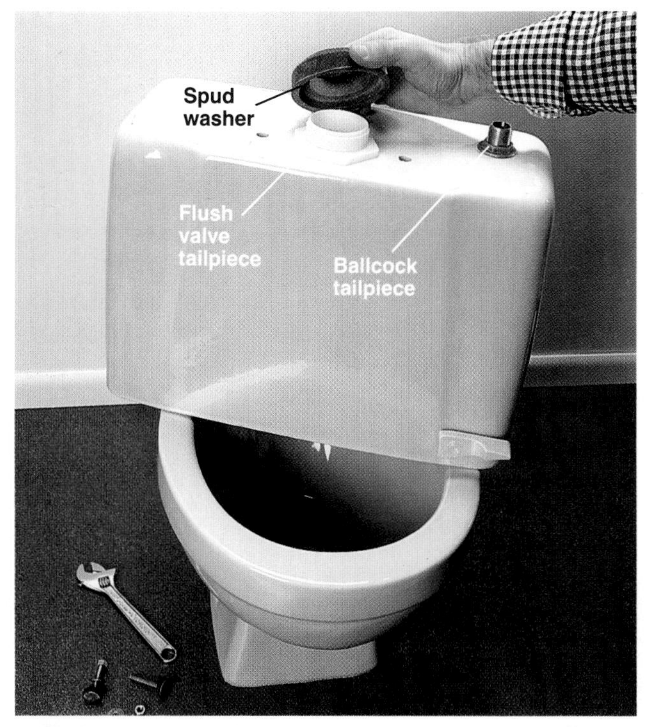

10 Prepare tank for installation. If necessary, install a handle (page 212), ballcock (page 216), and flush valve (page 217). Carefully turn tank upside down, and place soft spud washer over the flush valve tailpiece.

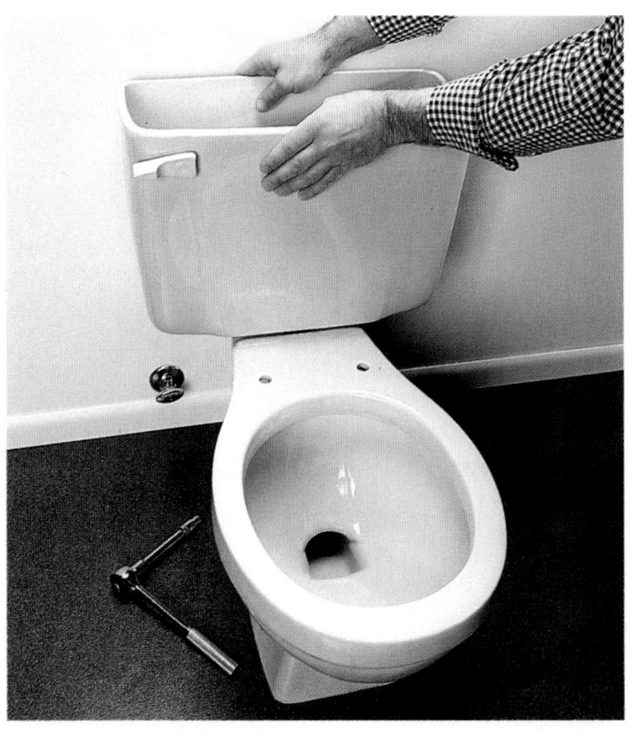

11 Turn tank right side up and position it on rear of toilet base so that spud washer is centered in water inlet opening.

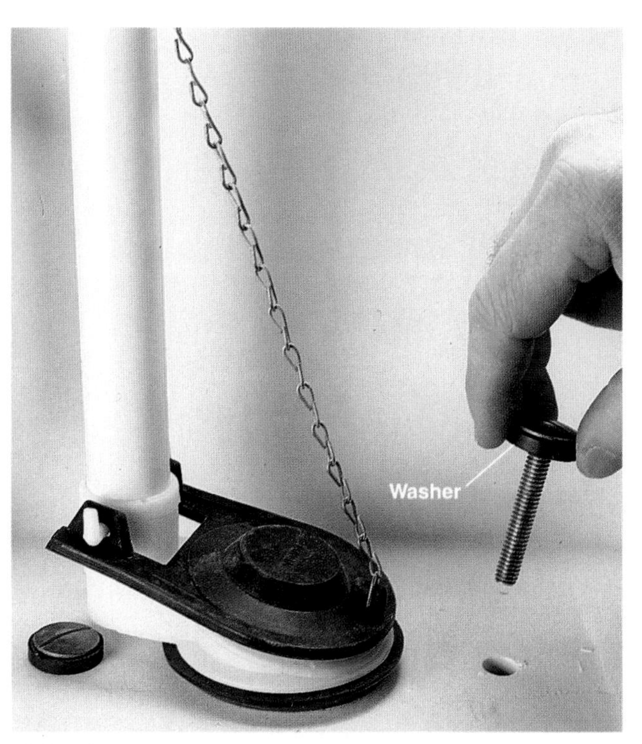

12 Line up the tank bolt holes with holes in base of toilet. Slide rubber washers onto the tank bolts and place the bolts through holes. From underneath the tank, thread washers and nuts onto the bolts.

13 Tighten nuts with ratchet wrench until tank is snug. Use caution when tightening nuts: most toilet tanks rest on the spud washer, not directly on the toilet base.

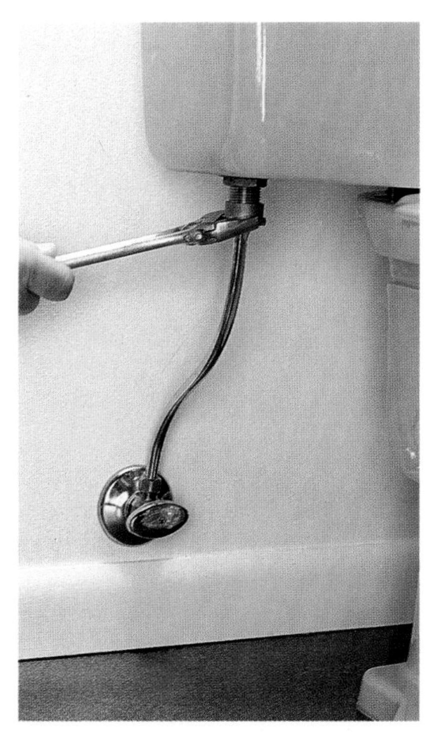

14 Attach the water supply tube to the ballcock tailpiece with an adjustable wrench (page 216). Turn on the water and test toilet. Tighten tank bolts and water connections, if necessary.

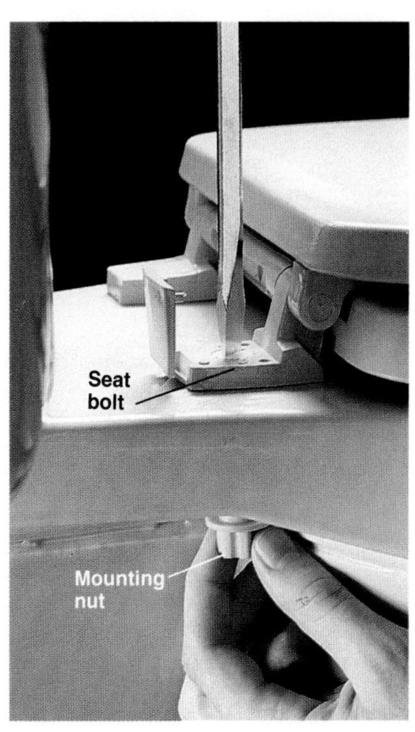

15 Position the new toilet seat, if needed, inserting seat bolts into mounting holes in toilet. Screw mounting nuts onto the seat bolts, and tighten.

Clearing Clogs & Fixing Drains

Clear a clogged drain with a plunger, hand auger, or blow bag. A plunger breaks up clogs by forcing air pressure into the drain line. Because a plunger is effective and simple to use, it should be the first choice for clearing a clog.

A hand auger has a flexible steel cable that is pushed into the drain line to break up or remove obstructions. An auger is easy to use, but for best results the user must know the "feel" of the cable in the drain line. A little experience often is necessary to tell the difference between a soap clog and a bend in the drain line (pages 226 to 227).

A blow bag hooks to a garden hose and uses water pressure to clear clogs. Blow bags are most effective on clogs in floor drains (page 233).

Use caustic, acid-based chemical drain cleaners only as a last resort. These drain cleaners, usually available at hardware stores and supermarkets, will dissolve clogs, but they also may damage pipes and must be handled with caution. Always read the manufacturer's directions completely.

Regular maintenance helps keep drains working properly. Flush drains once each week with hot tap water to keep them free of soap, grease, and debris. Or, treat drains once every six months with a non-caustic (copper sulfide- or sodium hydroxide-based) drain cleaner. A non-caustic cleaner will not harm pipes.

Occasionally, leaks may occur in the drain lines or around the drain opening. Most leaks in drain lines are fixed easily by gently tightening all pipe connections. If the leak is at the sink drain opening, fix or replace the strainer body assembly (page 225).

Clearing Clogged Sinks

Every sink has a drain trap and a fixture drain line. Sink clogs usually are caused by a buildup of soap and hair in the trap or fixture drain line. Remove clogs by using a plunger, disconnecting and cleaning the trap (page 224), or using a hand auger (pages 226 to 227).

Many sinks hold water with a mechanical plug called a *pop-up stopper.* If the sink will not hold standing water, or if water in the sink drains too slowly, the pop-up stopper must be cleaned and adjusted (page 224).

Everything You Need:

Tools: plunger, channel-type pliers, small wire brush, screwdriver.

Materials: rag, bucket, replacement gaskets.

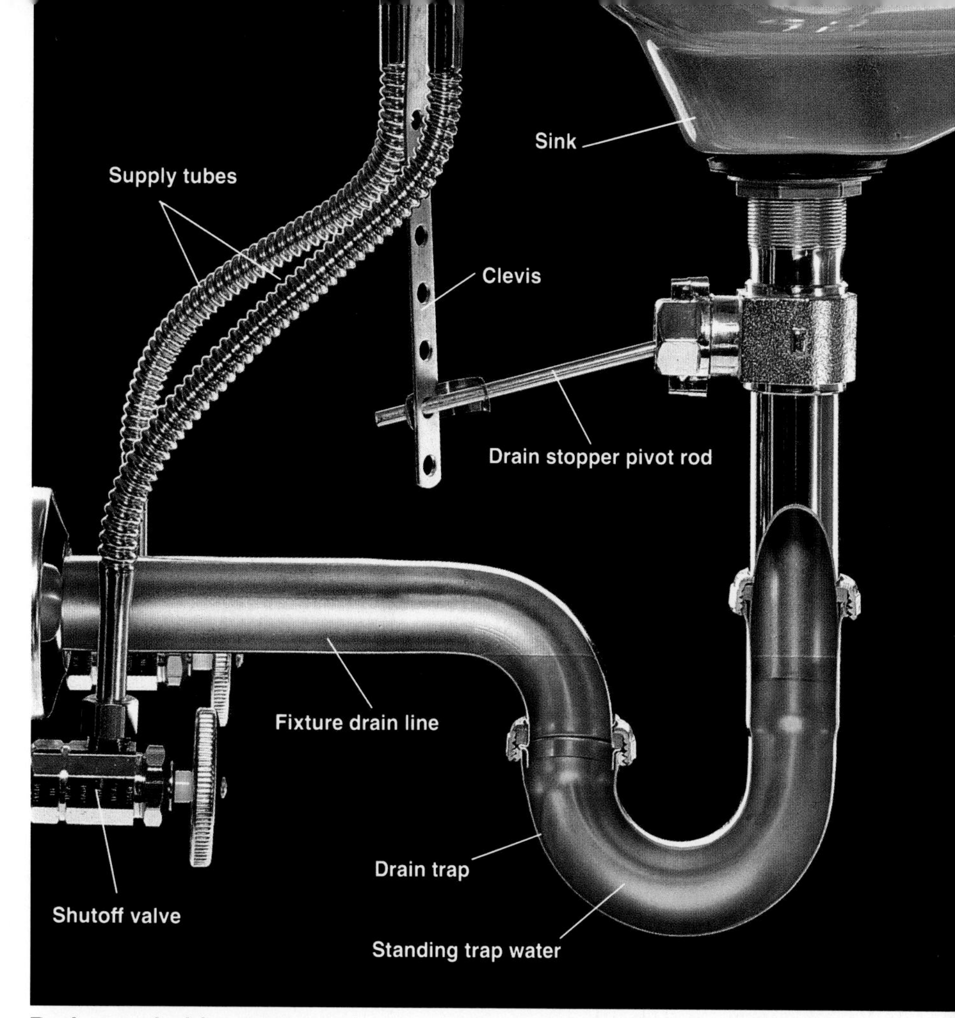

Drain trap holds water that seals the drain line and prevents sewer gases from entering the home. Each time a drain is used, the standing trap water is flushed away and replaced by new water. The shape of the trap and fixture drain line may resemble the letter "P," and sink traps sometimes are called P-traps.

How to Clear Sink Drains with a Plunger

1 Remove drain stopper. Some pop-up stoppers lift out directly; others turn counterclockwise. On some older types of stoppers, the pivot rod must be removed to free the stopper.

2 Stuff a wet rag in sink overflow opening. Rag prevents air from breaking the suction of the plunger. Place plunger cup over drain and run enough water to cover the rubber cup. Move plunger handle up and down rapidly to break up the clog.

How to Clean & Adjust a Pop-up Sink Drain Stopper

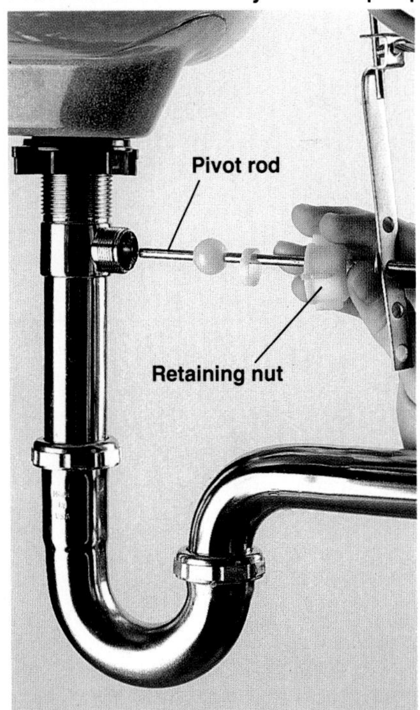

1 Raise stopper lever to full upright (closed) position. Unscrew the retaining nut that holds pivot rod in position. Pull pivot rod out of drain pipe to release stopper.

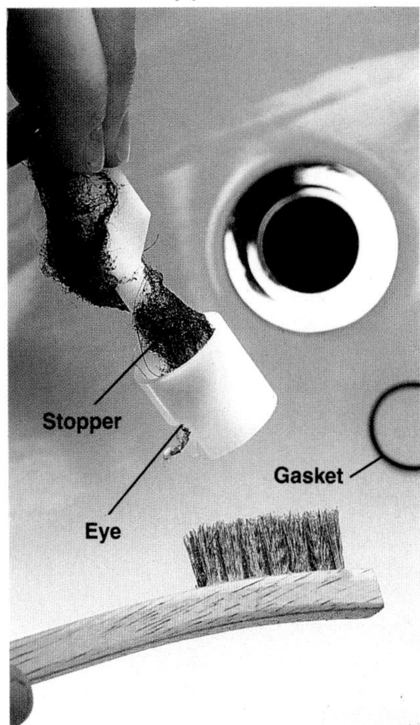

2 Remove stopper. Clean debris from stopper, using a small wire brush. Inspect gasket for wear or damage, and replace if necessary. Reinstall stopper.

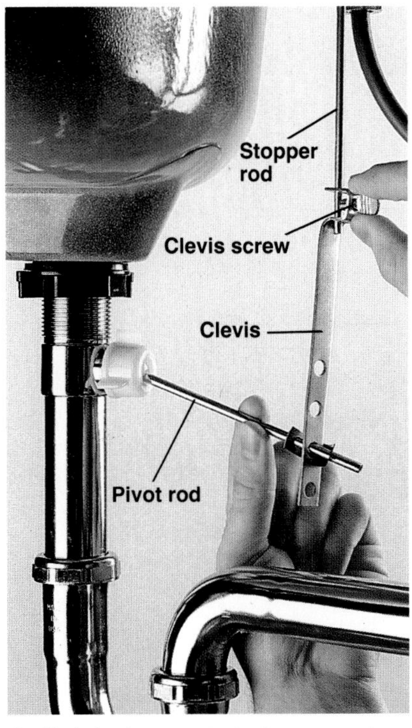

3 If sink does not drain properly, adjust clevis. Loosen clevis screw. Slide clevis up or down on stopper rod to adjust position of stopper. Tighten clevis screw.

How to Remove & Clean a Sink Drain Trap

1 Place bucket under trap to catch water and debris. Loosen slip nuts on trap bend with channel-type pliers. Unscrew nuts by hand and slide away from connections. Pull off trap bend.

2 Dump out debris. Clean trap bend with a small wire brush. Inspect slip nut washers for wear, and replace if necessary. Reinstall trap bend, and tighten slip nuts.

Fixing Leaky Sink Strainers

A leak under a sink may be caused by a strainer body that is not properly sealed to the sink drain opening. To check for leaks, close the drain stopper and fill sink with water. From underneath sink, inspect the strainer assembly for leaks.

Remove the strainer body, clean it, and replace the gaskets and plumber's putty. Or, replace the strainer with a new one, available at home centers.

Everything You Need:

Tools: channel-type pliers, spud wrench, hammer, putty knife.

Materials: plumber's putty, replacement parts (if needed).

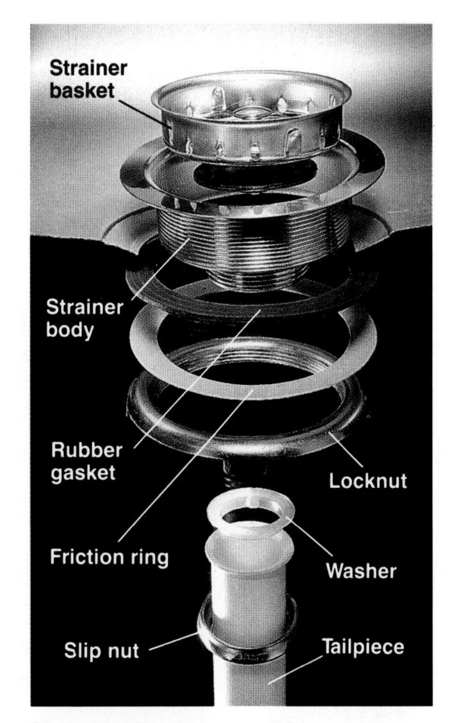

Sink strainer assembly connects the sink to the drain line. Leaks may occur where the strainer body seals against the lip of the drain opening.

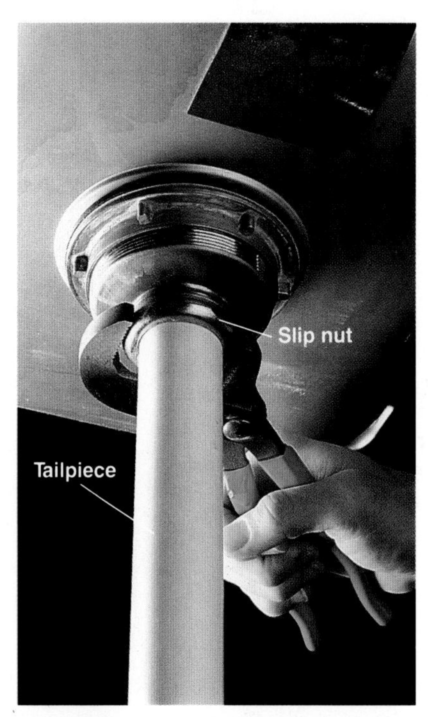

1 Unscrew slip nuts from both ends of tailpiece, using channel-type pliers. Disconnect tailpiece from strainer body and trap bend. Remove tailpiece.

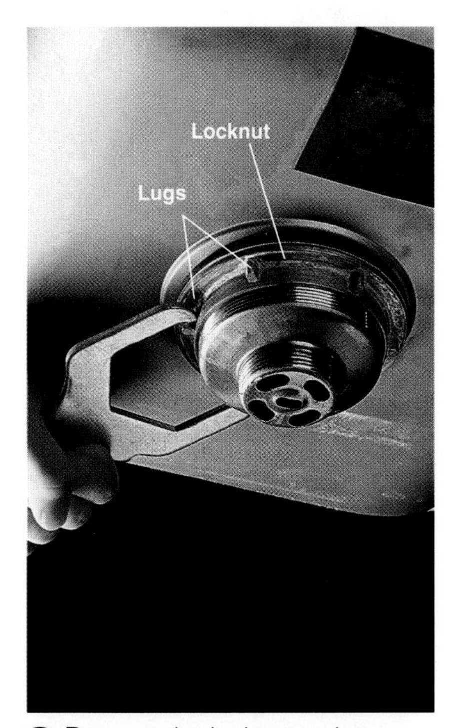

2 Remove the locknut, using a spud wrench. Stubborn locknuts may be removed by tapping on the lugs with a hammer. Unscrew the locknut completely, and remove the strainer assembly. If necessary, cut the locknut.

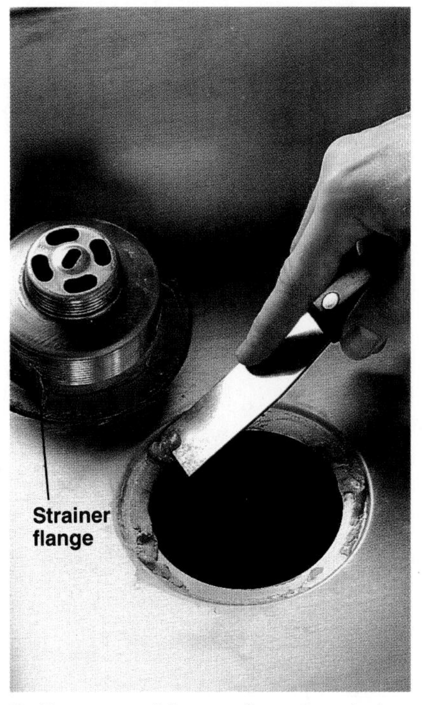

3 Remove old putty from the drain opening, using a putty knife. If reusing the old strainer body, clean off old putty from under the flange. Old gaskets and washers should be replaced.

4 Apply a bead of plumber's putty to the lip of the drain opening. Press strainer body into drain opening. From under the sink, place rubber gasket, then metal or fiber friction ring, over strainer. Reinstall locknut and tighten. Reinstall tailpiece.

How to Clear a Fixture Drain Line with a Hand Auger

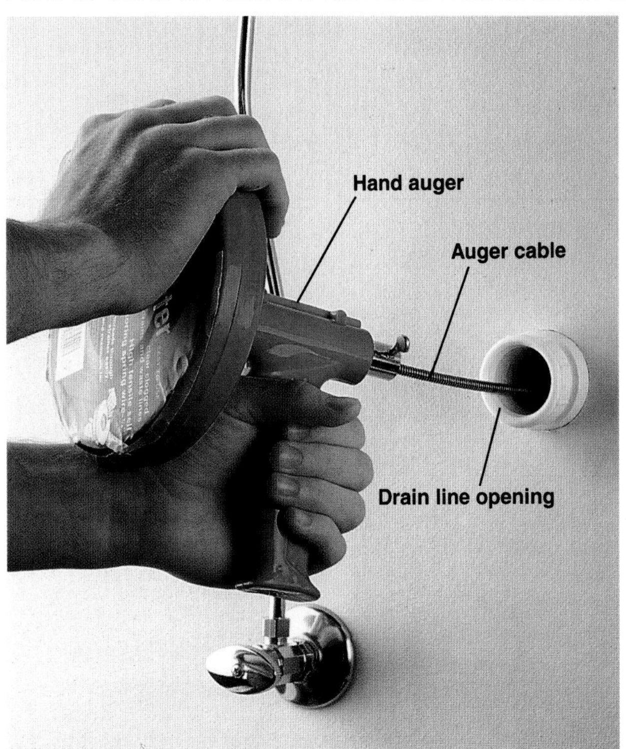

1 Remove trap bend (page 224). Push the end of the auger cable into the drain line opening until resistance is met. This resistance usually indicates end of cable has reached a bend in the drain pipe.

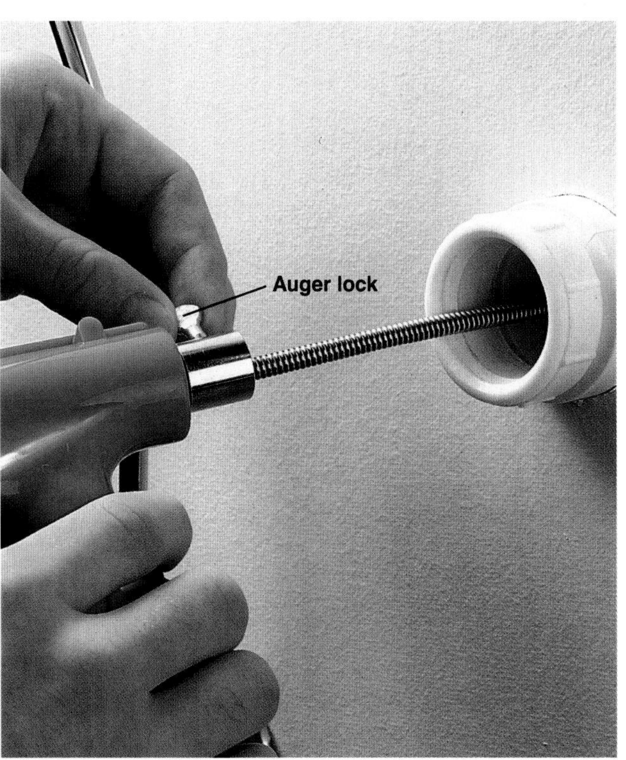

2 Set the auger lock so that at least 6" of cable extends out of the opening. Crank the auger handle in a clockwise direction to move the end of the cable past bend in drain line.

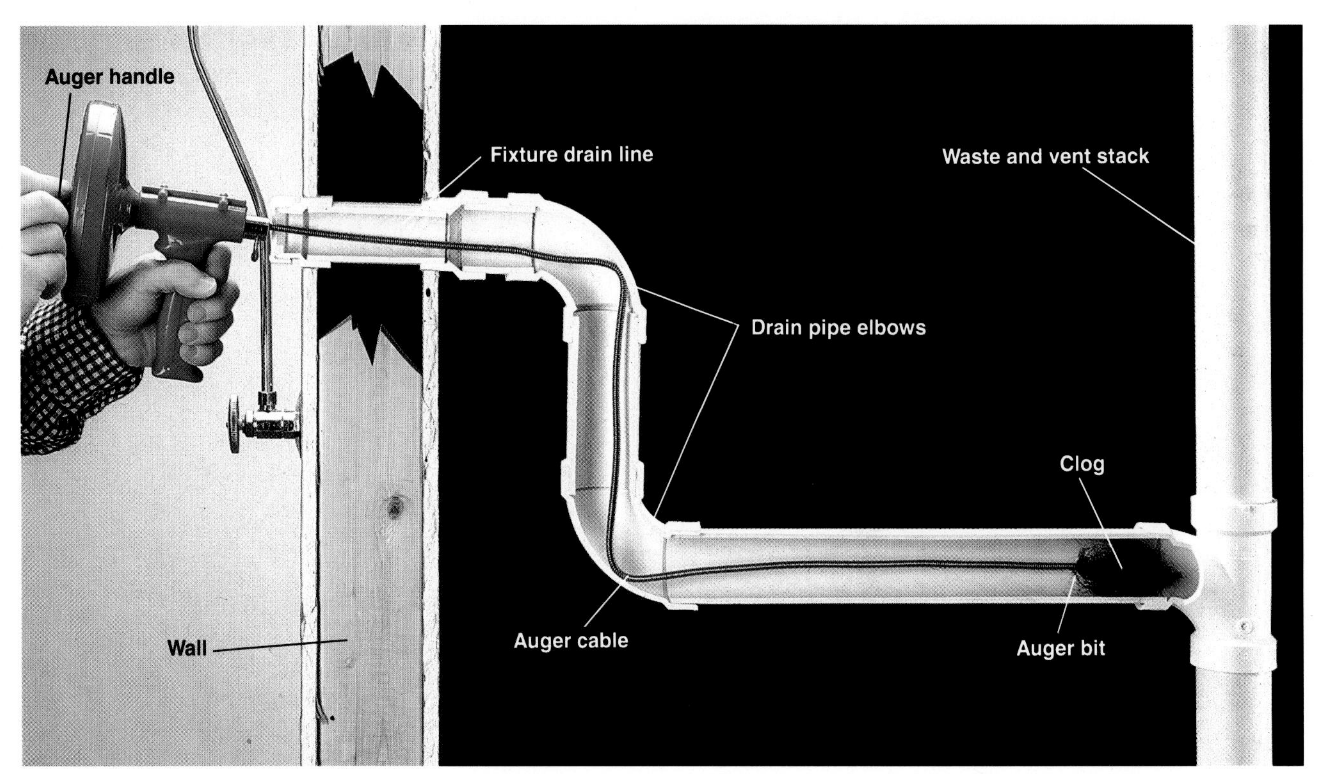

3 Release the lock and continue pushing the cable into the opening until firm resistance is felt. Set the auger lock and crank the handle in a clockwise direction. Solid resistance that prevents the cable from advancing indicates a clog. Some clogs, such as a sponge or an accumulation of hair, can be snagged and retrieved (step 4). Continuous resistance that allows the cable to advance slowly is probably a soap clog (step 5).

4 Pull an obstruction out of the line by releasing the auger lock and cranking the handle clockwise. If no object can be retrieved, reconnect the trap bend and use the auger to clear the nearest branch drain line or main waste and vent stack (pages 234 to 235).

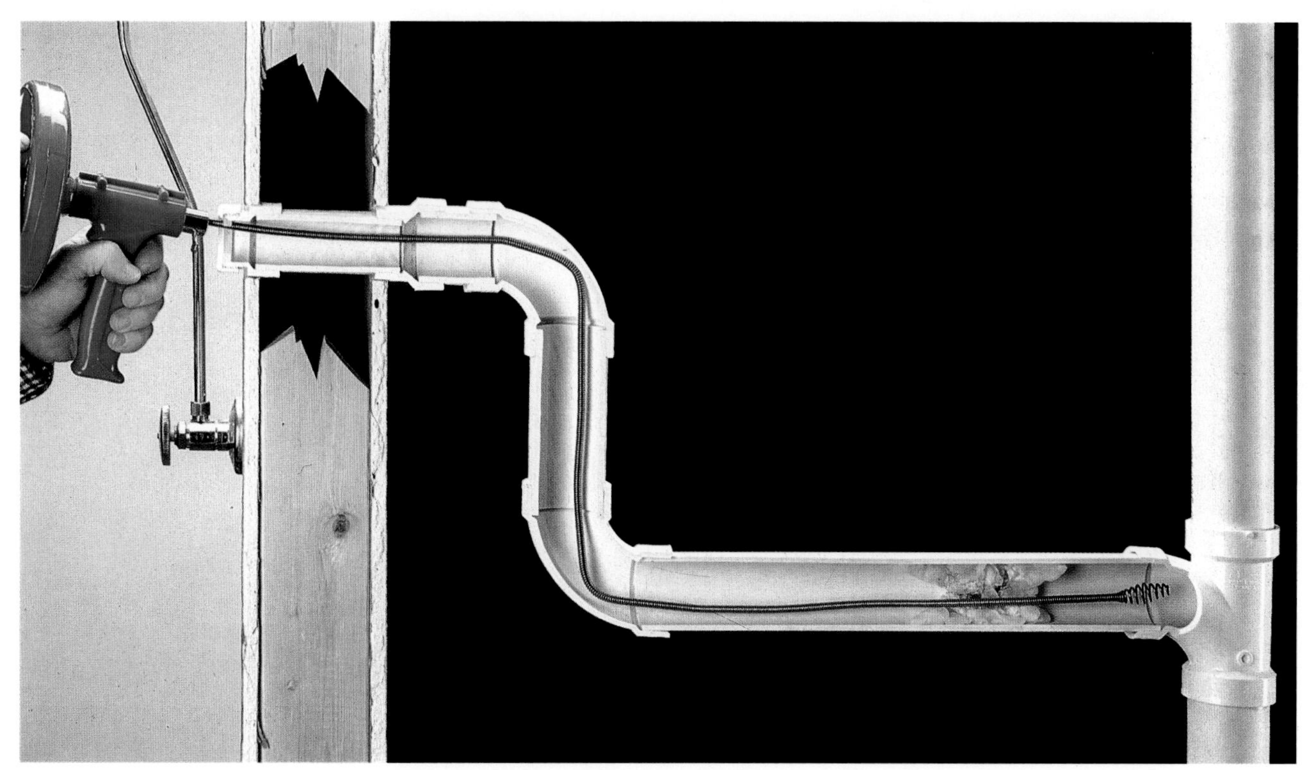

5 Continuous resistance indicates a soap clog. Bore through the clog by cranking the auger handle clockwise while applying steady pressure on the hand grip of the auger. Repeat the procedure two or three times, then retrieve the cable. Reconnect the trap bend and flush the system with hot tap water to remove debris.

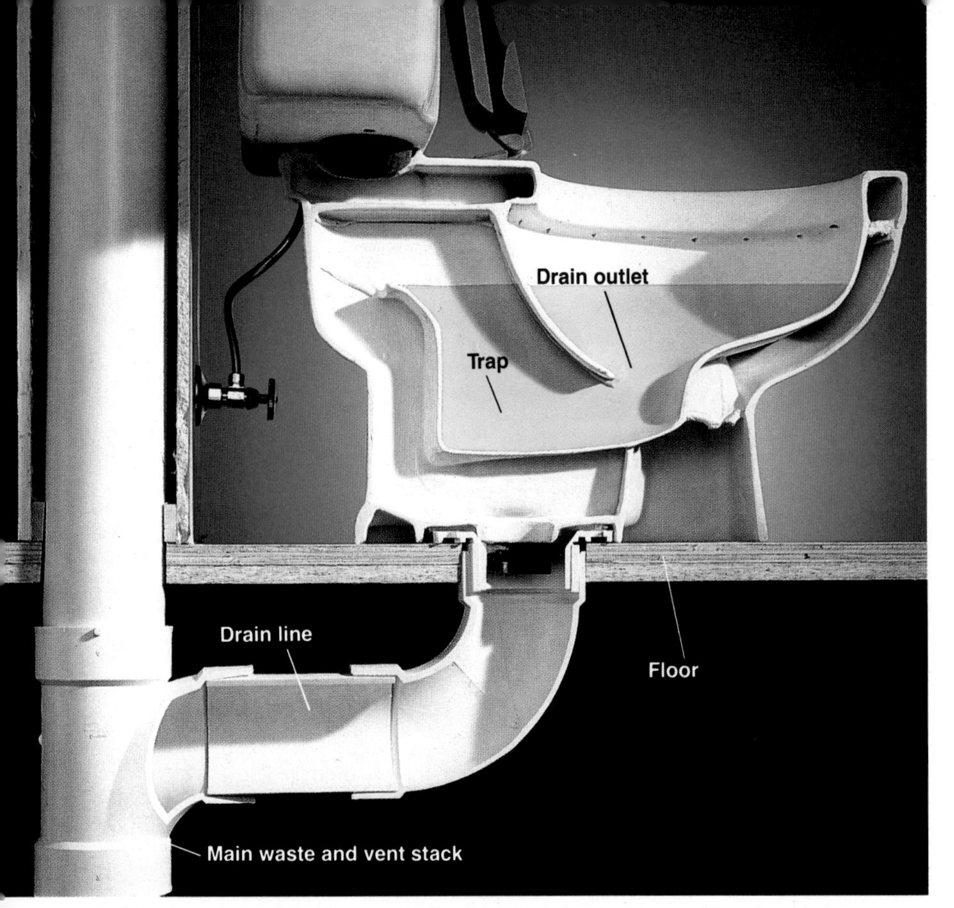

Toilet drain system has a drain outlet at the bottom of the bowl and a built-in trap. The toilet drain is connected to a drain line and a main waste and vent stack.

Clearing Clogged Toilets

Most toilet clogs occur because an object is stuck inside the toilet trap. Use a flanged plunger or a closet auger to remove the clog.

A toilet that is sluggish during the flush cycle may be partially blocked. Clear the blockage with a plunger or closet auger. Occasionally, a sluggish toilet flush indicates a blocked waste and vent stack. Clear the stack as shown on page 235.

Everything You Need:

Tools: flanged plunger, closet auger.

Materials: bucket.

How to Clear a Toilet with a Plunger

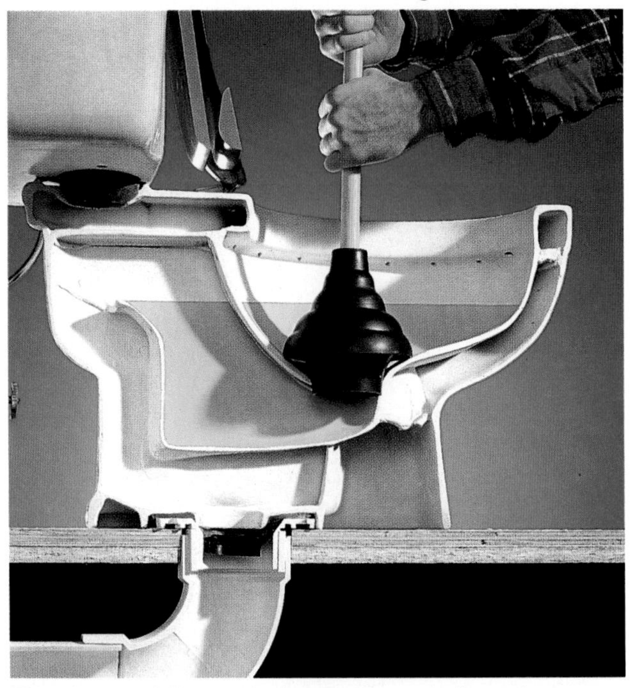

Place cup of flanged plunger over drain outlet opening. Plunge up and down rapidly. Slowly pour a bucket of water into bowl to flush debris through drain. If toilet does not drain, repeat plunging, or clear clog with a closet auger.

How to Clear a Toilet with a Closet Auger

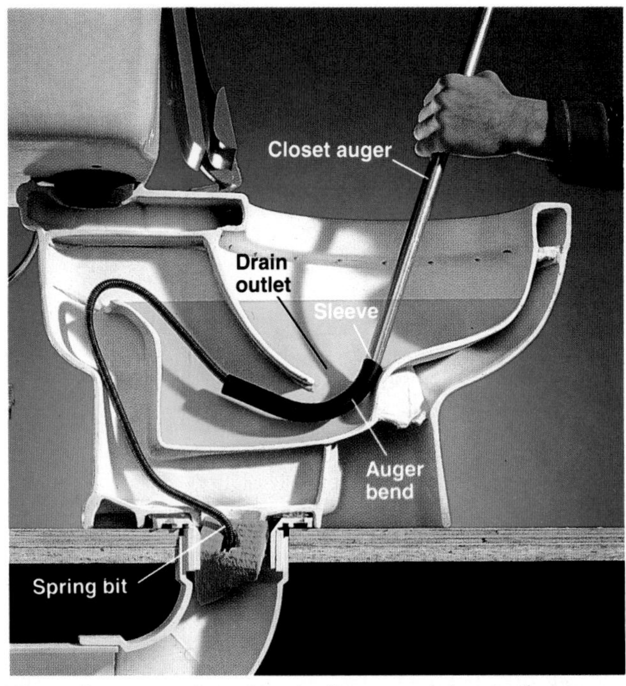

Place the auger bend in the bottom of the drain opening, and push the auger cable into the trap. Crank the auger handle in a clockwise direction to snag obstruction. Continue cranking while retrieving the cable to pull the obstruction out of the trap.

Clearing Clogged Shower Drains

Shower drain clogs usually are caused by an accumulation of hair in the drain line. Remove the strainer cover and look for clogs in the drain opening (below). Some clogs are removed easily with a piece of stiff wire.

Stubborn clogs should be removed with a plunger or hand auger.

Everything You Need:

Tools: screwdriver, flashlight, plunger, hand auger.

Materials: stiff wire.

Sloped floor
Drain opening
Floor
Drain line
Trap
Waste and vent stack

Shower drain system has a sloped floor, a drain opening, a trap, and a drain line that connects to a branch drain line or waste and vent stack.

How to Clear a Shower Drain

Check for clogs. Remove strainer cover, using a screwdriver. Use a flashlight to look for hair clogs in the drain opening. Use a stiff wire to clear shower drain of hair or to snag any obstructions.

Use a plunger to clear most shower drain clogs. Place the rubber cup over the drain opening. Pour enough water into the shower stall to cover the lip of the cup. Move plunger handle up and down rapidly.

Clear stubborn clogs in the shower drain with a hand auger. Use the auger as shown on pages 226 to 227.

Fixing Tub Drains

When water in the tub drains slowly or not at all, remove and inspect the drain assembly. Both plunger and pop-up type drain mechanisms catch hair and other debris that cause clogs.

If cleaning the drain assembly does not fix the problem, the tub drain line is clogged. Clear the line with a plunger or a hand auger. Always stuff a wet rag in the overflow drain opening before plunging the tub drain. The rag prevents air from breaking the suction of the plunger. When using an auger, always insert the cable down through the overflow drain opening.

If the tub will not hold water with the drain closed, or if the tub continues to drain slowly after the assembly has been cleaned, then the drain assembly needs adjustment. Remove the assembly, and follow the instructions on the opposite page.

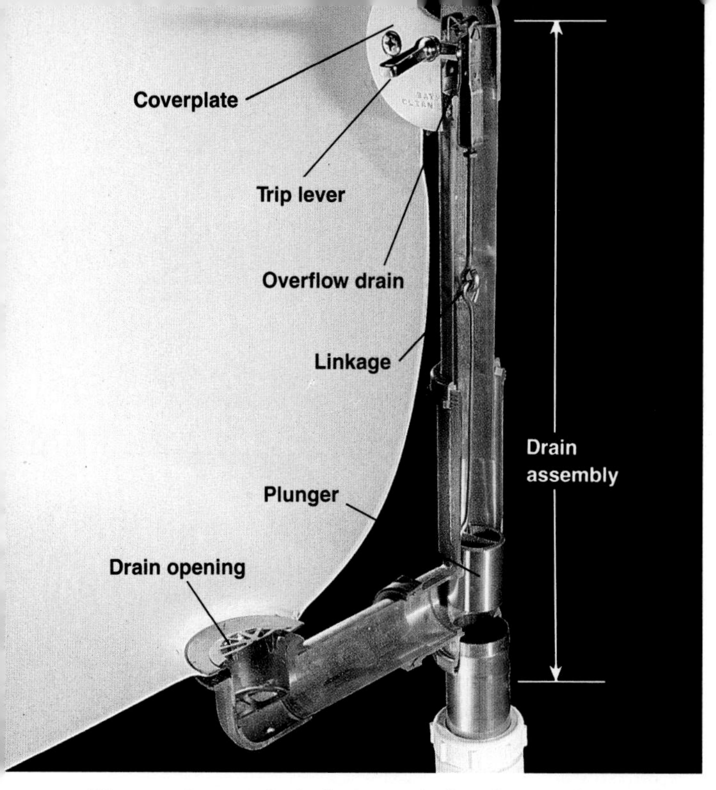

Plunger-type tub drain has a hollow brass plug, called a *plunger*, that slides up and down inside the overflow drain to seal off the water flow. The plunger is moved by a trip lever and linkage that runs through the overflow drain.

Everything You Need:

Tools: plunger, screwdriver, small wire brush, needlenose pliers, hand auger.

Materials: vinegar, heatproof grease, rag.

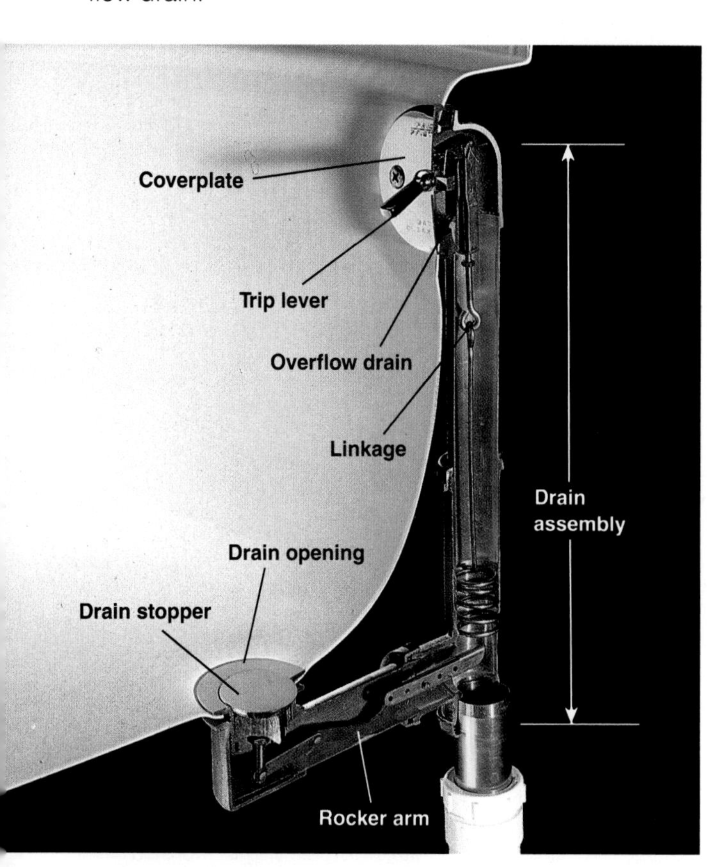

Pop-up tub drain has a rocker arm that pivots to open or close a metal drain stopper. The rocker arm is moved by a trip lever and linkage that runs through the overflow drain.

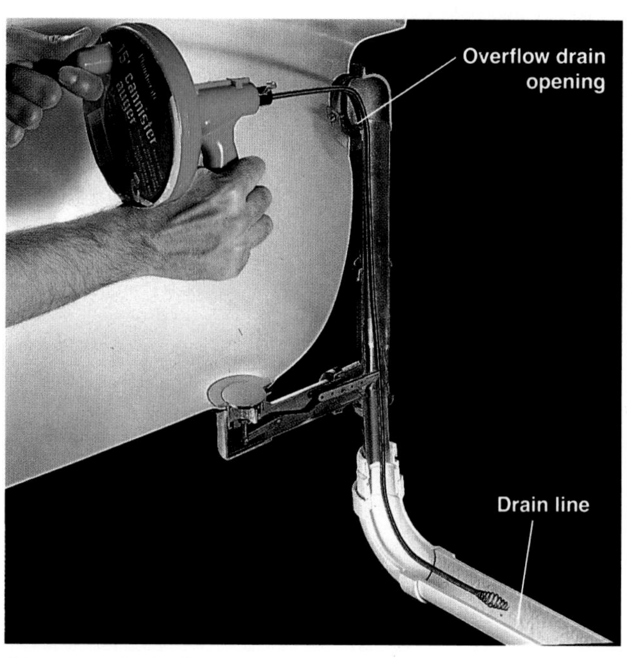

Clear a tub drain by running the auger cable through the overflow opening. First, remove the coverplate and carefully lift out the drain linkage (page opposite). Push auger cable into the opening until resistance is felt (pages 226 to 227). After using the auger, replace drain linkage. Open drain and run hot water through drain to flush out any debris.

How to Clean & Adjust a Plunger-type Tub Drain

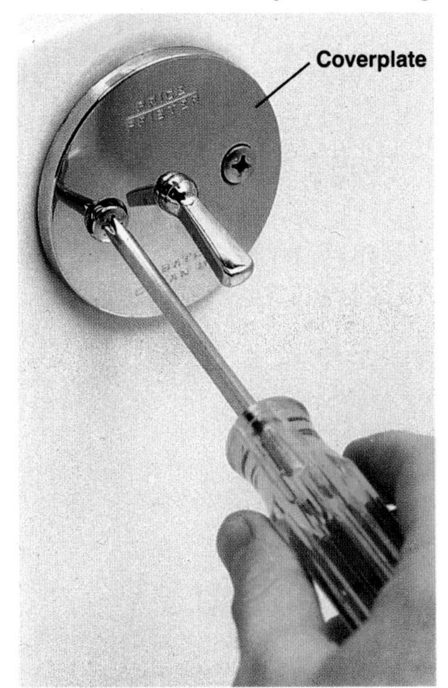

1 Remove screws on coverplate. Carefully pull coverplate, linkage, and plunger from the overflow drain opening.

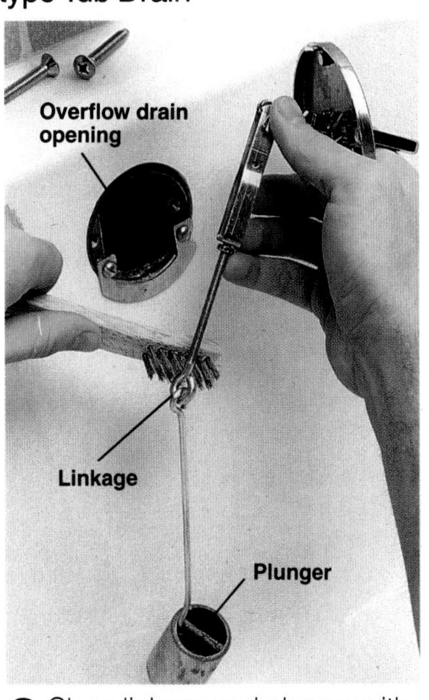

2 Clean linkage and plunger with a small wire brush dipped in vinegar. Lubricate assembly with heat-proof grease.

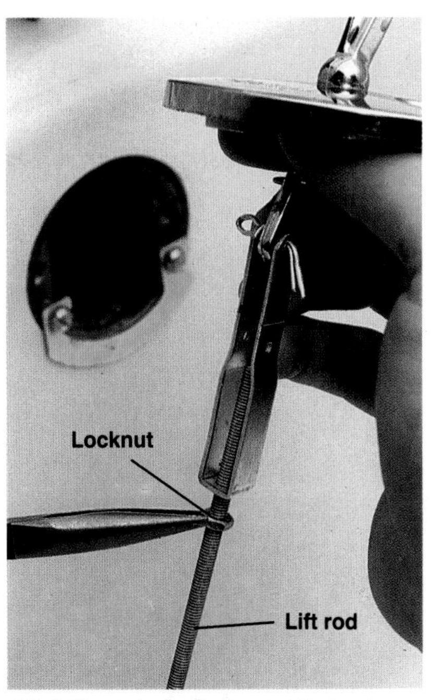

3 Adjust drain flow and fix leaks by adjusting linkage. Unscrew locknut on threaded lift rod, using needlenose pliers. Screw rod down about 1/8". Tighten locknut and reinstall entire assembly.

How to Clean & Adjust a Pop-up Tub Drain

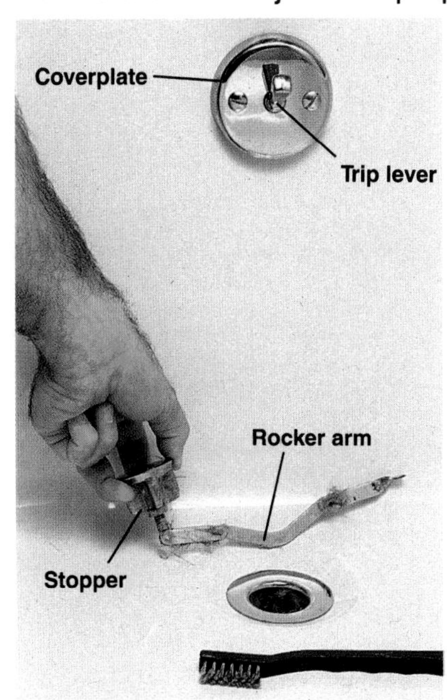

1 Raise trip lever to the full open position. Carefully pull stopper and rocker arm assembly from drain opening. Clean hair or debris from rocker arm with a small wire brush.

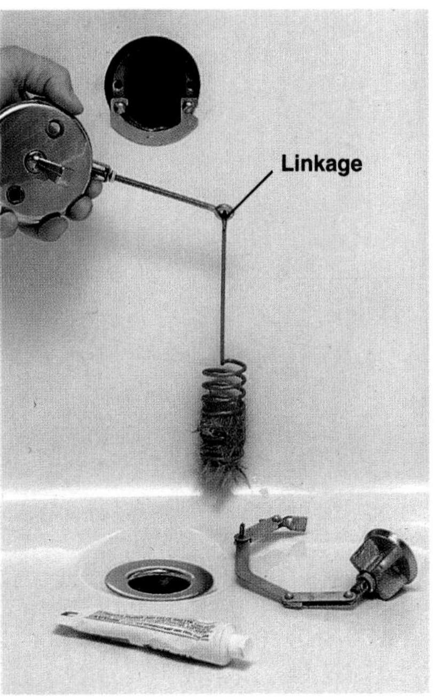

2 Remove screws from coverplate. Pull coverplate, trip lever, and linkage from overflow drain. Remove hair and debris. Remove corrosion with a small wire brush dipped in vinegar. Lubricate linkage with heat-proof grease.

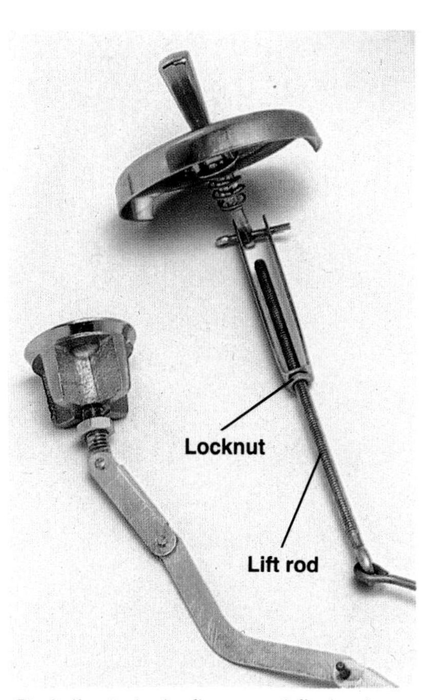

3 Adjust drain flow and fix leaks by adjusting the linkage. Loosen locknut on threaded lift rod and screw lift rod up about 1/8". Tighten locknut and reinstall entire assembly.

Clearing Clogged Drum Traps

A drum trap is a canister made of lead or cast iron. Usually, more than one fixture drain line is connected to the drum. Drum traps are not vented, and they are no longer approved for new plumbing installations.

In older homes, clogs in bathroom sinks or bathtubs may be caused by blockage in the drain lines connected to a drum trap. Remove the drum trap cover and use a hand auger to clear each drain line.

Drum traps usually are located in the floor next to the bathtub. They are identified by a flat, screw-in type cover or plug that is flush with the floor. Occasionally, a drum trap may be located under the floor. This type of drum trap will be positioned upside down so that the plug is accessible from below.

Everything You Need:

Tools: adjustable wrench, hand auger.

Materials: rags or towels, penetrating oil, Teflon tape.

How to Clear a Clogged Drum Trap

1 Place rags or towels around the opening of the drum trap to absorb water that may be backed up in the lines.

2 Remove the trap cover, using an adjustable wrench. Work carefully: older drum traps may be made of lead, which gets brittle with age. If cover does not unscrew easily, apply penetrating oil to lubricate the threads.

3 Use a hand auger (pages 226 to 227) to clear each drain line. Then wrap the threads of the cover with Teflon tape and install. Flush all drains with hot water for five minutes.

Clearing Clogged Floor Drains

When water backs up onto a basement floor, there is a clog in either the floor drain line, drain trap, or the sewer service line. Clogs in the drain line or trap may be cleared with a hand auger or a blow bag. To clear a sewer service line, see page 234.

Blow bags are especially useful for clearing clogs in floor drain lines. A blow bag attaches to a garden hose and is inserted directly into the floor drain line. The bag fills with water and then releases a powerful spurt that dislodges clogs.

Everything You Need:

Tools: adjustable wrench, screwdriver, hand auger, blow bag.

Materials: garden hose.

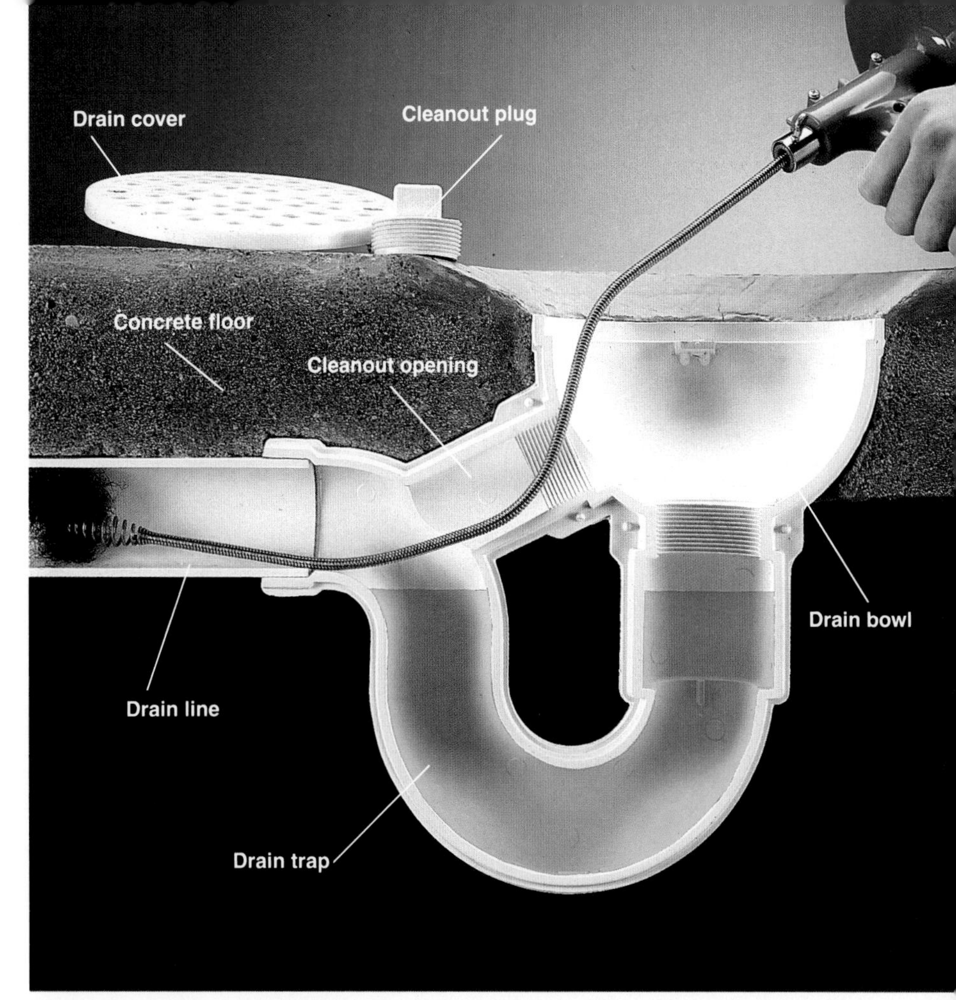

Clear clogged floor drains using a hand auger. Remove the drain cover, then use a wrench to unscrew the cleanout plug in the drain bowl. Push the auger cable through the cleanout opening directly into the drain line.

How to Use a Blow Bag to Clear a Floor Drain

1 Attach blow bag to garden hose, then attach hose to a hose bib or utility faucet.

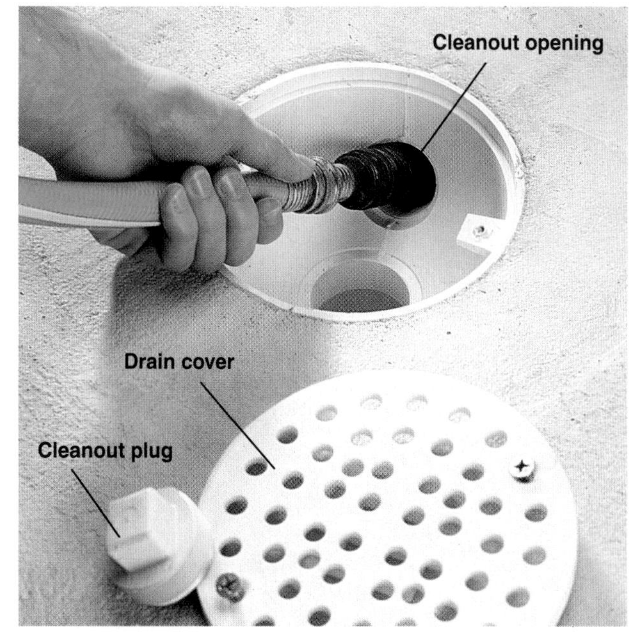

2 Remove drain cover and cleanout plug. Insert the blow bag completely into the cleanout opening and turn on water. Allow several minutes for the blow bag to work properly.

Clearing Clogs in Branch & Main Drain Lines

If using a plunger or a hand auger does not clear a clog in a fixture drain line, it means that the blockage may be in a branch drain line, the main waste and vent stack, or the sewer service line (see the photo on page 13).

First, use an auger to clear the branch drain line closest to any stopped-up fixtures. Branch drain lines may be serviced through the cleanout fittings located at the end of the branch. Because waste water may be backed up in the drain lines, always open a cleanout with caution. Place a bucket and rags under the opening to catch waste water. Never position yourself directly under a cleanout opening while unscrewing the plug or cover.

If using an auger on the branch line does not solve the problem, then the clog may be located in a main waste and vent stack. To clear the stack, run an auger cable down through the roof vent. Make sure that the cable of your auger is long enough to reach down the entire length of the stack. If it is not, you may want to rent or borrow another auger. Always use extreme caution when working on a ladder or on a roof.

If no clog is present in the main stack, the problem may be located in the sewer service line. Locate the main cleanout, usually a Y-shaped fitting at the bottom of the main waste and vent stack. Remove the plug and push the cable of a hand auger into the opening.

Some sewer service lines in older homes have a house trap. The house trap is a U-shaped fitting located at the point where the sewer line exits the house. Most of the fitting will be beneath the floor surface, but it can be identified by its two openings. Use a hand auger to clean a house trap.

If the auger meets solid resistance in the sewer line, retrieve the cable and inspect the bit. Fine, hair-like roots on the bit indicate the line is clogged with tree roots. Dirt on the bit indicates a collapsed line.

Use a power auger to clear sewer service lines that are clogged with tree roots. Power augers (page 37) are available at rental centers. However, a power auger is a large, heavy piece of equipment. Before renting, consider the cost of rental and the level of your do-it-yourself skills versus the price of a professional sewer cleaning service. If you rent a power auger, ask the rental dealer for complete instructions on how to operate the equipment.

Always consult a professional sewer cleaning service if you suspect a collapsed line.

Everything You Need:

Tools: adjustable wrench or pipe wrench, hand auger, cold chisel, ball peen hammer.

Materials: bucket, rags, penetrating oil, cleanout plug (if needed), pipe joint compound.

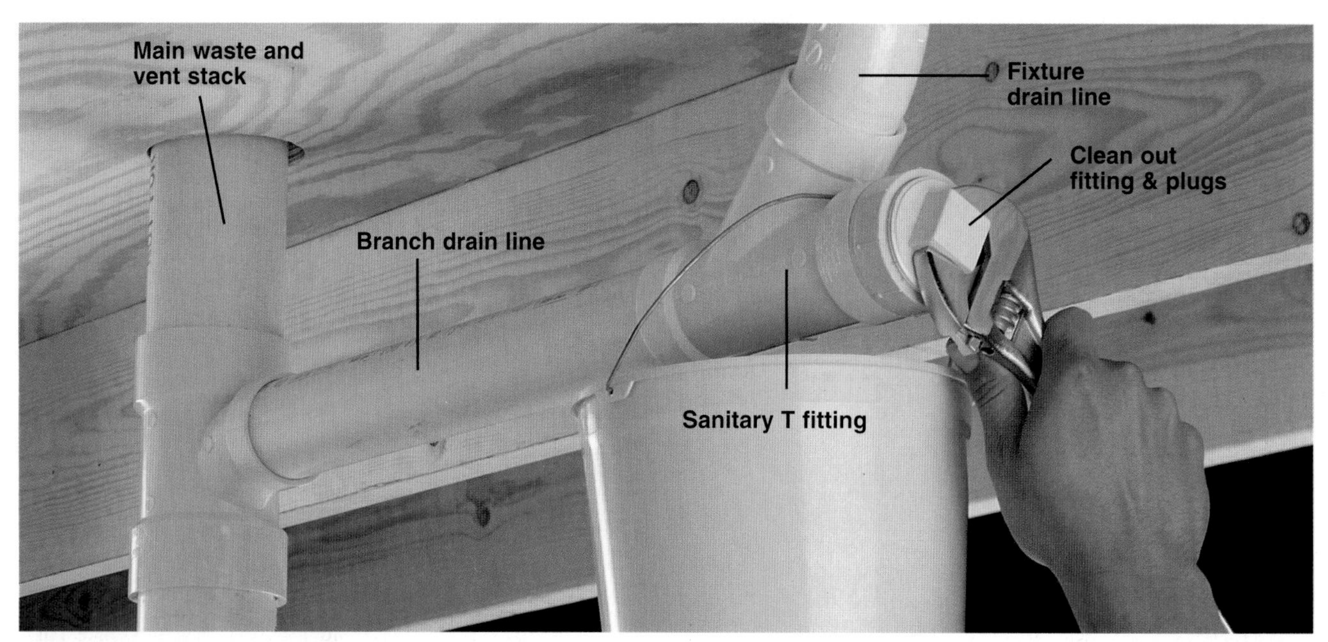

Clear a branch drain line by locating the cleanout fitting at the end of the line. Place a bucket underneath the opening to catch waste water, then slowly unscrew the cleanout plug with an adjustable wrench. Clear clogs in the branch drain line with a hand auger (pages 226 to 227).

Clear the main waste and vent stack by running the cable of a hand auger down through the roof vent. Always use extreme caution while working on a ladder or roof.

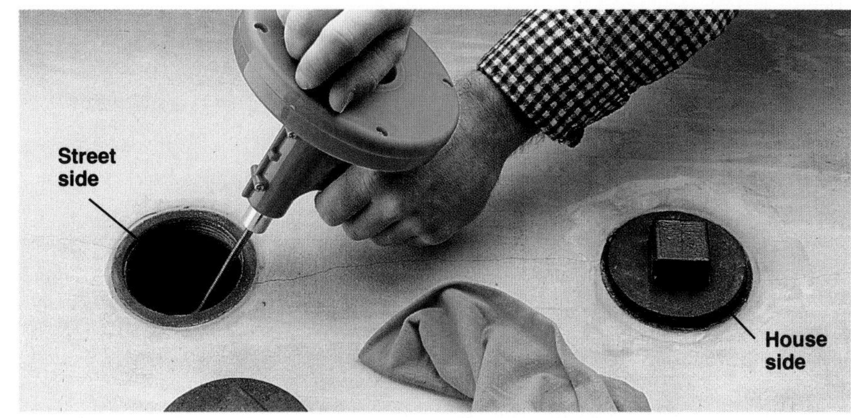

Clear the house trap in a sewer service line using a hand auger. Slowly remove only the plug on the "street side" of the trap. If water seeps out the opening as the plug is removed, the clog is in the sewer line beyond the trap. If no water seeps out, auger the trap. If no clog is present in the trap, replace the street-side plug and remove the house-side plug. Use the auger to clear clogs located between the house trap and main stack.

How to Remove & Replace a Main Drain Cleanout Plug

1 Remove the cleanout plug, using a large wrench. If plug does not turn out, apply penetrating oil around edge of plug, wait 10 minutes, and try again. Place rags and a bucket under fitting opening to catch any water that may be backed up in the line.

2 Remove stubborn plugs by placing the cutting edge of chisel on edge of plug. Strike chisel with a ball peen hammer to move plug counterclockwise. If plug does not turn out, break it into pieces with the chisel and hammer. Remove all broken pieces.

3 Replace old plug with new plastic plug. Apply pipe joint compound to the threads of the replacement plug and screw into cleanout fitting.

Alternate: Replace old plug with an expandable rubber plug. A wing nut squeezes the rubber core between two metal plates. The rubber bulges slightly to create a watertight seal.

gas water heater works: Hot water leaves tank through the **hot** **utlet (1)** as fresh, cold water enters the water heater through the **(2)**. As the water temperature drops, the **thermostat (3)** opens valve, and the **gas burner (4)** is lighted by pilot flame. Exhaust gases ted through **flue (5)**. When water temperature reaches preset tem-, the thermostat closes gas valve, extinguishing burner. The thermo-protects against gas leaks by automatically shutting off gas if pilot oes out. Anode rod protects tank lining from rust by attracting cor-ements in the water. Pressure-relief valve guards against ruptures by steam buildup in tank.

Fixing a Water Heater

Standard tank water heaters are designed so that repairs are simple. All water heaters have convenient access panels that make it easy to replace worn-out parts. When buying new water heater parts, make sure the replacements match the specifications of your water heater. Most water heaters have a nameplate (page 242) that lists the information needed, including the pressure rating of the tank, and the voltage and wattage ratings of the electric heating elements.

Many water heater problems can be avoided with routine yearly maintenance. Flush the water heater once a year, and test the pressure-relief valve. Set the thermostat at a lower water temperature to prevent heat damage to the tank. (Note: water temperature may affect the efficiency of automatic dishwashers. Check manufacturer's directions for recommended water temperature.) Water heaters last about 10 years on average, but with regular maintenance, a water heater can last 20 years or more.

Do not install an insulating jacket around a gas water heater. Insulation can block air supply and prevent the water heater from ventilating properly. Many water heater manufacturers prohibit the use of insulating jackets. To save energy, insulate the hot water pipes instead, using the materials described on page 250.

The pressure-relief valve is an important safety device that should be checked at least once each year and replaced, if needed. When replacing the pressure-relief valve, shut off the water and drain several gallons of water from the tank.

Problems	Repairs
No hot water, or not enough hot water.	1. **Gas heater**: Make sure gas is on, then relight pilot flame (page 247). **Electric heater**: Make sure power is on, then reset thermostat (page 249). 2. Flush water heater to remove sediment in tank (photo, below). 3. Insulate hot water pipes to reduce heat loss (page 250). 4. **Gas heater**: Clean gas burner & replace thermocouple (pages 238 to 239). **Electric heater**: Replace heating element or thermostat (pages 240 to 241). 5. Raise temperature setting of thermostat.
Pressure-relief valve leaks.	1. Lower the temperature setting (photo, below). 2. Install a new pressure-relief valve (pages 244 to 245, steps 10 to 11).
Pilot flame will not stay lighted.	Clean gas burner & replace the thermocouple (pages 238 to 239).
Water heater leaks around base of tank.	Replace the water heater immediately (pages 242 to 249).

Tips for Maintaining a Water Heater

Flush the water heater once a year by draining several gallons of water from the tank. Flushing removes sediment buildup that causes corrosion and reduces heating efficiency.

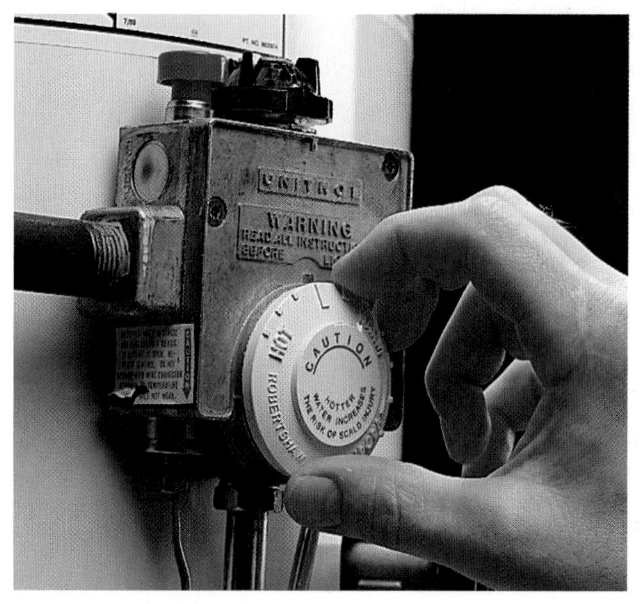

Lower the temperature setting on thermostat to 120°F. Lower temperature setting reduces damage to tank caused by overheating and also reduces energy use.

Fixing a Gas Water Heater

If a gas water heater does not heat water, first remove the outer and inner access panels and make sure the pilot is lighted. To relight a pilot, see steps 20 to 23, page 247. During operation, the outer and inner access panels must be in place. Operating the water heater without the access panels may allow air drafts to blow out the pilot flame.

If the pilot will not light, it is probably because the thermocouple is worn out. The thermocouple is a safety device designed to shut off the gas automatically if the pilot flame goes out. The thermocouple is a thin copper wire that runs from the control box to the gas burner. New thermocouples are inexpensive, and can be installed in a few minutes.

If the gas burner does not light even though the pilot flame is working, or if the gas burns with a yellow, smoky flame, the burner and the pilot gas tube should be cleaned. Clean the burner and gas tube annually to improve energy efficiency and extend the life of the water heater.

A gas water heater must be well ventilated. If you smell smoke or fumes coming from a water heater, shut off the water heater and make sure the exhaust duct is not clogged with soot. A rusted duct must be replaced.

Remember to shut off the gas before beginning work.

Everything You Need:

Tools: adjustable wrench, vacuum cleaner, needlenose pliers.

Materials: thin wire, replacement thermocouple.

How to Clean a Gas Burner & Replace a Thermocouple

1 Shut off gas by turning the gas cock on top of the control box to the OFF position. Wait 10 minutes for gas to dissipate.

2 Disconnect the pilot gas tube, the burner gas tube, and the thermocouple from the bottom of the control box, using an adjustable wrench.

3 Remove the outer and inner access panels covering the burner chamber.

4 Pull down slightly on the pilot gas tube, the burner gas tube, and thermocouple wire to free them from the control box. Tilt the burner unit slightly and remove it from the burner chamber.

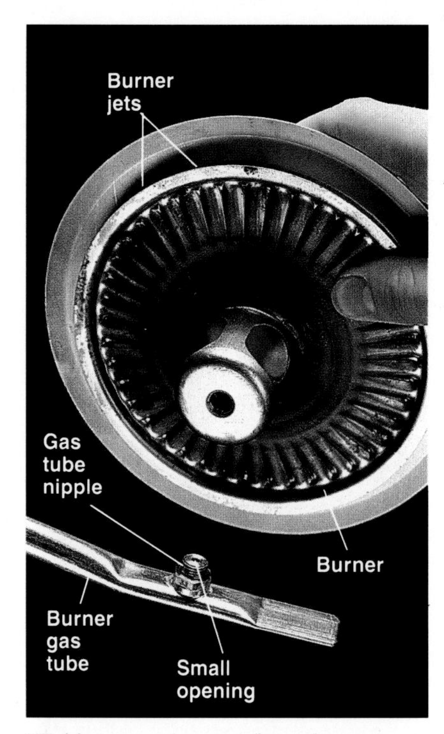

5 Unscrew burner from burner gas tube nipple. Clean small opening in nipple, using a piece of thin wire. Vacuum out burner jets and the burner chamber.

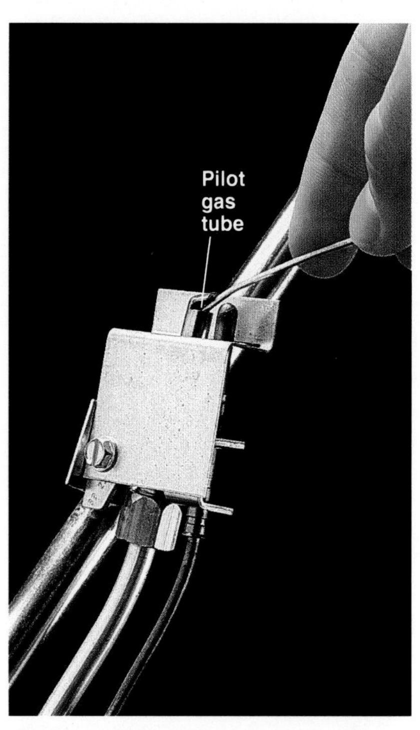

6 Clean the pilot gas tube with a piece of wire. Vacuum out any loose particles. Screw burner onto gas tube nipple.

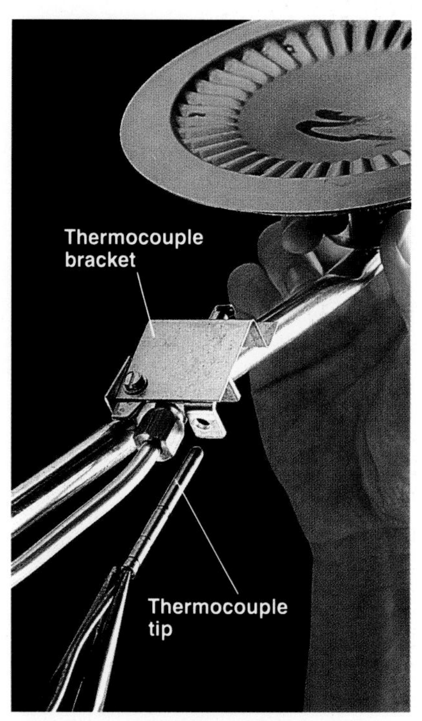

7 Pull the old thermocouple from bracket. Install new thermocouple by pushing the tip into the bracket until it snaps into place.

8 Insert the burner unit into the chamber. Flat tab at end of burner should fit into slotted opening in mounting bracket at the bottom of the chamber.

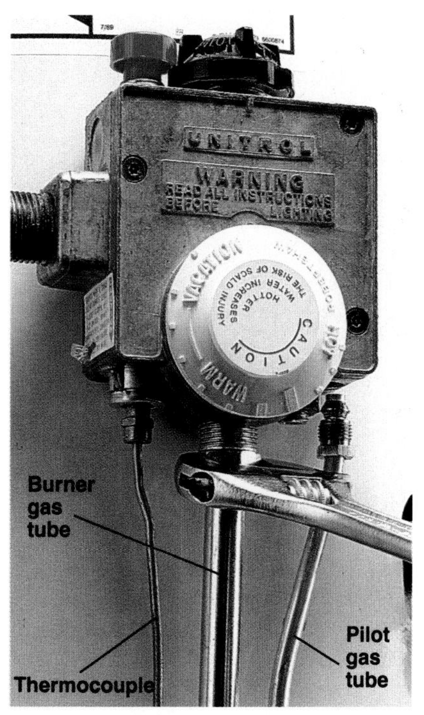

9 Reconnect the gas tubes and the thermocouple to the control box. Turn on the gas and test for leaks (page 246, step 19). Light the pilot (page 247, steps 20 to 23).

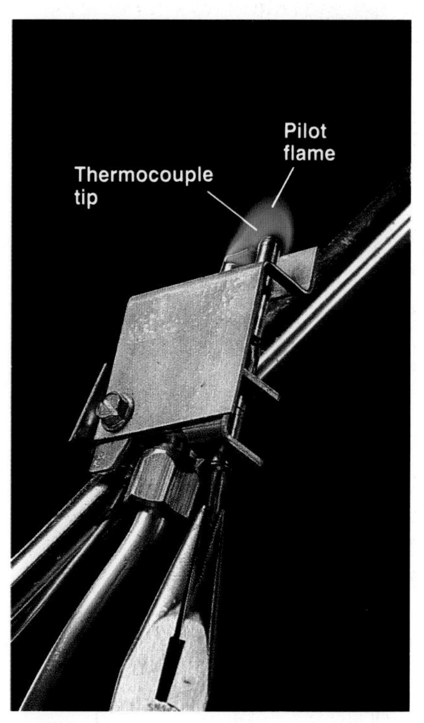

10 Make sure pilot flame wraps around tip of thermocouple. If needed, adjust thermocouple with needlenose pliers until tip is in flame. Replace the inner and outer access panels.

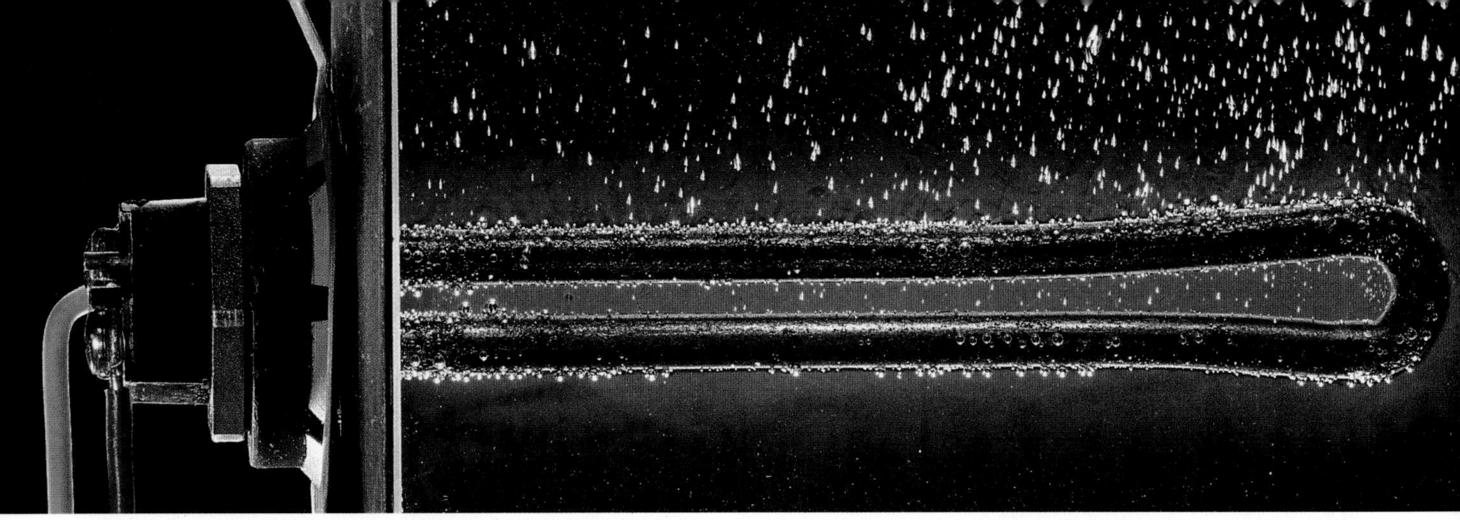

Electric water heater has one or two heating elements mounted in the side wall of the heater. Each element is connected to its own thermostat. When buying a replacement heating element or thermostat, make sure the replacement has same voltage and wattage rating as old part. This information is found on the nameplate (page 242).

Fixing an Electric Water Heater

The most common problem with an electric water heater is a burned-out heating element. Many electric water heaters have two heating elements. To determine which element has failed, turn on a hot water faucet and test the temperature. If the water heater produces water that is warm, but not hot, replace the top heating element. If the heater produces a small amount of very hot water, followed by cold water, replace the bottom heating element.

If replacing the heating element does not solve the problem, then the thermostat may need to be replaced. These parts are found under convenient access panels on the side of the heater.

Remember to turn off the power and test for current before touching wires (page 248, step 4).

Everything You Need:

Tools: screwdriver, gloves, neon circuit tester, channel-type pliers.

Materials: masking tape, replacement heating element or thermostat, replacement gasket, pipe joint compound.

How to Replace an Electric Thermostat

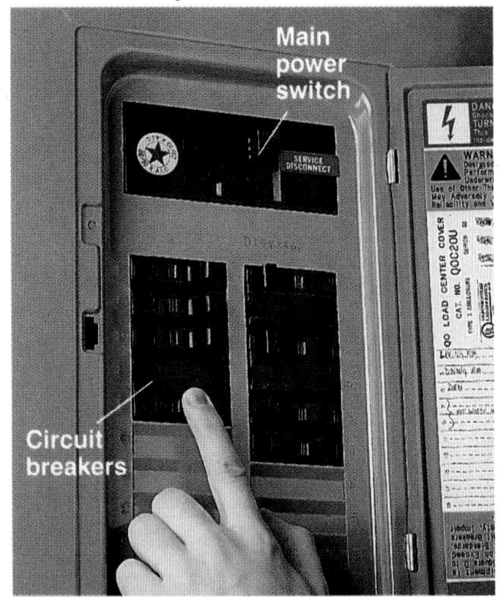

1 Turn off power at main service panel. Remove access panel on side of heater, and **test for current (page 248, step 4).**

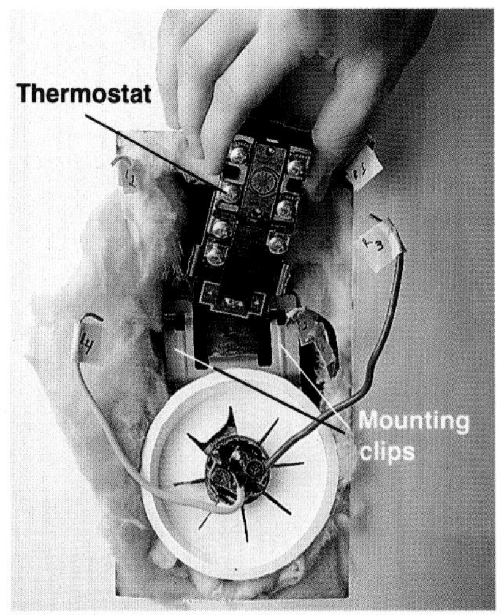

2 Disconnect thermostat wires, and label connections with masking tape. Pull old thermostat out of mounting clips. Snap new thermostat into place, and reconnect wires.

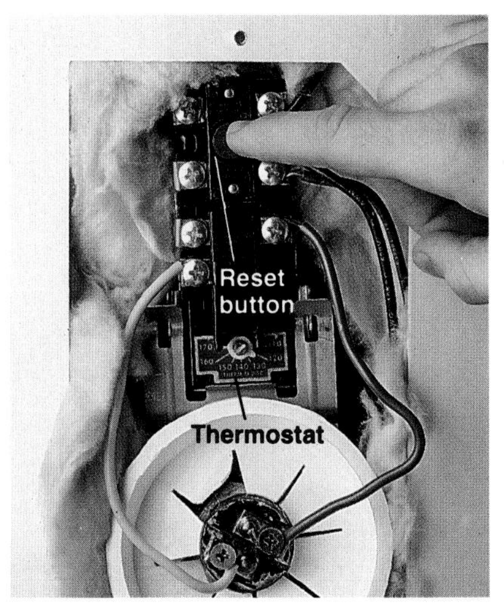

3 Press thermostat reset button, then use a screwdriver to set thermostat to desired temperature. Replace insulation and access panel. Turn on power.

How to Replace an Electric Heating Element

1 Remove access panel on side of water heater. Shut off power to water heater (page 248, step 1). Close the shutoff valves, then drain tank (page 243, step 3).

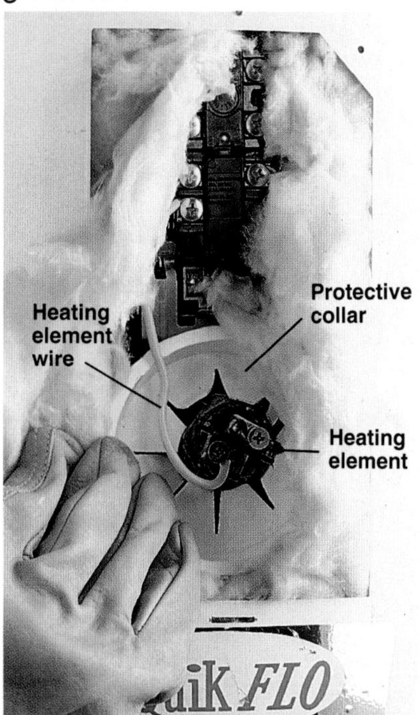

2 Wearing protective gloves, carefully move insulation aside. **Caution: test for current (page 248, step 4),** then disconnect wires on heating element. Remove protective collar.

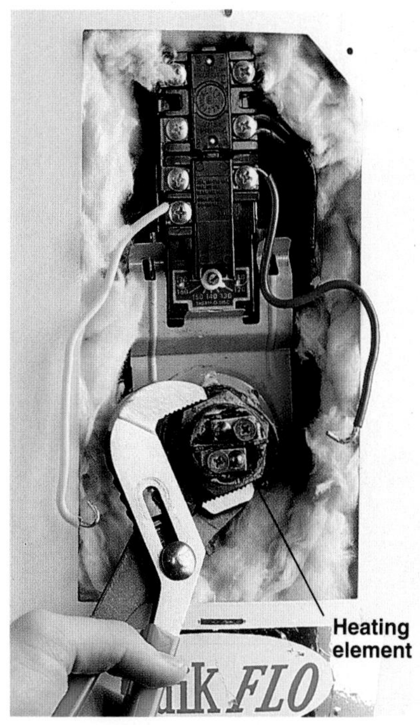

3 Unscrew the heating element with channel-type pliers. Remove old gasket from around water heater opening. Coat both sides of new gasket with pipe joint compound.

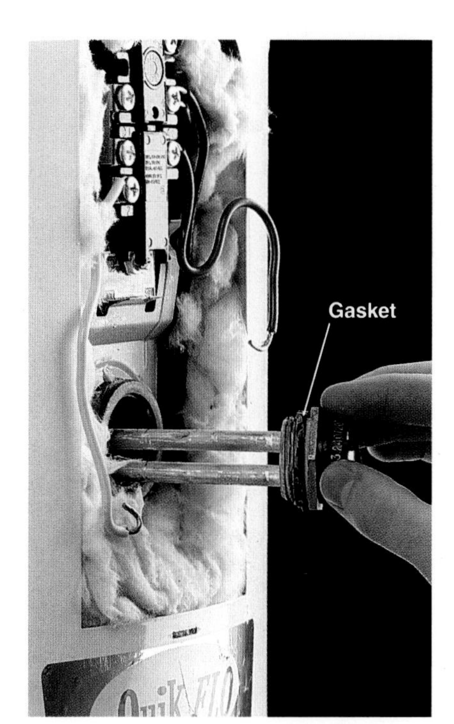

4 Slide new gasket over heating element, and screw element into the tank. Tighten element with channel-type pliers.

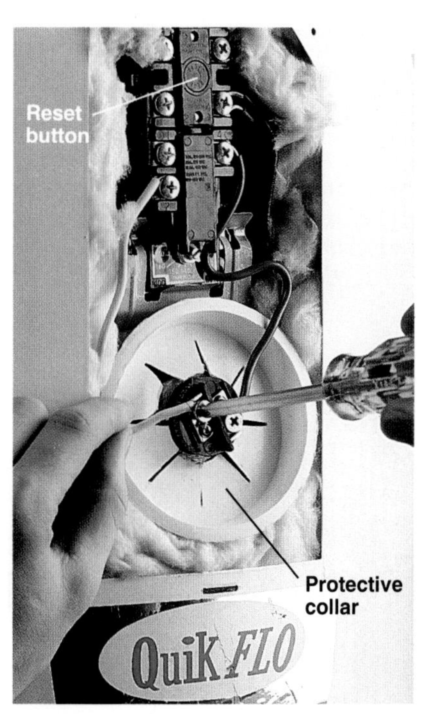

5 Replace protective collar, and reconnect all wires. Turn on hot water faucets throughout house, then turn on water heater shutoff valves. When tap water runs steadily, close faucets.

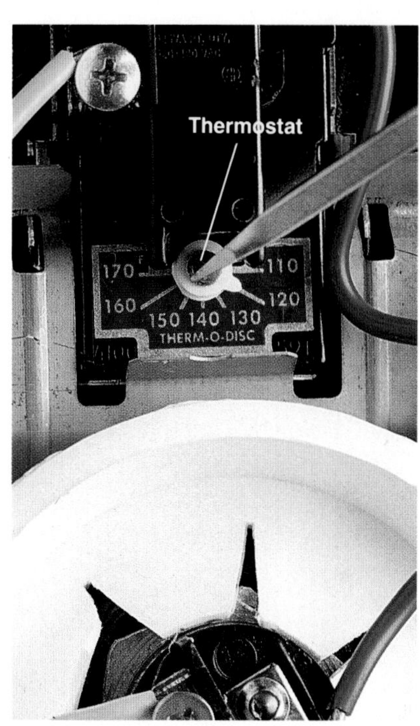

6 Use a screwdriver to set thermostat to desired temperature. Press thermostat reset buttons. Fold insulation over thermostat, and replace the access panel. Turn on power.

Replacing a Water Heater

A water heater that leaks should be replaced immediately to prevent expensive water damage. Leaks occur because the inner tank has rusted through.

When replacing an electric water heater, make sure the voltage of the new model is the same as the old heater. When replacing a gas water heater, maintain a clearance of 6" or more around the unit for ventilation. Water heaters are available with tank sizes ranging from 30 to 65 gallons. A 40-gallon heater is large enough for a family of four.

Energy-efficient water heaters have polyurethane foam insulation, and usually carry an extended warranty. These models are more expensive, but over the life of the water heater they cost less to own and operate. Some top-quality water heaters have two anode rods for extra corrosion protection.

The pressure-relief valve usually must be purchased separately. Make sure the new valve matches the *working pressure* rating of the tank (photo, left).

Nameplate on side of water heater lists tank capacity, insulation R-value, and working pressure (pounds-per-square-inch). More efficient water heaters have an insulation R-value of 7 or higher. Nameplate for an electric water heater includes the voltage and the wattage capacity of the heating elements and thermostats. Water heaters also have a yellow **energy guide label** (photo, top) that lists typical yearly operating costs. Estimates are based on national averages. Energy costs in your area may differ.

Everything You Need:

Tools: pipe wrenches, hacksaw or tubing cutter, screwdriver, hammer, appliance dolly, level, small wire brush, propane torch, adjustable wrench, circuit tester (electric heaters).

Materials: bucket, wood shims, #4 gauge 3⁄8" sheet-metal screws, pressure-relief valve, threaded male pipe adapters, solder, two heat-saver nipples, Teflon tape, flexible water connectors, 3⁄4" copper pipe, pipe joint compound, sponge, masking tape.

How to Replace a Gas Water Heater

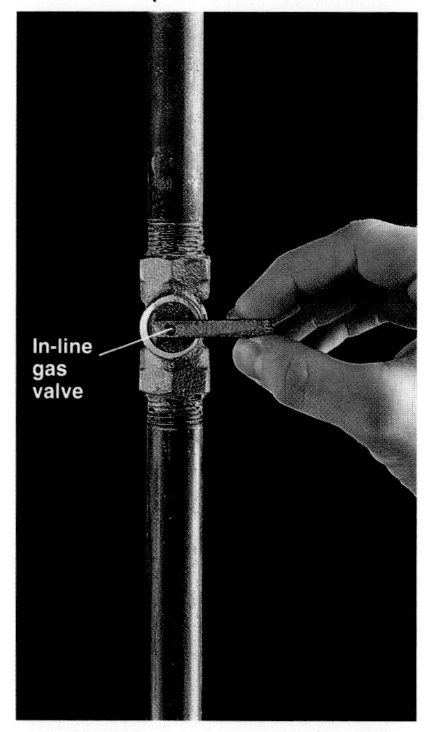

1 Shut off the gas by turning the handle of the in-line valve so it is perpendicular to gas line. Wait 10 minutes for gas to dissipate. Shut off the water supply at the shutoff valves (photo, below).

2 Disconnect gas line at the union fitting or at the flare fitting below shutoff valve, using pipe wrenches. Disassemble and save the gas pipes and fittings.

3 Drain water from the water heater tank by opening the hose bib on the side of the tank. Drain the water into buckets, or attach a hose and empty the tank into a floor drain.

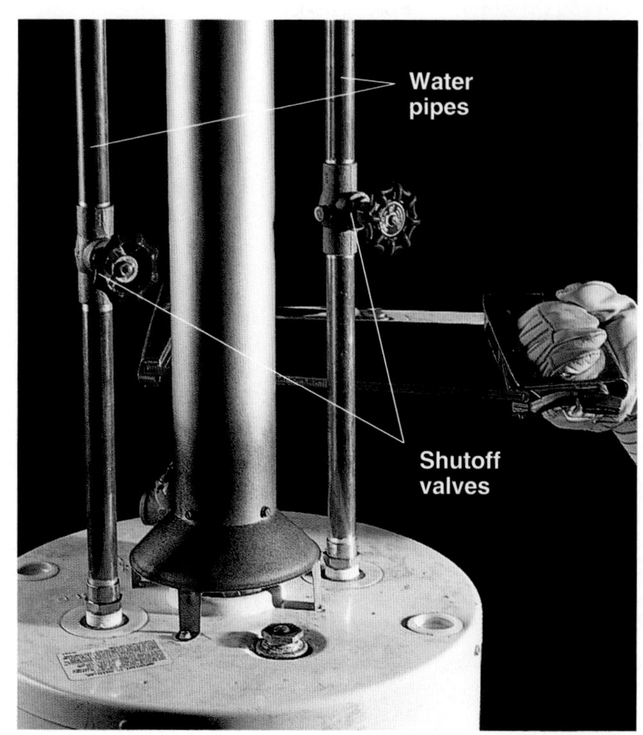

4 Disconnect the hot and cold water pipes above the water heater. If pipes are soldered copper, use a hacksaw or tubing cutter to cut through water pipes just below shutoff valves. Cuts must be straight.

5 Disconnect the exhaust duct by removing the sheetmetal screws. Remove the old water heater with a rented appliance dolly.

(continued next page)

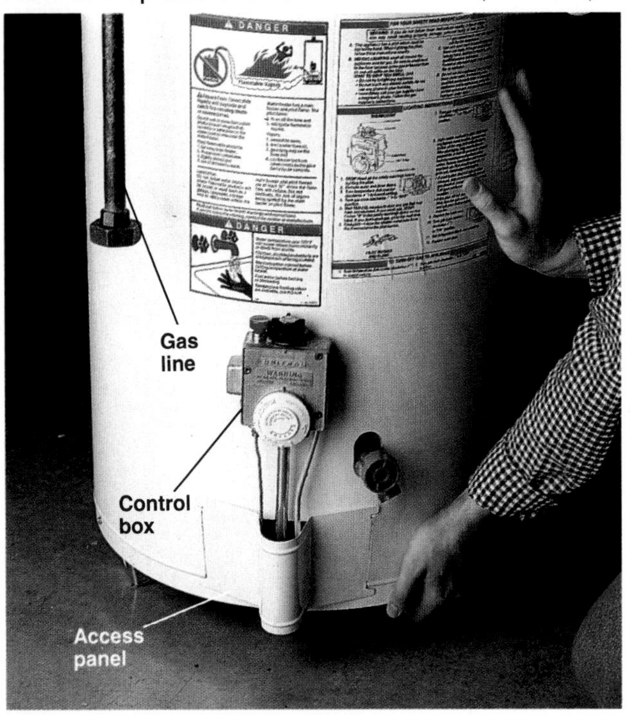

6 Position new heater so that control box is close to gas line, and access panel for burner chamber is not obstructed.

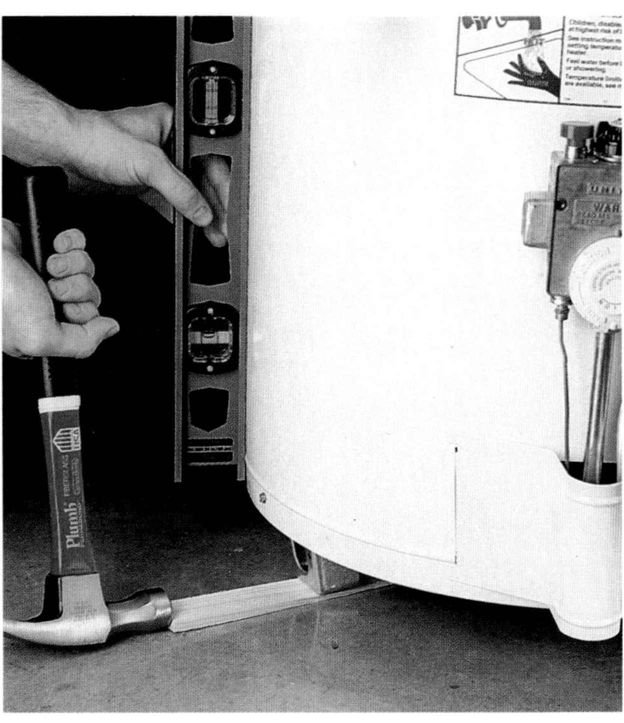

7 Level the water heater by placing wood shims under the legs.

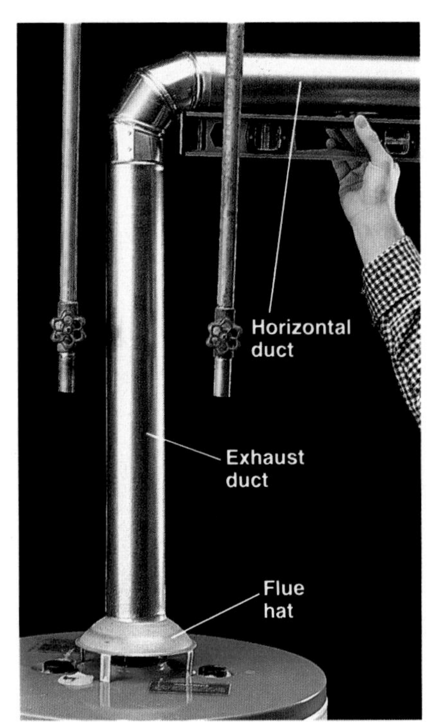

8 Position flue hat so legs fit into slots on water heater, then slip exhaust duct over flue hat. Make sure horizontal duct slopes upward ¼" per foot so fumes cannot back up into house.

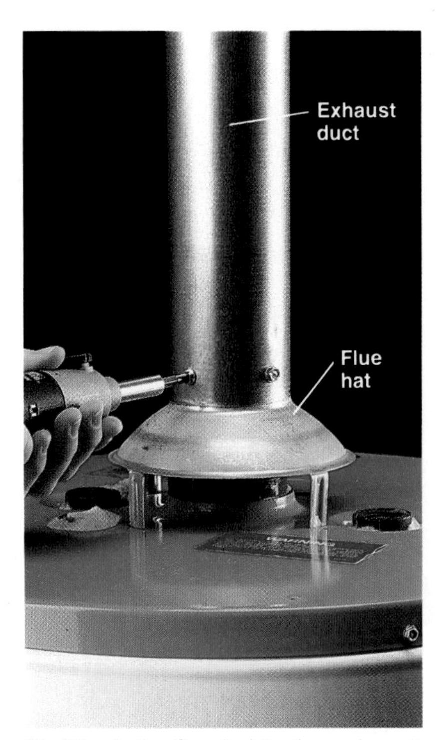

9 Attach the flue hat to the exhaust duct with #4 gauge ⅜" sheet-metal screws driven every 4".

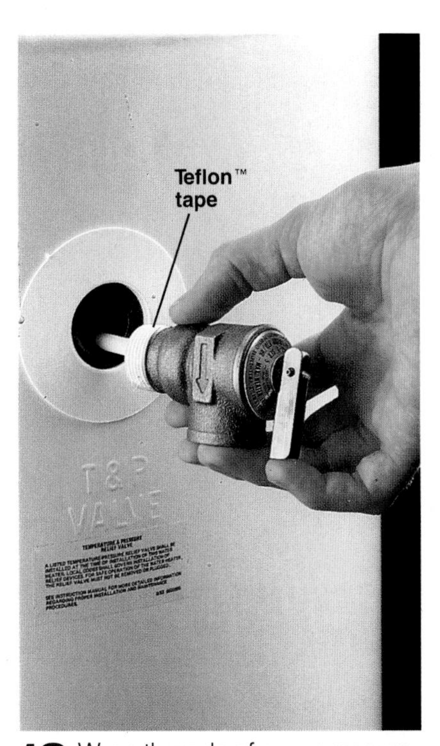

10 Wrap threads of new pressure-relief valve with Teflon tape, and a screw valve into tank opening with a pipe wrench.

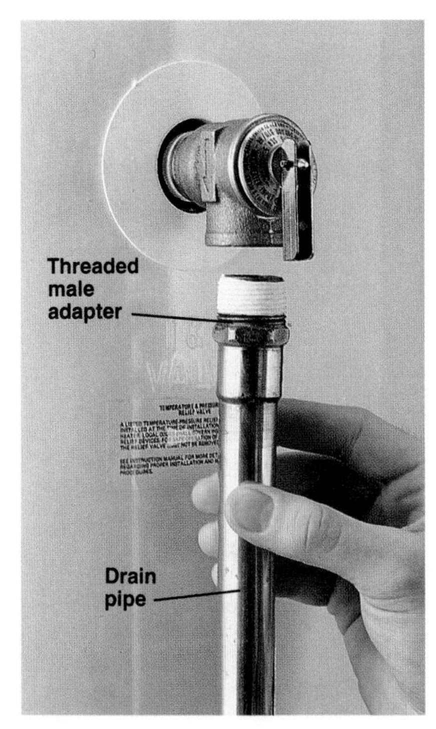

11 Attach a copper or CPVC drain pipe to the pressure-relief valve, using threaded male adapter (page 43). Pipe should reach to within 3" of floor.

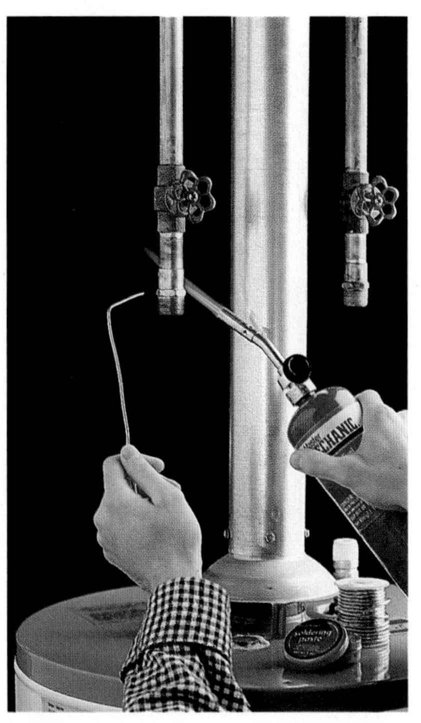

12 Solder threaded male adapters to the water pipes (pages 46 to 51). Let pipes cool, then wrap Teflon tape around threads of adapters.

13 Wrap Teflon tape around the threads of two heat-saver nipples. The nipples are color-coded, and have water-direction arrows to ensure proper installation.

14 Attach blue-coded nipple fitting to cold water inlet and red-coded fitting to hot water outlet, using a pipe wrench. On cold water nipple, water direction arrow should face down; on hot water nipple, arrow should face up.

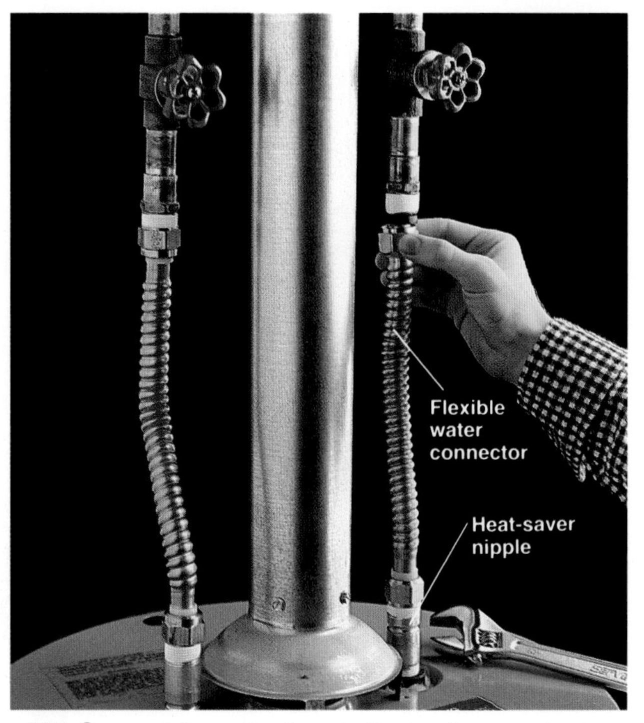

15 Connect the water lines to the heat-saver nipples with flexible water connectors. Tighten fittings with an adjustable wrench.

(continued next page)

How to Replace a Gas Water Heater (continued)

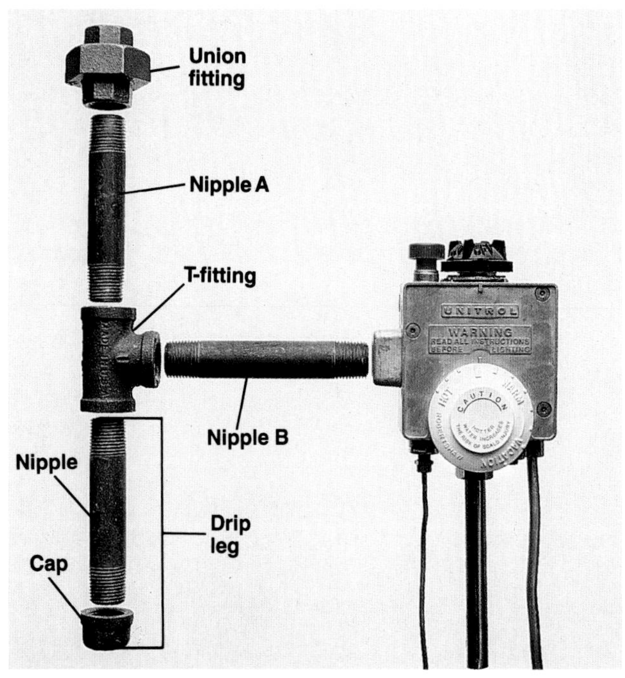

16 Test-fit gas pipes and fittings from old water heater (step 2). One or two new black-iron nipples (A, B) may be necessary if new water heater is taller or shorter than old heater. Use black iron, not galvanized iron, for gas lines. Capped nipple is called a drip leg. The drip leg protects the gas burner by catching dirt particles.

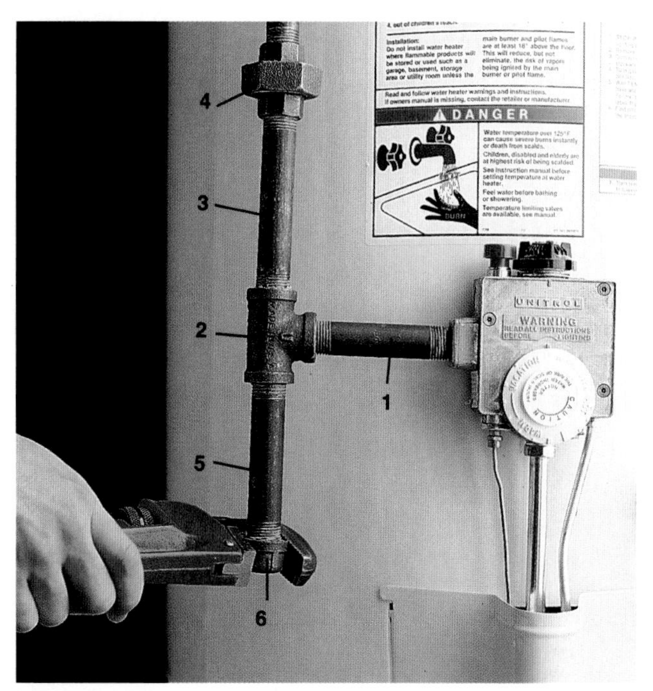

17 Clean pipe threads with a small wire brush, and coat the threads with pipe joint compound. Assemble gas line in the following order: control box nipple (1), T-fitting (2), vertical nipple (3), union fitting (4), vertical nipple (5), cap (6). (Black iron is fitted using same methods as for galvanized iron. See pages 64 to 67 for more information.)

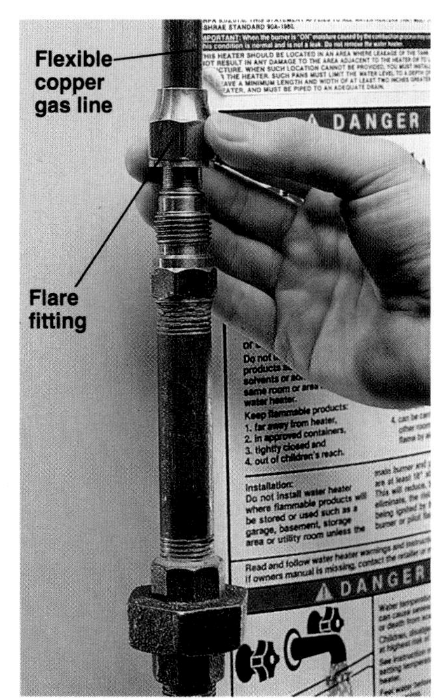

Alternate: if gas line is made of flexible copper, use a flare fitting to connect the gas line to the water heater. (For more information on flare fittings, see pages 54 to 55.)

18 Open the hot water faucets throughout house, then open the water heater inlet and outlet shutoff valves. When water runs steadily from faucets, close faucets.

19 Open the in-line valve on the gas line (step 1). Test for leaks by dabbing soapy water on each joint. Leaking gas will cause water to bubble. Tighten leaking joints with a pipe wrench.

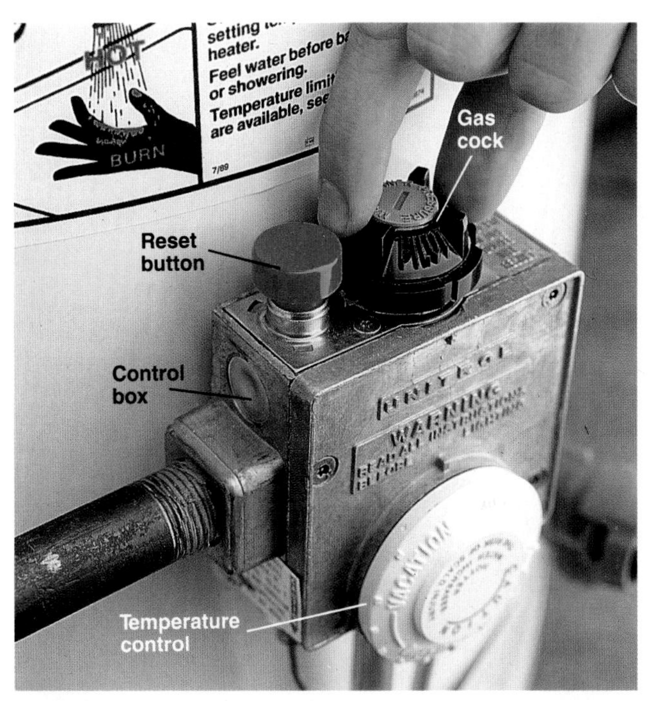

20 Turn the gas cock on top of control box to the PILOT position. Set the temperature control on front of box to desired temperature.

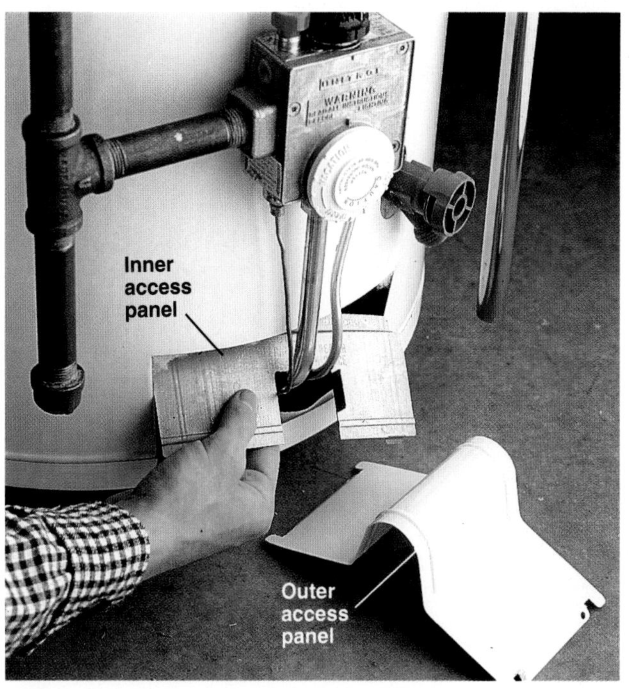

21 Remove the outer and inner access panels covering the burner chamber.

22 Light a match, and hold flame next to the end of the pilot gas tube inside the burner chamber. Be sure to keep your face away from the opening.

23 While holding match next to end of pilot gas tube, press the reset button on top of control box. When pilot flame lights, continue to hold reset button for one minute. Turn gas cock to ON position, and replace the inner and outer access panels.

How to Replace a 220/240-volt Electric Water Heater

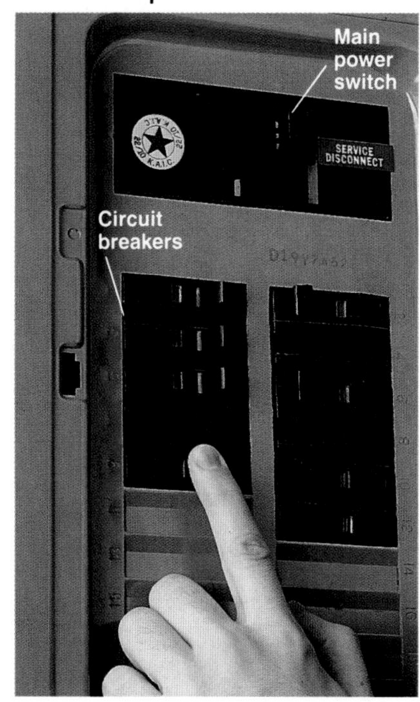

1 Turn off power to water heater by switching off circuit breaker (or removing fuse) at main service panel. Drain water heater and disconnect water pipes (page 243, steps 3 and 4).

2 Remove one of the heating element access panels on the side of the water heater.

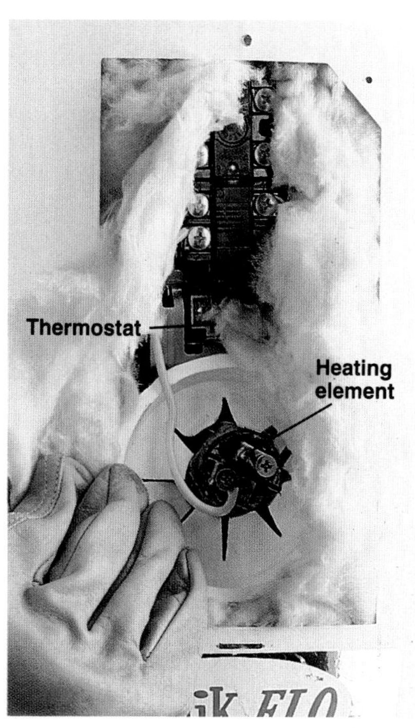

3 Wearing protective gloves, fold back the insulation to expose the thermostat. **Caution: do not touch bare wires until they have been tested for current.**

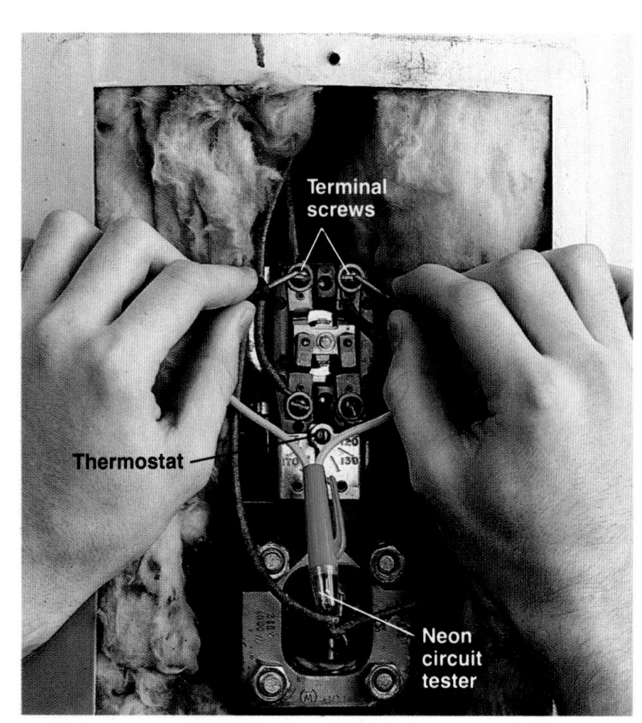

4 Test for current by touching probes of neon circuit tester to top pair of terminal screws on the thermostat. If tester lights, wires are not safe to work on; turn off main power switch and retest for current.

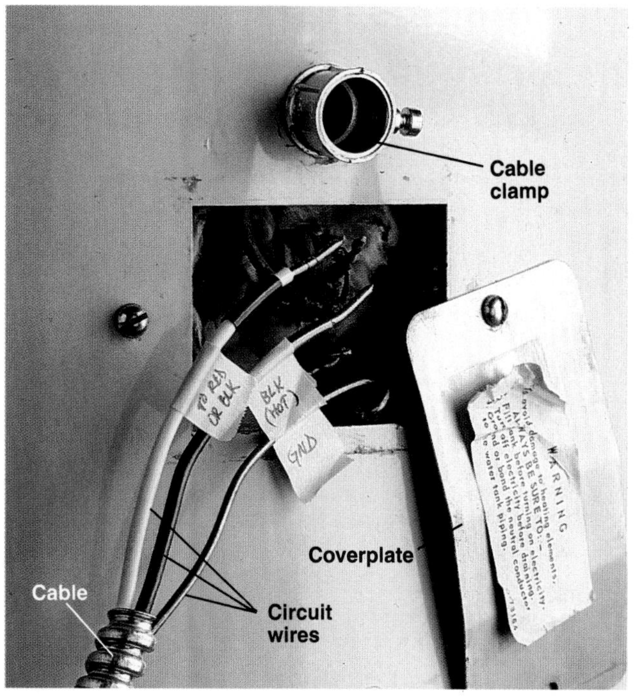

5 Remove coverplate on electrical box, found at side or top of water heater. Disconnect all wires, and label with masking tape for reference. Loosen cable clamp. Remove wires by pulling them through clamp. Remove old heater, then position new heater.

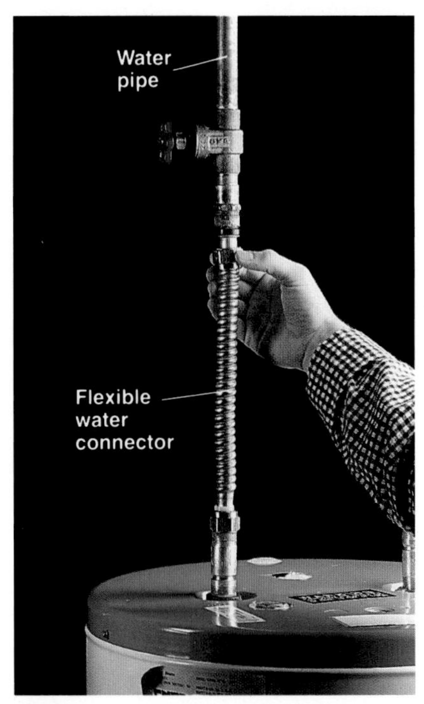

6 Connect water pipes and pressure-relief valve, following directions for gas water heaters (pages 244 to 245, steps 10 to 15). Open hot water faucets throughout house, and turn on water. When water runs steadily, turn off faucets.

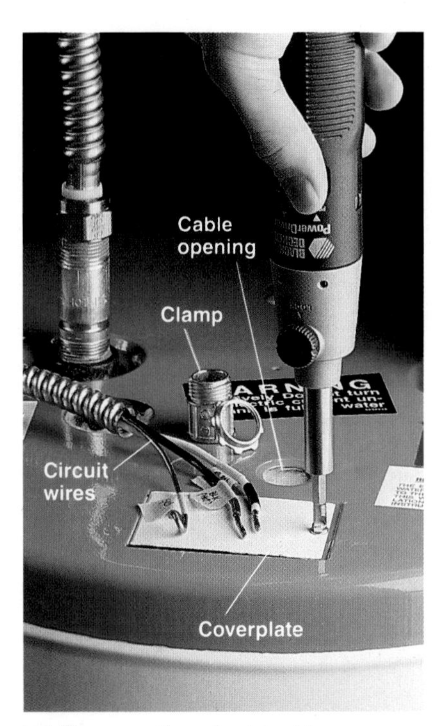

7 Remove the electrical box coverplate on new water heater. Thread the circuit wires through the clamp. Thread circuit wires through the cable opening on the water heater, and attach clamp to water heater.

8 Connect the circuit wires to the water heater wires, using wire nuts.

9 Attach bare copper or green ground wire to ground screw. Replace coverplate.

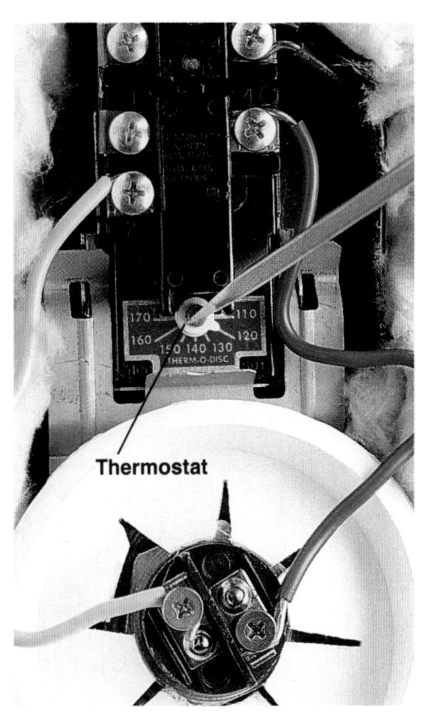

10 Remove access panels on side of water heater (steps 2 to 3), and use a screwdriver to set thermostats to desired water temperature.

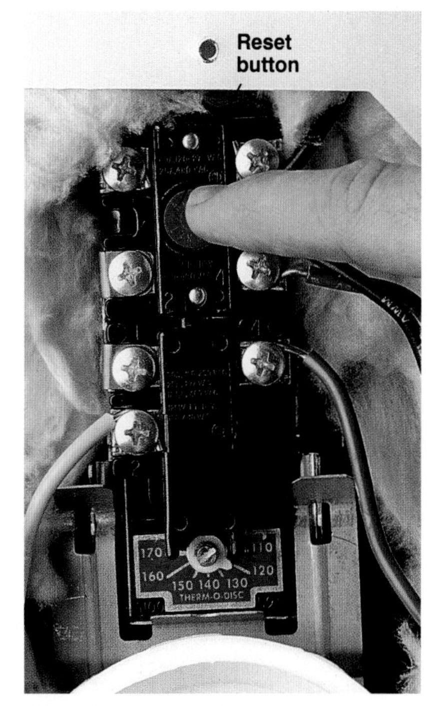

11 Press reset button on thermostats. Replace the insulation and access panels. Turn on power.

Fixing Burst or Frozen Pipes

When a pipe bursts, immediately turn off the water at the main shutoff valve. Make temporary repairs with a sleeve clamp repair kit (page opposite).

A burst pipe is usually caused by freezing water. Prevent freezes by insulating pipes that run in crawl spaces or other unheated areas.

Pipes that freeze, but do not burst, will block water flow to faucets or appliances. Frozen pipes are easily thawed, but determining the exact location of the blockage may be difficult. Leave blocked faucets or valves turned on. Trace supply pipes that lead to blocked faucet or valve, and look for places where the line runs close to exterior walls or unheated areas. Thaw pipes with a heat gun or hair dryer (below).

Old fittings or corroded pipe also may leak or rupture. Fix old pipes according to the guidelines described on pages 44 to 71.

Everything You Need:

Tools: heat gun or hair dryer, gloves, metal file, screwdriver.

Materials: pipe insulation, sleeve clamp repair kit.

Begin any emergency repair by turning off water supply at main shutoff valve. The main shutoff valve is usually located near water meter.

How to Repair Pipes Blocked with Ice

1 Thaw pipes with a heat gun or hair dryer. Use heat gun on low setting, and keep nozzle moving to prevent overheating pipes.

2 Let pipes cool, then insulate with sleeve-type foam insulation to prevent freezing. Use pipe insulation in crawl spaces or other unheated areas.

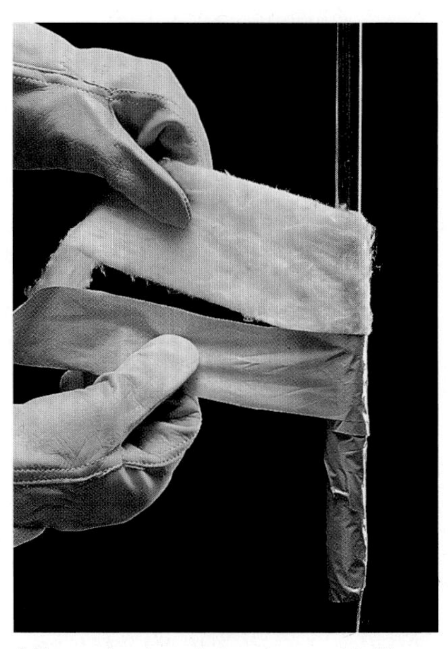

Alternate: Insulate pipes with fiberglass strip insulation and waterproof wrap. Wrap insulating strips loosely for best protection.

How to Temporarily Fix a Burst Pipe

1 Turn off water at main shutoff valve. Heat pipe gently with heat gun or hair dryer. Keep nozzle moving. Once frozen area is thawed, allow pipe to drain.

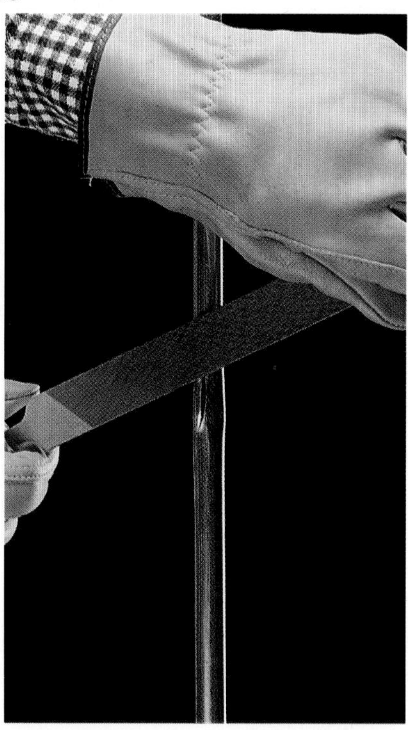

2 Smooth rough edges of rupture with metal file.

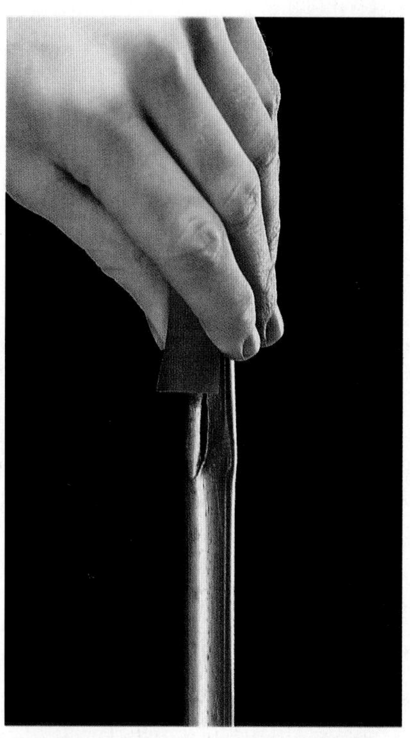

3 Place rubber sleeve of repair clamp around rupture. Make sure seam of sleeve is on opposite side of pipe from rupture.

4 Place the two metal repair clamps around rubber sleeve.

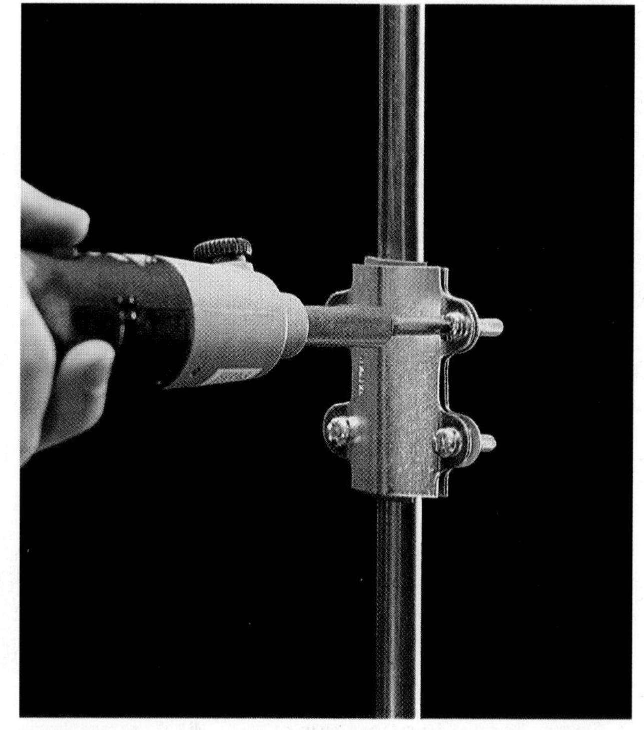

5 Tighten screws with screwdriver. Open water supply and watch for leaks. If repair clamp leaks, retighten screws. **Caution: repairs made with a repair clamp kit are temporary.** Replace ruptured section of pipe as soon as possible.

INDEX

E

F

G

H

I

K

L

T

U

V

W

Y

Z

Creative Publishing international, Inc.
offers a variety of how-to books.
For information write:
Creative Publishing international, Inc.
Subscriber Books
5900 Green Oak Drive
Minnetonka, MN 55343